The Molecular Vision of Life

Monographs on the History and Philosophy of Biology

RICHARD BURIAN, RICHARD BURKHARDT, JR.,
RICHARD LEWONTIN, JOHN MAYNARD SMITH
EDITORS

The Molecular Vision of Life

*Caltech, The Rockefeller Foundation,
and the Rise of the New Biology*

Lily E. Kay

Program in Science, Technology, and Society
Massachusetts Institute of Technology

New York   Oxford
OXFORD UNIVERSITY PRESS

Oxford University Press

Oxford   New York
Athens   Auckland   Bangkok   Bombay
Calcutta   Cape Town   Dar es Salaam   Delhi
Florence   Hong Kong   Istanbul   Karachi
Kuala Lumpur   Madras   Madrid   Melbourne
Mexico City   Nairobi   Paris   Singapore
Taipei   Tokyo   Toronto

and associated companies in
Berlin   Ibadan

Copyright © 1993 by Oxford University Press, Inc.

First published in 1993 by Oxford University Press, Inc.
198 Madison Avenue, New York, NY 10016

First issued as an Oxford University Press paperback, 1996.

Oxford is a registered trademark of Oxford University Press

Library of Congress Cataloging-in-Publication Data
Kay, Lily E.
The molecular vision of life : Caltech, the Rockefeller
Foundation, and the rise of the new biology / by Lily E. Kay,
p.   cm.—(Monographs on the history and philosophy of biology)
Includes bibliographical references and index.
ISBN 0-19-505812-7 (cloth)
ISBN 0-19-511143-5 (paper)
1. Molecular biology—History.
2. California Institute of
Technology—Research—History.
3. Rockefeller Foundation—Research—
—History.
I. Title. II. Series.
QH506.K39   1993   574.8'8'09—dc20
92-3460

1 3 5 7 9 8 6 4 2

Printed in the United States of America
on acid-free paper

To my teachers
and to the memory of my parents

Acknowledgments

My interest in molecular biology, the science and its history, go back a decade and a half, to my days at the Salk Institute for Biological Studies in La Jolla, California. This inspiring intellectual and physical space, and the support of friends and colleagues, particularly Ted Melnechuk and Tony Hunter, initiated a search for answers beyond the scope of the laboratory. As I embarked on my training in the history of science, the spirited guidance of the late Derek de Solla Price ensured the laboratory a central place in my historical reconstructions of the production of scientific knowledge. My raw ideas about the workings of sciences were challenged, refined, and molded into historical scholarship by my teachers and colleagues at the Department of the History of Science, The Johns Hopkins University. The rigorous training and warm support of Robert Kargon, Owen Hannaway, and especially my dissertation adviser, Sharon Kingsland, have sustained this work well beyond the era of graduate school.

A Smithsonian Fellowship contributed to this project in many ways. Financial support and other institutional resources have aided my travels and research. Ramunas Kondratas, Jon Eklund, and Nathan Reingold continuously encouraged my efforts; Paul Forman provided critical sources and insights that had a deep influence on the direction of this project. A grant from the National Science Foundation has aided the writing, and an Andrew W. Mellon Foundation Fellowship provided an exposure to the wealth of manuscript sources in the history of life science at the American Philosophical Society Library, broadening significantly the book's interpretive perspectives. I am very grateful to the Library's director Edward C. Carter II and staff for their cheerful help. I am also indebted to the Rockefeller Archive Center for their research grants, and especially to Thomas Rosenbaum for his expert guidance and generous cooperation during the entire duration of the work. I also thank Clifford Mead of the Oregon State University Archive for his attentive assistance, and the archivists of the California Institute of Technology Archives for providing me with access to many of the requested records. These sources were further augmented by the insights of several scientists who shared with me their personal experiences in molecular biology: Linus Pauling, Joshua Lederberg, James Bonner, Norman Horowitz, and Ray Owen. The various forms of support from members of the Program in Science, Technology, and Society at M.I.T. helped sustain my momentum in the final stretch of this book; and the care of the editors at Oxford University Press ensured its best possible presentation.

Many friends and colleagues have enriched this book by sharing with me their worlds and ideas; those mentioned here represent merely a fraction of these contributions. Peter Kuznick's scholarship and friendship have unfailingly nurtured my labors. Charles Weiner, Richard Burian, Diane Paul, Philip Pauly, Alan Trachtenberg, Evelyn Fox Keller, Jacqueline Broekhuysen, and colleagues at Penn and the University of Chicago have offered comments, criticisms, and encouragements. And in countless ways, Herbert Gottweis has provided emotional and intellectual sustenance which has fostered a quest for levels of knowing that my endeavors have yet to reach.

Boston, Massachusetts   Lily E. Kay
August, 1992

Contents

The Molecular Vision of Life

Introduction

During the 1930s a new biology came into being that by the late 1950s was to endow scientists with unprecedented power over life. These three decades culminated in the elucidation of the self-replicating mechanisms of DNA and an explanation of its action in terms of information coding, representations that laid the cognitive foundations for genetic engineering. Scientists could now manipulate genes on the most fundamental level and attempt to control the course of biological and social evolution; they laid claim to "the secret of life."

The new biology, which became known as molecular biology, emerged as a dominant disciplinary trend. Its molecular vision of life promised to function as Occam's razor, paring down the convoluted explanations offered by traditional biological fields. This new science did not just evolve by natural selection of randomly distributed disciplinary variants, nor did it ascend solely through the compelling power of its ideas and its leaders. Rather, the rise of the new biology was an expression of the systematic cooperative efforts of America's scientific establishment—scientists and their patrons—to direct the study of animate phenomena along selected paths toward a shared vision of science and society.

The aim of this book is to understand the historical process that propelled molecular biology to its dominant disciplinary status by uncovering the motivations and mechanisms empowering its ascent. It does so by focusing on two key institutions: the Rockefeller Foundation and the California Institute of Technology. As has been well documented, the Rockefeller Foundation served as the principal patron of molecular biology from the 1930s to the 1950s[1]; Caltech, a primary site for implementing the Foundation's project, became the most influential international center for research and training in molecular biology. Why did the Rockefeller Foundation launch and sustain with massive support a new biology program at that moment in history? From the entire range of contemporary biological in-

3

terests, why did scientists and their patrons privilege and promote a molecular study of life? Why was Caltech selected as a primary site, and what accounted for its remarkable influence? Studied together, the trails leading to the answers reveal a synergy between intellectual capital and economic resources, a potent convergence of scientific goals and social agendas, shaped initially by the cultural imperatives of the interwar period and later modified by the experience of World War II.

Some preliminary directions would be helpful before embarking on these trails. Specifically, a clarification of terms and meanings accompanies the delineation of the constitutive elements of the primary argument.

Molecular Biology (A New Biology?)

The term molecular biology and the authentic novelty of the field have been a subject of some debate among scientists and historians. Coined in 1938 by Warren Weaver, the director of the Rockefeller Foundation's natural science division, the term was intended to capture the essence of the Foundation's program: its emphasis on the ultimate minuteness of biological entities.[2] However, if by "molecular biology" one means the program sponsored by the Rockefeller Foundation, this definition would include many practitioners of life science (e.g., biophysics and immunochemistry) not typically associated with the term. Current parlance tends to equate molecular biology with DNA molecular genetics, a definition that would exclude most of the research in life science during the 1930s and 1940s. To complicate matters further, it appears that some scientists who would never have identified themselves as molecular biologists were retroactively, in light of the high status of molecular biology during the 1960s, all too eager to reconstruct their careers as part of the molecular biology success story. Others, especially biochemists, resisted the sweeping of their parent disciplines into the whirlwind of hybridization.[3]

In light of these divergent meanings of the term molecular biology, it is necessary to outline a set of criteria that together explain and justify use of the term in this study. Even though separately some of these structural features of molecular biology were not novel, assembled and amplified in a single program they eventually did constitute a coherent intellectual and institutional framework that departed sharply from traditional modes of biological research.

1. Programmatic statements about the newness of the new biology were not limited to its patrons. The new biology, as geneticist Thomas Hunt Morgan put it in 1928, stressed the unity of life phenomena common to all organisms rather than the diversity. Thus the new biology would concentrate, for example, on respiration or on reproduction as a central biological (in contradistinction to biochemical) problem, regardless of whether the object of study was a mammal or a bacterium.[4]

2. Based on this rationale, it became far more convenient to study fundamental vital phenomena on their minimalist levels. Thus the new biology increasingly employed simple biological systems—primarily bacteria and viruses—as phenomenological probes or as conceptual models (the line of thinking that led to

Jacques Monod's notorious dictum that what is true for the bacterium is true for the elephant).

3. By cleaving life processes from their host organisms, the molecular biology program aimed to discover general physicochemical laws governing vital phenomena. In so doing it distanced its concerns from emergent properties, from interactive processes occurring within higher organisms, between organisms (e.g., symbiosis), and between organisms and their environments, thus bracketing out of biological discourse a broad range of phenomena generally subsumed under the term "life." Concomitantly, this physicochemical approach bypassed historical explanations in biology and developmental and evolutionary accounts of life processes—the arrow of time. The new biology generally acknowledged only mechanisms of upward causation, ignoring the explanatory role of downward causation.

4. It would borrow methods not only from physics, mathematics, and chemistry but also from other fields of life science—genetics, embryology, physiology, immunology, microbiology. The new biology aimed to transcend disciplinary boundaries and employ whatever tools the problem at hand demanded. Although the transfer of techniques between fields was certainly not new, the design of a large-scale program based on interdisciplinary research encompassing several disciplines was unprecedented.

5. By defining life in terms of fundamental physicochemical mechanisms, molecular biology ultimately narrowed its principal focus to macromolecules; and until the mid-1950s it meant primarily the "giant protein molecules." Molecular biology was based on the protein paradigm, the premise that the salient features of life—reproduction, growth, neural function, immunity—could be explained through the structures and functions of proteins. In fact, guided by the protein paradigm, research on antibodies occupied a key position within the new biology. This important chapter, however, has been written out of the history of molecular biology.

6. Molecular biology thus defined the locus of life phenomena principally at the submicroscopic region between 10^{-6} and 10^{-7} cm. That this region was the main functional domain of the new biology had immense consequences for the form and content of research.

7. This domain could be investigated primarily with complex and sophisticated apparatus, specifically designed to investigate life at this range of dimensions. Whereas only a couple of decades earlier biology laboratories housed mainly microscopes, petri dishes, and autoclaves, the new biology laboratories displayed an imposing technological landscape. Electron microscopes, ultracentrifuges, electrophoresis, spectroscopy, x-ray diffraction, isotopes, and scintillation counters became the sine qua non of biological research.

8. The cognitive focus on the molecular level also shaped the social structure of research. Partly because of the interdisciplinary nature of the problems to be investigated and partly because of the complexities and costs of the new apparatus and the intricacies of the techniques, research problems were often defined by the instruments designed to examine them and were increasingly addressed as team projects. Molecular biology studies thus entailed structural changes in the organization of departments and laboratories and the prizing of cooperation as an institutional strategy and a personal ethos.

One important disciplinary consequence of these characteristics of molecular biology was the loosening of the traditional grip of medicine over biological research. Whereas in the past biological research was shaped by the extent of its service role to medical schools (and, to a lesser extent, agriculture), the new biology, which focused on fundamental physicochemical explanations, microorganisms, and submicrosocopic processes, had only indirect links to medicine.[5] As it turned out, the medical connection did affect the growth of molecular biology (even at Caltech) but in a convoluted and variable manner. During the 1930s, however, the designers of molecular biology carved for it a spacious, medicine-free niche within the disciplinary ecology of life science. They intended to create a new science of life, a science whose tributaries eventually converged on the molecular study of the gene.

Rockefeller Foundation: Knowledge and Cultural Hegemony

The cognitive and structural reconfigurations of molecular biology were greatly facilitated through the powerful resource base of the Rockefeller Foundation. During the years 1932–1959 the Foundation poured about $25 million into the molecular biology program in the United States, more than one-fourth of the Foundation's total spending for the biological sciences outside of medicine (including, from the early 1940s on, enormous sums for agriculture). Although there was a great deal of fluctuation in the annual expenditures for science during that 25-year period, the pre-World War II level of Rockefeller Foundation support for molecular biology amounted on average to about two percent of the entire federal budget for scientific research and development. This figure gains significance when we consider that the lion's share of government support for the life sciences went to agricultural research. When we also include the indirect effects of the Foundation's support for molecular biology in Europe and its massive support for biomedical research, the financial resources for molecular biology become even more impressive. It is clear that the Rockefeller Foundation was in a strong position to shape fields in life science that fell outside the scientific domain of the federal government.[6]

The Foundation's power to shape life science transcended the dollar amount of its investment; its effectiveness lay in creating and promoting institutional mechanisms of interdisciplinary cooperation through extensive systems of grants and fellowships, and in systematically fostering a project-oriented, technology-based biology. As the correspondence files, reports, and diaries of the officers reveal, during the 1930–1950s the Rockefeller projects became densely interwoven with the scientific agenda of those universities that were heavily supported through the Foundation's molecular biology program. The Foundation's network permeated their academic infrastructures; a significant number of Rockefeller trustees held top administrative positions in the universities. The Foundation officers cultivated advisers and contacts in nearly every discipline and had detailed knowledge of the academic traffic; the officers became central to what in effect became an informal peer review system. It was not uncommon for scientists and administrators to consult the Foundation's grapevine regarding academic appointments, reputations, personalities, travel, and potential projects. Weaver was even invited to sit in on faculty meetings; and within the limits of professional discretion and philanthropic etiquette, the officers were pleased to be closely involved on many interlocking levels of the scientific enterprise.

One of the strongest motives for redirecting academic practice and for creating institutional mechanisms compatible with the Foundation's design was the drive toward cooperation, a key feature in the molecular biology program. Cooperation did not mean just collaboration—the grass-roots, spontaneous activity of scientists sharing theoretical and experimental tools. Interdisciplinary cooperation, with its emphasis on group projects, was a long-range strategy and an overarching philosophy. Inaugurated in 1933, the molecular biology program was planned as part of a joint venture between the divisions of natural, medical, and social sciences. Within the natural science division, the molecular biology program itself was cooperative on several levels: among biological disciplines and between biology and the physical sciences. Again and again, as one reads through the Rockefeller reports, one is struck by the constant references to various projects as "the isotope group," "the protein group," or the "*Neurospora* group." One is also struck by the special attention to what the Foundation referred to as "cooperative individualists," men whose intellectual enterprise included a managerial temperament.

The term "cooperation" had an even broader meaning. It was more than an institutional strategy for fostering interdisciplinarity. As a modification of the extremes of laissez-faire, cooperation was a political and economic ideology of the evolving corporate structures of post-World War I America, specifically science, industry, and business. The reorganization of biology around the standard of cooperation reflected the broader reorganization of American science during the Great War and the restructuring of social relations in corporate America. Mirroring its industrial and business sponsors, the new scientific enterprise no longer extolled the virtuosity of the individual. Just as the multiunit business structures depended on the team player and the coordinating manager, so the new science relied on management and group projects directed toward interdisciplinary cooperation. There was a remarkable commensurability between levels: between the goals of the new program and the institutional structures that supported it, between ideology and form.

Similarly, the infusion of massive and sophisticated apparatus into biology must also be understood on several levels of significance. Instruments are not mere devices for discovering objective reality but complex processes of intervention for representing nature, processes that alter nearly all aspects of scientific practice. The new technologies not only raised the cost of research, they brought about structural changes. With large-scale commercial equipment almost unknown before World War II, new instruments often had to be constructed in situ. Workshops and special rooms had to be built to house the new equipment, thereby greatly expanding not only the budget but also the physical space of biological laboratories. These technologies also demanded a type of technical expertise that allied biology more closely with the physical sciences and engineering; this alliance too was part of the Foundation's goal.

More fundamentally, the reification of the molecular level as the essential locus of life, with the attendant reorientation of laboratory practice, altered the epistemological foundations of biological research, making the representation of life contingent upon technological intervention. Conceived at the twilight of an era characterized by its faith in technology and business, the design of the new biology not only reflected the particular bias of its principal architects—physicist Max Mason and mathematician Warren Weaver—but the more general bias of a technocratic elite who had dominated American culture during the 1920s. Simon Flexner, Rockefeller Foundation trustee and director of the Rockefeller Institute for Medical Research, in 1934 disapproved of the new program. Not only did he question the soundness of a biology managed by physicists, he doubted that a collection of instruments and techniques constituted a new biology. In ways that Flexner had not envisioned, the program did just that.[7]

The force of the Foundation's molecular biology program, and especially the effective management of Warren Weaver, have been amply acknowledged and debated. Caltech geneticist and Nobel Laureate George Wells Beadle observed that during the dozen years following 1953 (the elucidation of DNA structure) Nobel prizes were awarded to 18 scholars for research into the molecular biology of the gene, and all but one were either fully or partially sponsored by the Rockefeller Foundation under Weaver's guidance. Not only did Weaver help shape Beadle's scientific path, he influenced the careers of hundreds of others, inside and outside the field of molecular biology.[8]

Historians of science have offered divergent interpretations of the pervasive influence of the molecular biology program and Weaver's influential role. They have ranged from celebration to condemnation: The program was fruitful and innovative; the program was elitist and conservative, merely a process of technology transfer; it was a subversion of biological knowledge and of the academic order. All have concurred, however, that the program did have a profound impact on life science.[9] Although several works have examined the structures, mechanisms, and effects of the Foundation's molecular biology program, only scant attention has been paid to the broader intellectual and social agenda within which the program was nested. Little has been said about the inscribed cultural and ideological premises or the historical forces underlying the development of a molecular framework for the study of animate phenomena.

This book addresses this lacuna by situating the molecular biology program within the Rockefeller Foundation's "Science of Man" agenda, thereby viewing the program as both a scientific and a cultural enterprise. The motivation behind the enormous investment in the new agenda was to develop the human sciences as a comprehensive explanatory and applied framework of social control grounded in the natural, medical, and social sciences. Conceived during the late 1920s, the new agenda was articulated in terms of the contemporary technocratic discourse of human engineering, aiming toward an endpoint of restructuring human relations in congruence with the social framework of industrial capitalism. The support for life science must be seen within that larger investment in the human sciences. Within that agenda, the new biology (originally named "psychobiology") was erected on the bedrock of the physical sciences in order to rigorously explain and eventually control the fundamental mechanisms governing human behavior, placing a particularly strong emphasis on heredity.

This conjunction of cognitive and social goals had a strong historical connection to eugenics, to its promise and perils. By 1930 the Rockefeller Foundation had supported a number of eugenically directed projects. By the time of the inauguration of the "new science of man," however, the goal of social control through selective breeding had suffered severe setbacks. As an intellectual program, eugenics guided by the crude principles of Charles B. Davenport had lost much of its force; and as a social movement it carried the stigma of racial prejudice and political propaganda. Eugenics as such became a scientific liability. The quest for rationalized human reproduction, however, never quite lost its intuitive appeal (even when it was later modified by the Nazi experience). For the architects and champions of a science-based technological utopianism, human engineering through controlled breeding remained a compelling social vision.

Thus one of the subarguments of this book is that eugenic goals played a significant role in the conception and design of the molecular biology program, an argument that hinges on the politics of meaning. Precisely because the old eugenics had lost its scientific validity, a space was created for a new program that promised to place the study of human heredity and behavior on rigorous grounds. A concerted physicochemical attack on the gene was initiated at the moment in history when it became unacceptable to advocate social control based on crude eugenic principles and outmoded racial theories. The molecular biology program, through the study of simple biological systems and the analyses of protein structure, promised a surer, albeit much slower, way toward social planning based on sounder principles of eugenic selection. Time was seldom a deterrent for the visionaries of the Rockefeller Foundation. As Wickliffe Rose, head of the International Education Board, used to remind his pragmatic colleagues, "Remember we are not in a hurry." Moreover, as Raymond Fosdick, trustee and later president of the Rockefeller Foundation, acknowledged during the 1920s, in the quest for enlightened social control, "There is no royal road to the millennium, no short cut to the Promised Land."[10]

Indeed, support of the human sciences by the Rockefeller philanthropies did not begin during the early 1930s with their "Science of Man" agenda. By then they had supported biology for about 15 years and had been the main force behind the development of the social sciences in America. Scholars have generally concurred that the Rockefeller philanthropies have played a leading role in shaping the human sciences, and that their projects were in large measure an effort to build a base of technical expertise in order to lay a rational foundation for social reform. However, the meanings and interpretations of the Foundation's motives and its impact have diverged. The political significance of such terms as "technical expertise," "rational foundation," and "social reform" have been contested; and studies of the motivations behind the wholesale promotion of specific intellectual programs and institutional structures span a wide spectrum of approaches. They range from Marxian analyses to liberal perspectives, from arguments of economic determinism to claims of intellectual autonomy.[11]

Circumventing these interpretive dichotomies, this book examines the directed autonomy of leading practitioners of molecular biology sponsored by the Rockefeller program. We shall see how the social goals of the private sector—as individuals and collectively—interacted with the research goals of these life scientists in a mutually enhancing manner, as a nuanced process of consensus formation. The reference to the private sector encompasses industry, foundations, and individual donors (usually of industrial fortunes). My analysis does not place the Rockefeller philanthropies outside the ideological framework of the business ventures that gave rise to them. The corporate structure of the philanthropic enterprise mirrored the structure of the business corporation; and the visions of the Foundation's trustees, leaders of business and industry, reflected their ideologies and social world. True, legal distinctions such as tax laws and nonprofit status did differentiate these quasipublic institutions from their parent corporations, but they are proximate mechanistic causes.

Programmatic statements, reports, and memoranda from the various Rockefeller divisions attest that on social and ideological levels there was no fundamental separation of purpose between the heads of corporations and the leadership of the Foundation. Drawn primarily from the business sector, the Rockefeller trustees exerted their ideological and economic influence on general policies and on specific grants (trustees' approval of academic projects was a continuous concern for the Foundation's officers).[12] Animated by a potent conjunction of Protestant values and technocratic visions, the Foundation's civic missions were formulated within the dominant cultural categories of race, class, and gender, as well as within a socioeconomic framework that defined norm and deviance for individuals and groups. The Rockefeller philanthropies cultivated scientific and managerial elites in order to address the root causes of social dysfunction: culturally specific and historically contingent forms of maladjustment. Their projects aimed to restructure human relations and to develop social technologies commensurate with the material and ideological imperatives of industrial capitalism.

These observations are not Machiavellian attributions or pronouncements of academic subversion and cooptation. The complex set of relations of scientists to patrons and of intellectual programs to the social agenda can be better explained within an analytic framework of cultural hegemony: through the explicit and tacit constitutive processes of consensus formation. Within that framework, "power" includes intellectual, cultural, political, and economic power; and mental life is not a mere shadow of material life. From this perspective the maintenance of hegemony does not require active commitment by an academic constituency (or by subordinates) to legitimate elite rule. Rather, the two reinforce each other in a circular manner to form a "hegemonic bloc" sustained by formal and informal systems of incentives and power sharing, particularly through half-conscious modes of complicity. Hence this viewpoint does not regard hegemony as a form of subtle coercion or top–down social control but, rather, as an interactive process between different social groups vying for power.[13]

Within this theoretical framework it does not matter that many of the scientists funded through the Rockefeller Foundation's molecular biology program were unaware of their social function within the "Science of Man" agenda. As we shall see, the leaders of American life science, many of them acting as scientific advisers to the Foundation, did understand the larger picture; but even they did not always share all the goals of their patrons. The rise of the new biology was a process of consensus building among interdependent though not identical professional constituencies with common as well as separate goals. In search of patronage, however, most American leaders of pure science did argue for their service role. Their pleas and pledges often projected conflicting images and contradictory purposes: pure research as disinterested knowledge; research as investment in economic growth, human betterment, and political power; objectivity as a mark of professionalization; relevance as a measure of social worth; science as a democratic institution; science as an elite enterprise. Perhaps no scientific figure had embodied and reinforced these contradictions more pointedly than the German émigré Jacques Loeb, America's emblem of pure *wissenschaft*, who, while cleansing biology from the taint of pill-pushing, articulated his intellectual mission in terms of the technological control of life.[14]

Thus consensus can form without active complicity. Whether practitioners of molecular biology valued their works as contributions to a broad social agenda, or through perceived intellectual autonomy they deployed the appropriate rhetoric to fund their basic research, is secondary to the process of consensus formation. As the evidence shows, however, at the highest levels of management there was a substantial degree of agreement. As members of a national elite with similar social, religious, and educational backgrounds, the leaders of science and the Foundation's officers and trustees were cut from the same cultural cloth. This resonance facilitated the emergence of a hegemonic bloc. What is of primary importance, however, is that through their authority scientists did empower the Rockefeller program. By offering expertise they supplied an instrumental rationality that not only legitimated their own enterprise but also validated the cultural objectives of their patrons.

California Institute of Technology: Engineering and Consensus

No single institution exemplified this cultural resonance between patronage and science better than the California Institute of Technology. No other academic enterprise so clearly expressed the powerful conjunction of intellectual capital, economic resources, and institutional structures. The selection of an exemplary institution affords an opportunity for a detailed examination of these trends and operations, a close-up view that would be obstructed by a wider perspective on the rise of molecular biology. At the same time, there is a limit to extrapolating from the experience of one institution. Within these polarities of density and diffusion, the focus on Caltech represents a strategic optimization between the textures gained from a microstudy and the generality lost without macroanalysis, because on several counts Caltech serves as a paradigmatic case study of the rise of molecular biology. In a single center we can fruitfully examine the development of some of the major cognitive trends: the focus on physicochemical genetics, the emphasis on instruments, and the dominance of the protein paradigm, with its heavy emphasis on immunology. We are able to follow institutional and social mechanisms in action: group projects and interdisciplinary cooperation. To be sure, there were some hurdles and conflicts, but we generally witness the process of consensus formation on social goals, representations of life, and the future shape of biological research.

A cursory glance at Caltech during the 1930–1950s reveals that the Institute nurtured some of the most important "founding fathers" of American molecular biology. All supported by the Rockefeller Foundation, these future Nobel laureates built at Caltech influential research schools based on the cooperative model. The premier geneticist Thomas Hunt Morgan was engaged in 1928 to head the new biology division—and not only because of his Mendelian virtuosity on *Drosophila*. No stranger to business fortunes and foundations' politics, he was sought for his effective scientific management, which had propelled genetics from a disciplinary marginality to the vanguard. In his cooperative new division—comprising genetics, biochemistry, biophysics, embryology, and physiology— researchers were expected not only to build bridges between these biological fields but to forge links with physics and chemistry.

The German physicist Max Delbrück formed some of the earliest links between genetics, physics, and mathematics, establishing at Caltech during the late 1930s the foundations of the "phage school." This research program, which employed bacterial viruses as conceptual models of gene action, has been generally recognized as one of the most fruitful approaches to the gene problem and a principal turning point in the history of molecular biology. His intellectual leaps did not fully account for his ascendancy, however. When he received a professorship in biology at Caltech in 1946, it was only partly for his personal scientific output. His strength lay in having established a socially cohesive program and a collaborative network spanning several disciplines, institutions, and countries.

A similar vignette emerges around George W. Beadle. A brilliant geneticist in his own right and founder of American biochemical genetics, he demonstrated his interdisciplinary flexibility early in his career at Caltech. Moving from corn to fruit flies to fungi, by the early 1940s he was managing an impressive *Neurospora* project at Stanford. Groomed for leadership by the Rockefeller Foundation and by Caltech's establishment, Beadle returned to head the biology division after World War II with a track record of cooperative projects spanning university, industry, and the military.

The career of Linus Pauling displays a similar pattern. Director of Caltech's chemistry division and one of the principal architects of molecular biology, Pauling towered as both scientist and manager. His studies of the chemical bond and of protein structure revolutionized concepts of the architecture of living matter and were central to America's prominence in x-ray crystallography. Propelled by Dionysian forces far stronger than any of his colleagues', Pauling's intellectual ambition was reinforced by bold managerial maneuvers that placed him and Caltech at the forefront of Rockefeller support and of the production of molecular knowledge.

These research schools, in turn, attracted during the 1930s–1950s visiting scholars and collaborators from around the world. The 1930s to the mid-1940s were challenging times for Caltech's program, but the mid-1940s were a watershed. Delbrück predicted in 1947 that "Caltech in the coming years will be to biology what Manchester was to physics in the 1910s."[15] Indeed, like Manchester, Caltech became an international training ground, a nursery for a new science. Nearly every leading molecular biologist by the 1960s had had some connection to Caltech. Hundreds of young life scientists—graduate students, postdoctoral fellows, research associates—passed through the Kerckhoff and Gates Laboratories during the formative period of molecular biology, most of them supported by Rockefeller money. Propagating the new techniques and research strategies, this second generation spread the molecular vision of life to other research and teaching centers.

Caltech, of course, was not the sole center for the new biology. The developments at Caltech, however, become even more remarkable from a comparative perspective, when viewed in light of the Rockefeller Foundation's support of molecular biology at other universities. A survey of the Foundation's annual reports from 1930 to 1955 reveals that the Foundation supported molecular biology projects in scores of elite institutions but invested the largest sums in six. In accord with its long-standing policy of supporting the strong—"making the peaks higher"—the University of Chicago, Caltech, Stanford, Columbia, Harvard, and the University of Wisconsin most consistently received grants for various projects in molecular biology. Of these institutions, the Foundation regarded Chicago and Caltech (each receiving about $5 million) and Stanford (about $1 million) as (in that order) the most promising centers for developing a unified program of molecular biology. Among these three, Chicago presented "the most convincing case in this country and the most important program opportunity."[16] Columbia (about $1 million), Harvard, and Wisconsin (less than $500,000) occupied favorite positions within the Rockefeller academic network, but these pockets of excellence did not evolve into coherent programs in molecular biology.

Why did these institutions lag so far behind Caltech? To understand these developments we must not only count dollars (grants, after all, were a vote of confidence) but see the molecular biology projects within the ecology of knowledge—in terms of the goodness-of-fit between cognitive activities and their institutional contexts. From this perspective, the Wisconsin case is quite easy to explain. The university did enjoy special status—both Weaver and Foundation president Max Mason spent most of their academic lives there. Fields central to the molecular biology program did thrive at Wisconsin, especially biochemistry and biophysics, endocrinology, microbiology, and genetics. The dominant influence of the food and drug industries, however, tended to guide research projects in these fields primarily toward agricultural, veterinary, and medical processes and products. Harvard's case is straightforward as well. Biology at Harvard had always lived in the shadow of medicine. By 1940, with genetics at Harvard nearly extinct, the principal projects in physicochemical biology were conducted primarily in the medical school. In both places, Wisconsin and Harvard, a new biology could hardly develop an identity independent of its traditional service roles.[17]

The Columbia situation is somewhat more challenging to explain, as the university made significant contributions to molecular biology. In addition to work in protein chemistry, there was promising research under the direction of physicist Harold C. Urey on the biological effects of heavy hydrogen, the work on radioactive isotopes by Rudolph Schoenheimer, and of course Erwin Chargaff's groundbreaking studies of nucleic acids composition. This research, however, was conducted mainly in the biochemistry department at Columbia's College of Physicians and Surgeons, with little interaction with the zoology department or with the perspectives of genetics. Under the direction of geneticist L. C. Dunn, the zoology department (formerly towering with the stature of Morgan and E. B. Wilson) emerged from a state of partial eclipse only during the early 1940s. Zoology possessed neither the will nor the capacity for interdisciplinary partnership with the medical school. Rockefeller grants from the mid-1940s to the mid-1950s to the zoology department aimed to restore Columbia's eminence by supporting primarily mammalian and population genetics. These institutional and intellectual dynamics were unfavorable for fostering a unified molecular biology program.[18]

At Stanford, though a relatively young institution, two trends had solidified by 1930: the strong presence of the medical school, and the lack of interdisciplinary cooperation. Through the trustees' directives and the presidency of Ray L. Wilbur (former dean of the medical school and later Rockefeller Foundation trustee), medicine came to occupy a dominant fiscal and disciplinary position at Stanford. The School of Biological Sciences was medically oriented, including departments of anatomy, bacteriology, physiology, botany, and zoology (and the Hopkins Marine Laboratory); but until Beadle's appointment in 1937 there was no research in genetics. Interesting projects developed at Stanford: ultracentrifugation studies of macromolecules; research of bioelectric phenomena, radiation, developmental mechanics, bacterial chemistry, cell metabolism, cell biology, and viruses; and during the early 1940s biochemical genetics. For structural and personal reasons there was little common planning and collaboration. In fact the lack of cooperation between chemistry and biology was one of the motives behind Beadle's departure in 1946. The strong medical influence and the sluggish response to the calls for cooperation were detrimental to the emergence of a unified interdisciplinary program with a distinctive biological signature.[19]

The University of Chicago offers the most fruitful comparative perspective on Caltech and the other four universities. Its institutional network was densely intertwined with the Rockefeller hierarchy; and the Foundation's support of Chicago equaled that of Caltech. Through the energetic leadership of Frank R. Lillie (adviser to the Foundation and director of the Marine Biological Laboratory at Woods Hole), the Chicago program during the early 1930s exemplified both the tradition and the vanguard of American biology. In fact Lillie played a decisive role in shaping the intellectual design of the molecular biology program. With a large group working on the biology of sex, with luminaries such as Sewall Wright in genetics, F. C. Koch in biochemistry, Karl Lashley in psychobiology, and later biologist Paul Weiss, Chicago's life sciences were at their peak. Given the remarkable strength of the social sciences, and the strong interest in psychiatry at the medical school, the University of Chicago should have rightly emerged as the premier center for implementing a program that sought to address the genetic, developmental, and chemical processes of behavior in cooperation with the social and medical sciences.

Great traditions do have costs, however. They are grounded in history and tend to impede alternative paths. By the late 1930s a large number of prominent life scientists at Chicago reached retirement age. The nature of new appointments would hinge on administrative reshufflings. The division of biology had been integrated into the medical curriculum for three decades at that point, and its biochemistry, physiology, pharmacology, bacteriology, anatomy, and psychology departments were regarded as preclinical fields. Lillie's directorship had sustained the division's distinctive biological identity, but with the replacement of Lillie as dean of the division of biology by immunologist W. H. Taliaferro, the medical school came to have greater influence on biology during the 1940s, tending to orient it toward clinical research. Furthermore, Chicago had a strong tradition of evolutionary biology and natural history nested within the university's formidable intellectual framework of progressive evolution. The evolutionary standpoint was fundamentally incompatible with the mechanistic conception of life. That Chicago's genetics had a strong focus on populations and that under the leadership of Paul Weiss biological research addressed developmental and interactive processes from organismic and holistic perspectives attested to the viability of older traditions at Chicago, traditions that competed with the new disciplinary goals during the formative years of the molecular biology program.[20]

Caltech had none of these competing disciplinary traditions: no evolutionary biology or natural history, no agricultural mandate, and above all no medical school. Despite intermittent pressures from the 1920s to the early 1950s to add medical research to the life sciences, Morgan's original plan to resist the medical imperative prevailed (plant physiology and agricultural interests experienced similar tensions). Similarly, the strong presence of Theodosius Dobzhansky during the 1930s did not orient the division toward evolutionary biology. The new biology was implanted into a disciplinary matrix that diverged sharply from that of the other institutions. Caltech's biology program explicitly aimed to depart from established biological traditions in order to create a new science of life based on cooperation with the physical sciences and engineering; the engineering setting would be especially compatible with a technology-based biology. In terms of curriculum, new linkages were formed on the undergraduate and graduate levels, producing after a decade a generation of biologists trained in the physical sciences.

This redirection of biology was accomplished through Caltech's unique institutional structure. Unlike older universities, Caltech had already distinguished itself during the mid-1920s as an elite institution that championed new fields grounded in interdisciplinary cooperation. The appropriate institutional mechanisms were created from the start in order to encourage a problem-oriented approach to science, structures that were guided by corporate models of management and inspired by the cooperative projects of World War I. Although geographically peripheral and an academic neophyte, by the late 1920s Caltech formed the hub of America's scientific establishment. Its illustrious leadership glided smoothly through the corridors of power that linked academe with industry and the philanthropic foundations. These leaders also forged a formidable alliance with Southern California's business elite. Their scientific and ideological commitment to the region's industrialization situated Caltech squarely within the local political economy, thereby garnering enormous community support, which in turn stimulated the flow from foundations' coffers. This convergence of institutional and cognitive strategies in the absence of competing biological traditions, buttressed by enormous social and economic resources, explains why—of all the well-endowed universities—Caltech emerged as the premier center for molecular biology. At Caltech we witness a remarkable resonance of means and ends between scientists and their patrons, the formation of a consensus on scientific and social goals based on the primacy of the molecular vision of life.

Molecular Vision of Life

Now that the constitutive elements of the argument are explained, we are in a position to extract the levels of significance embedded in the primary thesis, "the molecular vision of life." Above all, we can explore the linkages between the form and content of biological knowledge and the directionality and modality of seeking knowledge. Viewed for the moment strictly on the cognitive level of science, the question arises whether life is molecular or it is only the vision that is molecular. By studying macromolecules, do we study the salient attributes of life, or only a molecular representation of life, one of many possible representations of animate nature? To put it differently, the title of this book indirectly addresses the epistemological tension between various knowledge claims competing for the appropriation of "the secret of life," interpretations that, in turn, give rise to divergent social and medical prescriptions.

Ultimately the resolution of this tension hinges on the currently unfashionable, question "What is life?"—a centuries-old query that, although by no means intellectually bankrupt, has proved too challenging for philosophers and is deemed unproductive by most practitioners of life science. My own point of departure is that there are multiple biological realities within which life is systematically explained. Various research schools in life science during either different or the same historical periods have privileged different scientific representations of life; with representations referring to the totality of biological practices that include the choice of biological system, methods, theoretical frameworks, and interpretations of results. Each program has labored under the conviction that its own representational practice grasps the essence of life.

Thus evolutionary biology, in addressing the quintessential property of living entities—their ability to evolve—has given primacy to time and environment as principal arbiters of the structures and functions of organisms, species, and populations (and more recently biomolecules). Various schools of ecology, though differing in their approaches, would agree on the premise that some of the most important processes subsumed under the term life cannot be properly understood outside the interactive models that account for profound reciprocal changes in organisms and their environments. On the organismic level, the followers of Walter Cannon, for example, would give primacy to homeostasis as a salient attribute of organismic life, thereby stressing holistic or integrative approaches to vital processes. Practitioners in these diverse biological fields might value molecular explanations but would surely disagree that molecular representations derived primarily from microorganisms supply an overarching explanatory framework for animate phenomena.

The disagreement has deep historical roots. Since the Aristotelian beginnings of the debate of teleology versus mechanism, the question of how much we can learn about life by examining only its building blocks has not been resolved. Mechanisms of upward causation have been remarkably effective for a finite range of biological phenomena but have not effectively explained some of the primary characteristics of life. Neither the Cartesian program, the cell theory, the supposed victory of mechanism over vitalism around 1850 (which was actually a confrontation between mechanism and teleology), nor the DNA double helix have successfully accounted for emergent properties of life such as differentiation, growth, evolution, or human consciousness. On the other hand, even for the subcellular level, mechanisms of downward causation have revealed properties and processes of cellular components that are not manifested in their unassembled state. Few would contest that the packaging of viruses or the hardware of bacteria offers only minimal insights into the biological organization of mammals. The abundance of rigorous quantitative antireductionist models that have developed during the second half of the twentieth century attests to the limits of the mechanistic and physicochemical approach for solving problems of biological organization.[21]

This broad spectrum of biological fields and the multiplicity of biological realities grounded in divergent conceptions of life make it clear that a number of viable biological programs existed that could have been singled out and promoted by the Rockefeller Foundation during the 1930s. The evolutionary, ecological, and organismic standpoints spotlighted many secrets of life to be unraveled. They supplied different kinds of knowledge about the human body and mind as well as alternative paths to understanding social and environmental maladjustments. In short, there were different possible human sciences. Why then did the Rockefeller Foundation's "Science of Man" agenda privilege a molecular vision of life? The answer to this question is embedded in the matrix that linked the particular forms of social control sought by that agenda with the specific kinds of control supplied by the new biology.

The control of animate and inanimate nature, of course, was not a twentieth century project. Viewed diachronically the interlocking of "representing" and "intervening" was inaugurated with the Baconian program, with the birth of an autonomous experimental tradition whose primary aim was to manipulate and control nature for the utility of man and to collapse the dichotomy of the natural and the artificial.[22] From a synchronic perspective, the particular political, economic, and social configurations in early twentieth century America gave this conjunction of knowing and doing historically specific meanings. Greatly influenced by Jacques Loeb's project, which had adopted the engineering standpoint toward the control of life, the Rockefeller Foundation officers and their scientific advisers sought to develop a mechanistic biology as the central element of a new science of man whose goal was social engineering. In the social sciences the emphasis on controlling human behavior gave rise to mechanistic conceptions of behavior, personality, and socialization; interventionist strategies were inscribed into these scientific formulations. The life sciences aimed to map the pathways in the molecular labyrinth of the human soma and psyche in order to control biological destiny.

A biology governed by faith in technology and in the ultimate power of upward causation is far more amenable to strategies of control than a science of downward causation, where elements cannot be fully understood apart from the whole. There is seductive empowerment in a scientific ideology in which the complexities of the highest levels can be fully controlled by mastering the simplicity of the lowest. The rise of molecular biology, then, represented the selection and promotion of a particular kind of science: one whose form and content best fitted with the wider, dominating patterns of knowing and doing. The molecular vision of life was an optimal match between technocratic visions of human engineering and representations of life grounded in technological intervention, a resonance between scientific imagination and social vision.

*   *   *

The organization of this book is primarily chronological, the narrative interweaving between projects of the biology and chemistry divisions.

The first chapter traces the origins of the science of social control, positioning the conception and design of the new biology within the Rockefeller philanthropies' promotion of the human sciences during the early part of the twentieth century. The second chapter focuses on the emergence of life science at Caltech, situating the new biology within the Rockefeller Foundation's program and within Southern California's social nexus and political economy. Chapter 3 explores the intellectual and institutional dynamics in Morgan's new biology division, with a particular focus on the ecology of the new biological knowledge.

Chapter 3 is followed by an Interlude, which traces Caltech's commitment to the protein paradigm back to the intellectual trends that had dominated genetics and biochemistry since the beginning of the twentieth century. The Interlude provides the necessary background for subsequent chapters, especially Chapter 4.

Chapter 4 explores some of the influential American researches in physiological genetics: that of Jack Schultz, George Beadle and Boris Ephrussi, and Max Delbrück. It focuses on the linkages originating at Caltech between genetics and embryology, biochemistry, and biophysics, which contributed to the primacy of proteins in the nascent molecular biology program.

Chapter 5 traces the emergence of the chemistry division under Pauling as a major center for molecular biology—a joint effort between Caltech, the Rockefeller Foundation, and Pasadena's business community. Chapter 6 examines the development of immunochemistry under Pauling as a key feature of the protein paradigm and a decisive advantage for war-related research. Chapter 7 reconstructs the intellectual and social history of Beadle's *Neurospora* program during World War II, activities that later took him to Caltech as head of the biology division.

A second Interlude surveys some of the effects of the dislocations of the war on the policies of the Rockefeller Foundation and the impact these shifts in power had on the molecular biology program at Caltech. The final chapter examines the consolidation of the protein-based molecular biology program under the leadership of Beadle and Pauling, placing Caltech at the vanguard of the nascent discipline. The epilogue recounts the paradigm shift from protein to DNA and raises questions of continuity and change in the intellectual and social goals of postwar molecular biology.

This book is not a microstudy of the American path to the double helix. On the contrary, the "wrong turns" have been given considerable attention in order to capture the cognitive thrust of a quarter of a century of the molecular biology program: the protein paradigm. Morover, these chapters are not, by any means, a history of biology at Caltech; important projects been left out, such as Dobzhansky's work in evolutionary biology and many of the activities in plant physiology and neurophysiology. The first two were peripheral to the Rockefeller program and the third, though programmatically central, for historical reasons played only a secondary role in the rise of molecular biology at Caltech.

More importantly, the narrative in this book is not "The Story" of the rise of the new biology, as there are other possible narratives, including the versions of those bracketed out of the consensus. Rather, it is "a story" told mostly from the select perspective of the "winners," who not only shaped historical events but, through their documentation projects influenced the historical reconstruction of these events. Like all record repositories, the archives of the Rockefeller Foundation and the California Institute of Technology (the two principal sources informing this study) carry the inscriptions of their own historical perspectives and biases. The following pages, then, provide a particular historical reconstruction of those activities that together shaped a distinctive kind of biology centered around the physicochemical study of the gene. It is a critical study of the story of the "winners" who joined intellectual and social forces, propelling Caltech to the vanguard of a new biology that came to dominate life science during the second half of the twentieth century.

Notes*

1. Raymond B. Fosdick, *The Story of the Rockefeller Foundation*, (New York: Harper & Brothers, 1952), pp. 156–189; R. E. Kohler, *Partners in Science: Foundations and Natural Scientists 1900–1945* (Chicago: University of Chicago Press, 1991); R. E. Kohler, "The Management of Science: The Experience of Warren Weaver and the Rockefeller Foundation Programme in Molecular Biology," *Minerva, 14* (1976), pp. 249–293; Edward J. Yoxen, "Giving Life a New Meaning: The Rise of the Molecular Biology Establishment," in N. Elias, H. Martins, and R. Whitly, eds., *Scientific Establishments and Hierarchies: Sociology of the Sciences, IV*, (Dordrecht: D. Reidel, 1982), pp. 123–143; Pnina Abir-am, "The Discourse of Physical Power and Biological Knowledge in the 1930s: A Reappraisal of the Rockefeller Foundation's 'Policy' in Molecular Biology," *Social Studies of Science, 12* (1982), pp. 341–382.

2. Warren Weaver, "Molecular Biology: Origins of the term," *Science, 170* (1970), pp. 591–592.

3. See for example, John Kendrew, "Some Remarks on the History of Molecular Biology," *Biochemical Society Symposia, 30* (1970), pp. 5–10; Seymour Cohen, "The Biochemical Origins of Molecular Biology: Introduction," *Trends in Biochemical Sciences, 9* (1984) pp. 334–336; Robert C. Olby, "Biochemical Origins of Molecular Biology: A Discussion," *TIBS, 11* (1986), pp. 303–305; Erwin Chargaff, "Preface to the Grammar of Biology," *Science, 172* (1971), pp. 637–642; Erwin Chargaff, "Building the Tower of Babel," *Nature, 248* (1974), pp. 776–779; Erwin Chargaff, *Heraclitean Fires: Sketches from a Life Before Nature* (New York: Rockefeller University Press, 1978).

---

*The key to coded archival sources appears at the end of the book.

4. Thomas H. Morgan, "Study and Research in Biology," *Bulletin of the California Institute of Technology, 36* (1928), p. 87.

5. Philip J. Pauly, "The Appearance of Academic Biology in Late-Nineteenth Century America," *Journal of the History of Biology, 17* (1984), pp. 369–397.

6. Warren Weaver, "A Quarter Century in the Natural Sciences," *Rockefeller Foundation Annual Report* (1958), pp. 28–34; and A. Hunter Dupree, *Science in the Federal Government* (Baltimore: Johns Hopkins University Press, 1957/1986), Chs. 17 and 18.

7. Flexner's criticisms were cited in Robert E. Kohler, "Warren Weaver and the Rockefeller Foundation Program in Molecular Biology: A Case Study in the Management of Science," in Nathan Reingold, ed., *The Sciences in the American Context: New Perspectives* (Washington, DC: Smithsonian Institution Press, 1979), pp. 236–270.

8. George W. Beadle, "Foreword," in Warren Weaver, *Science and Imagination* (New York: Basic Books, 1967), pp. vii-xiv.

9. See Note 1.

10. RAC, RG3.1, 900/Pro-38, Box 23.174, "Comments on Personalities Instrumental in Developing Original Programs and Policies of the RF," Alan Gregg, September 1945, p. 14; and R. B. Fosdick, *The Old Savage and the New Civilization* (New York: Charles Scribner's Sons, 1930), p. 187.

11. For a discussion on these debates see Barry D. Karl and Stanley M. Katz, "Foundations and Ruling Class Elites," *Daedalus, 20* (1987), pp. 1–40.

12. On assessments and prescriptions, RAC, RG3, 900/Org-31a, Box 19.141, "Memorandum on the Relation of the Trustees to the Policy and Control of the Rockefeller Foundation," Gerome D. Green, October 28, 1935.

13. The original concepts of cultural hegemony and the role of intellectuals in society were formulated by Antonio Gramsci, *Selection from the Prison Notebooks* (1929–1935), Q. Hoare and G. N. Smith, eds. and trans. (New York: International Publishers, 1971). For an elaboration of the Gramscian framework and its usage in American history see T. J. Jackson Lears, "The Concept of Cultural Hegemony: Problems and Possibilities," *American Historical Review, 90* (1985), pp. 567–393, and six papers on the scope and limits of Gramscian hegemony, "A Round Table: Labor, Historical Pessimism, and Hegemony," *Journal of American History, 75* (1988), pp. 115–162.

14. Loeb's popular image is embodied in Professor Max Gottlieb in Sinclair Lewis, *Arrowsmith* (New York: Harcourt, Brace, 1924). For interpretations see Charles Rosenberg, *No Other Gods* (Baltimore: Johns Hopkins University Press, 1976), pp. 123–132; and Philip J. Pauly, *Controlling Life: Jacques Loeb and the Engineering Ideal in Biology* (New York: Oxford University Press, 1987), Ch. 8.

15. CIT, Delbrück Papers, Box 1.7, Delbrück to Bohr, January 11, 1947. See also Robert H. Kargon, *Science in Victorian Manchester* (Baltimore: Johns Hopkins University Press, 1977), pp. 1–3.

16. RAC, RG1.1, 205D, Box 8.103, Grant to Stanford University Biology, April 5, 1939, p. 3.

17. RAC, RG1.1, 200D, Box 140.1725–27, 1936–1937, Harvard; Boxes 162.1994, 162.1997–98, and 164.2013, Wisconsin, 1934–1945; and Allocations in Experimental Biology, *Rockefeller Foundation Annual Reports*, 1933–1955. Charles Rosenberg, "Toward an Ecology of Knowledge: On Discipline, Context, and History," in Alexandra Oleson and John Voss, eds., *The Organization of Knowledge in Modern America* (Baltimore: Johns Hopkins University Press, 1976), pp. 440–455; Robert E. Kohler, *From Medical Chemistry to Biochemistry* (Cambridge: Cambridge University Press, 1982); and Pauly "The Emergence of Academic Biology," (see Note 5).

18. RAC, RG1.1, 200D, Boxes 130.1604, 130.1607–08, 131.1609, and 132.1630–31, support of Columbia's biochemistry and genetic projects, 1930s–1950s.

19. RAC, RG1.1, 205D, Boxes 8.103–108 and 9.122–126, support for Stanford biology 1930s–1950s. See also Roger L. Geiger, *To Advance Knowledge: The Growth of American Research Universities, 1900–1940* (New York: Oxford University Press, 1986), pp. 17, 89–90; and Kohler, *Partners in Science*, pp. 213–214 (see Note 1).

20. RAC, RG1.1, 216D, Box 8.103–113, support of biology projects at Chicago 1930s–1950. On the relation of Chicago to Rockefeller philanthropies see R. L. Gieger, *To Advance Knowledge,*

pp. 196–200 (see Note 19). On Chicago's intellectual traditions, see P. J. Pauly, *Controlling Life* (see Note 14).

21. For a thoughtful discussion on this issue see Timothy Lenoir, *The Strategy of Life: Teleology and Mechanics in Nineteenth-Century German Biology* (Chicago: University of Chicago Press, 1982), Introduction.

22. Ian Hacking, *Representing and Intervening* (Cambridge: Cambridge University Press, 1983), pp. 130–146; and Carolyn Merchant, *The Death of Nature* (San Francisco: Harper & Row, 1980), Chs. 7–9.

CHAPTER 1

"Social Control": Rockefeller Foundation's Agenda in the Human Sciences, 1913–1933

Salvation through Experts

In 1894 the young sociologist Edward Alsworth Ross rejoiced in his "great new discovery in Sociology." Having jotted down 33 ways in which society exercised social control, he proceeded to develop these preliminary thoughts into the organizing principle of sociology. A newcomer to the discipline from economics, Ross intended to resolve the social and political dilemmas of modernity confronting a society whose national self-conception of historical uniqueness had been formed early in the century as a reaction against forces of modernization sweeping the Old World. Through the concept of social control, Ross intended to empower sociology by supplying it with an analytical framework grounded in liberal premises and sociopsychological sciences. The series of articles published beginning in 1896 in the *American Journal of Sociology* and his book *Social Control* (1901) not only revolutionized the discipline of sociology but also shaped the other human sciences.[1] In the social sciences it had a profound impact on the direction of political science, economics, and psychology; and in the biological sciences it influenced the course of genetics, eugenics, and psychobiology. In its later technocratic forms, social control would play a critical role in the conception and design of the molecular biology program during the 1930s.

Developed in the ideologically charged context of the debate between capitalism and socialism, Ross's vision of social control was an argument for a new liberalism that accepted the inevitability of class conflict and the social inequality of capitalism. The idea of "social control" was premised on the fundamental dissonance between the interests of the individual and those of society. To maintain itself, society had to modify individual feelings, ideas, and behavior to conform to social interests; the formal and informal constitutive processes of exercised power—condign (force), compensatory (money), and conditioned power (persuasion), to use J. K. Galbraith's typology—Ross called social control. His prescriptive discourse, however, concerned itself mainly with the subtle mechanisms of conditioned power appropriate to a democratic republic.[2]

Soft and meditative in comparison with the more coercive and technocratic versions of the 1920s and 1930s, the ideal of social control was inspired by Spencerian naturalistic philosophy and conceptualized through the dominant cultural categories and racialist doctrines of the Gilded Age. That discursive framework valued private enterprise, inward temperament, morality, and self-mastery as the innate drives that vaulted the Protestant Anglo-Saxon elite to world dominance. Although Ross's earlier writings contained only passing references to race, by 1900 his fear of Mediterranean immigrants in America and the resultant social flux had forced the problem of "moral varieties" into sharp relief. With the rapid retreat of the frontier with its "pitiless sifting," Ross's formulation of social control sought to erect a framework for preserving the virtue of the Aryan stock.[3] Ross observed:

> There is reason to believe that even to-day difference in race psychology leads people of the same development to adopt different measures of control. The anthropologists now put Europeans into two great races—the tall, long-skulled blonds and the shorter, broad-skulled brunets. This distinction corresponds roughly to the old divisions of Aryan and Celto-Slav, or Germanic and Latin peoples. It is agreed that the former are more enterprising and variative than the latter. They conquer, and constitute the upper caste in most countries. . . . [They are by temperament] of an inward-looking self-analysing bent, in contrast to the outward-looking, sensuous people of the South, they can be reached by such illusions as "moral law," "conscience," and "duty," which install the reflecting self in the judgment seat of the soul. With them Protestantism and moral philosophy have real power because they corroborate certain inner experience.

Beyond racial classification and social theorizing, Ross's analysis articulated a service role for the new discipline of sociology, arming it with tools for social action. Thus in the conclusion to his book Ross pinned the success of social control on the scientific expert, the professional sociologist, who "will address himself to those who administer the moral capital of society—to teachers, clergymen, editors, lawmakers, and judges, who yield the instruments of control; to poets, artists, thinkers and educators, who guide the human caravans across the waste."[4]

Ross's concept of social control struck a resonant chord with practitioners of the social sciences, designating a new liberal reform program. Social control of individual self-interest meant public control of the private sector, though the categories "private" versus "public" could not always be neatly drawn. With the rise of quasipublic entities such as large business corporations and foundations, the boundary between individual and corporate self-interest, between private and public control, would be increasingly blurred. In sociology "social control" assumed a striking disciplinary identity, becoming a key concept in the field. Sociologists began to direct their attention to psychosocial processes of social control, a trend that gave rise to the term "socialization" as a measure of personal adaptation. It was this emphasis on social psychology that focused attention on the individual's development within society; and it was the elaboration of the framework of social control that later channeled research efforts toward personality and behavior. Outside sociology, the term was embraced by economists and political scientists and assumed a broad range of social and political goals.[5] It also served as a rationale for studies of personality and behavior, linking it to the biological sciences.

In the Progressive Era's activist climate, reform agendas were frequently articulated in terms of control. The broad spectrum of problems associated with rationalizing human relations in a changing social order—education, work, class, race, sex, family and groups, public opinion, religion, and law—were subsumed under the term social control. Social control became a meeting ground for mutually enhancing academic and social interests. While the social and biological sciences raised their national status through their instrumental rationality, by supplying theoretical frameworks for utilitarian ends, reformers seeking to develop and implement programs of social control increasingly capitalized on the resources and the rising authority of the social, biological, and physical sciences. During the Progressive Era science emerged as a symbol of reason and efficiency, the fountainhead of objective knowledge and industrial prowess, a euphemism for technological and social progress that unified the disparate crusades and reforms, the icon of America's Babbitry. The founding of the Carnegie Corporation in 1911, the Rockefeller Foundation and the Bureau of Social Hygiene in 1913, and the Laura Spelman Rockefeller Memorial in 1918 expressed Americans' faith in the efficacy of research and in a science-based social intervention. These institutions would harness the expertise of the human sciences to stem what was perceived as the nation's social and biological decay and help realize the vision of America's destiny.[6]

By the Progressive Era, the concept of the American destiny, its self-conception of uniqueness, or "American exceptionalism," had evolved from a resistance to modernity to an accommodation to industrial capitalism.[7] Under the leadership of Theodore Roosevelt, the United States entered the twentieth century as an aspiring empire with political interests stretching from Latin America to Asia. The large business corporations, epitomized by Rockefeller's Standard Oil and Carnegie's U. S. Steel, had become primary vehicles for exerting economic and cultural influence. These corporations were by then international enterprises with continuously expanding world markets.[8]

"History is becoming more and more the story of industrial development," thundered industrialist Frank A. Vanderlip, vice president of the Rockefeller-owned National City Bank of New York in an address titled "The Americanization of the World," delivered before the Commercial Club of Chicago in February 1902.

> A nation's strength is measured by its wealth. Its position in the world's progress by its relative commercial growth. . . . We are now not only a billion dollar country, but a country of billion dollar corporations. . . . I believe in the great corporation. I believe there is no more effective way for us to impress ourselves on the trade situation of the world than through these great industrial units that can project into the world's markets the strength of their commercial position

with irresistible force. . . . I have the firmest conviction in America's ultimate destiny. The twentieth century is America's century.[9]

As he spoke, American industrial productivity was outdistancing national consumption, and Rockefeller's Standard Oil was being shipped around the globe.

Beyond economics, the vision of America's destiny was grounded in perceptions of moral and social ascendancy. Rockefeller business and philanthropic enterprises epitomized the conjunction of ideology, wealth, and power. The doctrine of predestination and the Calvinist concept of "calling," coupled with the moral imperative to fulfill one's obligation in daily affairs, sanctified devotion to a life of labor. Wealth was a proof of virtue and a symbol of self-control and mastery over nature, a tool for social action based on private initiative. In his self-perceived duty to amass and dispense fortunes, John D. Rockefeller realized the Protestant notion of stewardship and its dialectic of servitude and liberation. The work ethic legitimized the critical link between wealth and virtue, paving the road to salvation. If these values were shared by all members of society, prosperity and social stability would follow. The reality of industrial capitalism, however, fell short of the vision. Demographic dislocations fractured community and family structures, consumerism eroded spiritual values, social and economic conflicts pitted capital against labor; factory work bore little relation to the promise of salvation. The Protestant business establishment confronted a labor force swollen with foreign elements and was challenged by social ills even greater than those in industrialized Europe.[10]

The early immigrations to the East Coast brought in predominantly northern and western Europeans, America's cultural cognates; but by 1890 waves of immigration reached the Atlantic shores from southern and eastern Europe, importing foreign cultures, alien ethos, and incongruent work habits. These huddled masses (more than 18 million arrived between 1890 and 1919) aggregated mainly in urban centers, infusing factories and sweatshops with abundant cheap labor. At the same time, poor working and living conditions turned many of these ethnic pockets into breeding grounds of medical and social pathologies, problems that were later compounded by the "Negro problem"—the effects of large-scale migration from the rural South to northern cities.[11] These social ills seemed to support the Anglo-Saxon anxiety over racial inferiority, backward temperament, and mental deficiency—and over the general deterioration of American society.

The academic concept of social control thus struck a resonant chord with the perceived experience of those who, to use Edward Ross's words, "administered the moral capital of society." Social control as a system of knowledge and as a discursive practice supplied the articulations for various social reform projects of the Rockefeller philanthropies. No social policy can ever be independent of history and culture. Ross's formulation of social control offered culturally specific and historically contingent prescriptions for social stability—notions premised on the imperatives of industrial capitalism and Nordic privilege. The various projects of the Rockefeller philanthropies aimed to develop social control grounded in cultural and behavioral norms derived from American Protestantism. By guiding action and conduct, the Protestant ethic, with its material correlates, implicitly informed a range of attitudes and articulations of self and other, the control of nature and nurture.

To extract the full significance of the Rockefeller philanthropies' commitment to social reform, their projects must be viewed on two interconnected levels of commensurability: the economic and the ideological. On the materialistic and utilitarian level, the projects in the social and biological sciences were intended to foster favorable conditions for raising economic productivity and managing social stability—making the world safe for private enterprise. On the level of consciousness and ideology, the reform programs were intended to combat vice, raise moral standards, and improve human conduct. The life of labor, the practice of self-control, and the drive for prosperity formed the essential elements along the spiritual-material continuum of a social intervention project based on Protestantism, republican principles, and industrial capitalism.

The leaders of the Rockefeller Foundation articulated the primacy of conduct, control, and moral value with great conviction. It is clear that they regarded medicine, education, and public health as part of a larger process of enculturation leading to social control and economic stability. In his 1913 policy outline for the Rockefeller Foundation, trustee Harry Pratt Judson, president of the University of Chicago and first choice candidate for the Foundation's presidency, grouped plans for the welfare of humanity into two categories: uncontroversial ones, which accorded with human wants (medicine and education), and challenging ones, which conflicted with human desire. For the second kind of activity, he stressed that "the real hope of ultimate security lies in reinforcing the police power of the state by training of the moral nature so painstaking and so widespread as to restrict these unsocial wants and substitute for them a reasonable self control." Listing the problems of social hygiene—drug addiction, alcoholism, and sexual promiscuity—he cautioned that scientific knowledge of these social aberrations had only limited value. Avoiding these ills required the "practice of self-control."

Although there were limits to the Foundation's power against these enormous forces of evil, Judson thought that it could be an important mission of the Foundation. "But it is to human conduct, in the end, and that as affecting a great mass of mankind, that appeal must be made if there is to be any material change in the social situation."[12] This project of human betterment would draw on the armamentaria of the social and biological sciences: on the expertise associated with social control.

The captains of business and industry, notably Carnegie and the Rockefellers, were generally receptive to the promotion of basic research, though not in pure academic form and not as an end in itself, but as an investment in social reform. They looked to the human sciences as a means of reaching the root causes of social dysfunction. Because scientific management had proved its value at the workplace, it could also be applied to the wider social world. By the turn of the century, the Rockefeller charity had given rise to the University of Chicago (1892), The Rockefeller Institute for Medical Research (1901), and the General Education Board (1903). The founding of these institutions reflected a strong commitment to social, educational, and scientific development as an investment in human welfare, at a total price of about $200 million. Similarly, $10 million in United States Steel

Corporation bonds were transferred in 1902 to the newly created Carnegie Institution of Washington, with a similar amount slated for education and research. These philanthropic offerings supported research as a scientific and a cultural enterprise. As an expression of utility and ideology, the philanthropies accommodated a resonance between science and religion. The same notions of calling, work, service, efficiency, organization, order, rationality, and scientific management that animated the business corporations also permeated their charitable ventures; the visions that propelled their profit-making also guided their ameliorative actions.[13]

Several areas in the biological sciences addressed fundamental processes related to mental attributes and conduct. Of these areas, eugenics carried considerable intellectual and social authority, promising effective means of social control. The rediscovery of Mendel's laws in 1900 and the subsequent research on Mendelian inheritance supplied the Spencerian doctrine and its racialist pillars with new theoretical foundations and experimental tools. In the gene the goal of selective breeding as a corrective to the perceived swamping and degeneration of the Anglo-Saxon stock acquired a precise target. Thomas H. Huxley's protoplasmic theory of life (ca. 1870s) had already reduced the physical and mental attributes of life to protoplasm. By the turn of the century the proteinaceous substance had already been enshrined as the source of biological diversity and the locus of material and cognitive control. The subsequent identification of the gene with protoplasmic endowment became not only a potent scientific concept but a compelling cultural image, a synonym for the physiological site of social control.[14]

After 1904, with the zealous leadership of Charles B. Davenport, the support of the Harriman fortune, and under the aegis of the Carnegie Institution of Washington, the eugenic movement acquired a strong resource base and a stable headquarters at Cold Spring Harbor for its large constituency, predominantly the upper social echelons: business, Protestant churches, the professions, and the intelligentsia, including most American geneticists. Unlike the eugenic preoccupations after World War II, which would be articulated primarily through medical discourse, the early promoters of eugenics expressed relatively little concern over medical disorders. Their primary disciplinary allies were the social sciences rather than medicine, and their principal targets were mental attributes: temperament, personality, and, above all, intelligence. Often formulated in terms of social control, eugenic projects addressed problems of social dysfunction by focusing on psychological traits, traits constructed primarily within the dominant historical categories of race and class.[15]

By the time of the launching of the molecular biology program, the Rockefeller philanthropies had considerable experience with eugenics. Although the Rockefeller philanthropies did not establish a eugenic program per se—largely to avoid duplication of the Carnegie effort—they did support eugenic projects, such as the sterilization campaign of the National Committee for Mental Hygiene to restrict the breeding of the feeble-minded. The Rockefeller philanthropies also acted in the area of eugenics through the Bureau of Social Hygiene (BSH) and the Laura Spelman Rockefeller Memorial (LSRM). The BSH was incorporated in 1913 for the purpose of "the study, amelioration, and prevention of those social conditions, crimes and diseases which adversely affect the well being of society, with special reference to prostitution and evils associated therewith." Through its 30-year history, many of the topics covered by the BSH, such as sex education, maternal health, birth control, venereal diseases, and population control, were distinctly eugenic in conception or impact, drawing upon Davenport's and Henry Goddard's hereditarian frameworks of social deviance. Similarly, the LSRM—incorporated in 1918 with the purposes of gaining understanding of social problems concerning women and children and developing means for social control on an international scale—was informed by eugenic doctrines and practices.[16]

Thus by the end of the Progressive Era, even before the large-scale commitment to the "advancement of knowledge" spurred by World War I, the human sciences received considerable support from the large foundations. Their numerous projects and the unprecedented scope of their financial and institutional resources shaped the development of culture and the production of knowledge in the United States. Through education, public opinion, stimulation of specific research agenda, and the promotion of selective categories of knowledge and research, the Foundation played a key role in the creation of a hegemonic bloc; the resources and prestige flowing into those fields relevant to problems of social control were instrumental in the formation of consensus between social and political elites, on the one hand, and academic interests on the other.[17]

Criticism during the 1910–1920 decade recorded the dawning perception of the foundations' grip on culture. During the period of challenge to industrial exploitation, Rockefeller's "robber baron" reputation continuously offended public opinion. The Standard Oil trust was in the terminal phase of a losing battle against dismemberment for violating antitrust regulations. The "Ludlow Massacre" in 1913 at the Rockefeller's Colorado Fuel and Iron Company implicated John D. Rockefeller, Jr., a director of the Colorado Fuel and Iron and soon to be president of the newly established Rockefeller Foundation, in ruthless policies against labor. The Walsh Commission identified the Foundation as a thinly disguised capitalistic manipulation of the social order and challenged "the wisdom of giving public sanction and approval to the spending of a huge fortune through such philanthropies as that of the Rockefeller Foundation."[18]

There were echoes of minor dissent even within the academic community. Although during the 1910–1920 decade the Rockefeller and Carnegie fortunes supported university research only indirectly, they had already played a decisive role in shaping scientific knowledge and in higher education.[19] James McKeen Cattell, editor of *Science*, leveled his criticism in 1917 against their pervasive influence.

> The Carnegie and Rockefeller Foundations have undertaken to dictate educational affairs all over the country and all the way from the primary school to the university. The fact that their large resources enable them to employ able men is a dangerous aspect of the situation. So many institutions are now subsidized by one or both of these foundations, that many educational leaders are not free to express their real opinion or are not in a position to form unprejudiced opinions.[20]

Edwin B. Wilson of the Massachusetts Institute of Technology expressed similar concerns in 1918. "Sometimes I think it would be well to go in with them [Carnegie Corporation] and sometimes I feel that there is no use of putting good money after bad in the hands of dishonest people."[21] These voices of dissent, however, did not represent the majority of academic scientists, who kept knocking on the foundations' gates and underscoring the relevance of their biological and social research to the welfare of mankind. The efforts of this group came to fruition after World War I.

Taming the Savage

World War I catapulted science to dominance within new configurations of power. From the preparedness period in 1916, throughout the war effort, and into the "normalcy" of 1920, academic science, in cooperation with revitalized business and industry, played a central role in America's rise to world supremacy. The reciprocal effects of science and war transformed both enterprises. On the one hand, through the large scale mobilization of science and the diverse cooperative war projects it spawned—wireless communication, submarine detection devices, chemical warfare, pharmaceuticals, blood banks, and mental testing—science permanently altered the nature of warfare and grew indispensable to it. On the other hand, these projects shaped the organization of scientific knowledge by placing a premium on interdisciplinary cooperation and on a liaison with industry and business.[22]

An organized scientific community emerged (notably, a strong academic and industrial chemistry community) self-conscious of its weight in the political arena. A powerful scientific leadership—including physicists Max Mason and Robert A. Millikan; astrophysicist George E. Hale; chemist Arthur A. Noyes; life scientists Frank R. Lillie, Thomas H. Morgan, and Simon Flexner; and social scientists Lewis M. Terman and Robert M. Yerkes—played up the social importance of fundamental research. Ideologically ill-disposed to government control, the leaders of American science lobbied successfully for a substantial increase in financial support of science by the private sector, notably the Carnegie Corporation and the Rockefeller Foundation. The establishment of the National Research Council (NRC) in 1918 and the Social Science Research Council (SSRC) in 1923—both strongly backed by the Rockefeller Foundation—signaled the beginning of a steady and close cooperation between academic science and the private sector.

During the 1920s the amplified symbiosis between academe and the foundations resulted in a 100-fold jump in research grants to universities. During the first half of the decade the foundations still tended to support science mainly in research institutions, for example, the biological sciences in the Marine Biological Laboratory at Woods Hole, the Phipps Institute, and the Food Research Institute. By 1925 this trend substantially broadened to include university research. Frederick Keppel, president of the Carnegie Corporation, noted that foundations, inspired by the NRC's effective management of cooperative war projects, were investing large sums in graduate and postgraduate research; he estimated no fewer than 1500 fellowships a year. Both parties now converged on the view that academic research was not an end in itself. Scientists convinced foundations' trustees—mainly America's business leaders—that research need not be mere ivory-tower indulgence but a socially responsive activity with enduring political and cultural returns.[23]

These shared interests were not purely utilitarian. Some of the salient cultural and ideological features of the 1920s were inscribed into this coalition of science and the private sector, notably the values derived from the conjunction of technocracy, business, and religion that guided the missions of the philanthropic foundations. Science and business forged their formidable alliance during a period marked by the triumph of conservatism, presided over by the business-oriented Republican administrations of Warren Harding, Calvin Coolidge, and Herbert Hoover. During this decade of commercial growth, technocratic expansion, and the sweep of mass media, Americans generally shed their earlier mistrust of business, deifying private enterprise as a moral force and social service, a deification based on a synergy between religion and business.[24] Scientific management entered the churches, transforming both their organization and their sermons. The contemporary literary genre that portrayed heros of the Old and New Testaments as savvy businessmen attested to the success of the business ethos at the pulpit.[25]

As the flourishing churches attested, the alliance proved that godliness was profitable. Books and articles observed that Protestant churches teemed with successful people, whereas the poor floundered outside the gates of virtue. Studies derived from *Who's Who* and surveys of successful businessmen—among them J. Ogden Armour, Chicago's meat-packing king; George F. Baker of the First National Bank; Elbert H. Gary of U. S. Steel; E. M. Statler, the hotel tycoon; and Samuel Insull, President of the Commonwealth Edison—verified the link between the Protestant ethos and material success. A series of advertisements listed some 25 major industries headed by devout church members, notably Standard Oil, Packard Motor Car Co., General Electric, the Pennsylvania Railroad, Firestone Tire and Rubber, Proctor and Gamble, and Sears, Roebuck. These executives typified the business aristocracy comprising the board of trustees of the large foundations, whose ideology determined the foundations' conservative mode. Both church leaders and business executives trumpeted similar orations: The Protestant ethic of thrift and industry wedded to Christian qualities of charity, integrity, ambition, enterprise, and self-control materialized in worldly success.[26]

This synergy between Christian concepts and business culture animated the leadership of the Rockefeller Foundation in the 1920s. During the 1920s and throughout the 1930s, the boards of trustees of the Rockefeller and Carnegie philanthropies were drawn primarily from the business and social elite of the Northeast and, increasingly, the Middle West. "Heavily weighted toward conservatism" the trustees (and officers), as self-appointed custodians of national symbols, believed that they represented what they termed, "the American temperament."[27] A statistical survey attempting to establish a typical profile of the foundation trustee concluded in 1936 that he belonged to the "higher income-receiving class of the population." He was "respectable" and "associated with men of prestige, power, and influence."

His "intelligence" is ranked high by various institutions of higher learning from whom he has received signal honors. He receives his income primarily from profits and fees. In short, he is a member of that successful and conservative class which came into prominence during the latter part of the nineteenth and early twentieth century, the class whose status is based primarily upon pecuniary success.[28]

Such material success, Owen D. Young, head of General Electric and trustee of the Rockefeller Foundation, assured the congregation of the Rockefeller-built Riverside Church in New York, was inspired by religious values. Honesty and integrity simply made sense; morality and intelligence formed the backbone of big business.[29]

This Christian spirit permeated the leadership of the Rockefeller Foundation throughout the 1920s. Alan Gregg, director of the medical sciences division recalled the hypnotic power of the Baptist minister Frederick T. Gates, a key figure in the Rockefeller hierarchy. He remembered Mr. Gates around 1924 shaking his fist at a sedate but respectfully attentive Board, shouting:

> And when you die and come to approach the Judgement of Almighty God what do you think He will demand of *you*—yes, each one of you? Do you for an instant presume to think He will inquire into your petty failures, your trivial sins, your paltry virtues? NO! He will ask you just One Question: "WHAT did you do as a Trustee of the Rockefeller Foundation?"

No one in the audience moved, no amused looks followed the volcanic sermonizing, according to Gregg.[30] The premise that the Protestant temper and Christian values would spawn global rewards was implicit in the domestic and international missions of the foundations. As one of the Rockefeller Foundation trustees lamented some years later, "When the Foundation was founded, our own basic philosophy, largely influenced by Christian conceptions, was taken for granted and it was also pretty generally assumed that it was only a question of time before it would extend throughout the world."[31] The developments in the following decades would increasingly challenge these premises, but the trustees' republican ideology, Christian ethic, and sense of destiny continued to inform the Foundation's missions into the 1950s.[32]

No single figure exemplified this synergy of Christian values and science-based social mission better than Raymond B. Fosdick, Rockefeller trustee, counselor to J. D. Rockefeller, Jr., and later president of the Foundation. Raised in Buffalo, New York, during the 1880s in a home devoted to Protestantism, Raymond Fosdick and his prominent brother Harry Emerson Fosdick (Rockefeller trustee and pastor of Riverside Church) were suffused with the Calvinist ethos. Both later tempered their extreme evangelical training but retained the deep sense of calling, duty, and predestination in a modernized form, accommodating both science and social action. Whereas brother Harry fulfilled his calling through the social gospel of the ministry, reaching millions throughout America, Raymond ministered to society through law and social science.[33]

His childhood's political slumber violently shaken by the assassination in Buffalo of his hero President McKinley, Raymond Fosdick's commitment to social action eventually led him from a successful legal practice to the Bureau of Social Hygiene. There he distinguished himself during the 1920s through his studies of European and American police systems, authoritarian manifestations of the ideal of social

Figure 1 Raymond B. Fosdick, undated but probably ca. 1920s. Courtesy of the Rockefeller Archive Center.

control. After having played an active role in the war mobilization effort, he became an avid promoter of internationalism and a champion of the League of Nations.[34] The tributaries flowing through his various projects, speeches, and books converged as an anxious torrent over the future of Western civilization, underscoring the urgency for the success of social control measures that would subdue the irrational forces of technology, which he thought were subverting democracy. His writings during the late 1920s are a jeremiad on American society, a society symbolized as the old savage in a new civilization—"a naked Polynesian parading in top hat and spats"—a socially primitive culture enamored with gadgets it fails to comprehend, unable to control its play with dangerous technology.[35]

With the carnage of World War I still vivid—10 million dead soldiers, 13 million dead civilians, and millions wounded and devastated—Fosdick had little tolerance for the backlash of narrow nationalism and virulent xenophobia. Paradoxically, during the era of increased technological sophistication and cosmopolitanism, a regressive panic swept the country, often driven by eugenic arguments. The "big Red scare" produced a hysterical hunt for anarchists and Bolsheviks, and antipathy to the un-Americanism of Catholics, Jews, and organized labor. The Ku Klux Klan was revived, anti-Japanese hysteria raged in California, Henry Ford launched his notorious anti-Semitic campaign, and Ivy League colleges instituted quotas to restrict admission of Jews. A new edition of Madison Grant's *Passing of the Great Race* sold widely, as did William McDougal's *Is America Safe for Democracy*, questioning whether democracy would survive the mongrelization of a Nordic civilization. Between 1920 and 1923, Lathrop Stoddard's *The Rising Tide of Color Against World-Supremacy* went through 14 editions, and the movie industry contributed potent racist imagery to popular culture.[36]

As trustee of a foundation representing an interconnected world system with the United States at its center, an organization with economic and cultural interests around the globe, Fosdick opposed many of these trends, blaming them on democracy. He abhorred the nationalistic zeal and the populist sentiment behind it, loathed the mob mentality and its naive obsession with gadgetry, and deprecated the rampant Baalist consumerism with its orgy of self-destruction oblivious to civilization's spiritual decay. Democracy, Fosdick thought, amplified these social ills. His Nietzschean diatribe on democracy denounced "The pack instinct for solidarity [which] is reinforced by the enthusiasm of democracy for leveling human expression and imposing the measures of mediocrity. Democracy is the apotheosis of the commonplace, the glorification of the divine average." On the other side of the ideological divide hung the dread of the Hegelian fallacy: socialism. The state as collective good existing above the individual threatened the ethos of private initiative.[37]

Like other leaders in business and academia, Fosdick understood the exigencies of history. The 1920s business culture of cooperation, with its dependence on teamwork and management, could no longer subscribe to naive nineteenth-century laissez-faire or to the rugged individualism of a frontier society any more than it could surrender to Bolshevik oppression. Somewhere in the scheme of things, Fosdick believed, there had to be a place for social intervention based on neither populist prejudice and mob rule nor totalitarian coercion, but on principles of excellence, rationality, and science—the kind of system that would beget an Aristotle, a Newton, or a Darwin.[38] The role of the Rockefeller Foundation was to supply the resources and mechanisms for such a system of excellence, through processes of top–down democracy; science was its reservoir of theoretical knowledge and technical expertise.

Not all branches of science were equally relevant to the pressing problem of social control, however. During the 1920s, under the leadership of Wickliffe Rose, the International Educational Board directed its resources to the physical sciences— grants for the rehabilitation of European research centers and travel fellowships for chemists, physicists, and mathematicians. Fosdick, however, was critical. Guided in part by Frederick Soddy's fear of premature use of nuclear fission, Fosdick prophesied doom if the physical sciences proceeded unchecked and the savage was left unrestrained. Despite R. A. Millikan's vociferous objections, and even before the 1929 stock market crash and the renewed calls for a "science holiday," Fosdick questioned the wisdom of promoting pure physical knowledge. As a key figure in the 1928 reorganization of the Rockefeller philanthropies, his priorities would carry enormous weight. Fosdick's judgment that overinvestment in the physical sciences merely accelerated the pace of runaway technology and that underinvestment in the human sciences amplified the atavistic social response became the guiding principle in the Foundation's singular commitment to support the human sciences during the 1930s.[39]

Fosdick's "old savage" suffered from what Chicago sociologist William F. Ogburn defined as "cultural lag." Adviser to the Rockefeller Foundation and later member of President Hoover's influential Committee on Social Trends in the United States, Ogburn identified some of the alarming symptoms of social dysfunction—

growing divorce rates, delinquency, crime, mental deficiency, personality difficulties, immigrant assimilation, prostitution, alcoholism, and job instability—as
manifestations of cultural lag. They were described as large-scale maladjustments
due to society's inability to adapt to the dislocations of technological change.[40]
These problems would be addressed through a science of social control.

"We see the abyss upon the edge of which the race is standing. We see the
inevitable doom that lies ahead," Fosdick thundered, "unless we can achieve a
measure of social control far greater than any which we have hitherto exercised."
Pure knowledge was no guarantee of sanity and rationality: Germany's rarefied
wissenschaften nearly annihilated Western civilization. The purpose of knowledge
is to understand and control destructive trends. There was an urgent need for "the
same kind of fearless engineering in the social field that in the realm of physical
science has pushed out so widely the boundaries of human understanding." The
human sciences—biology, eugenics, and the social sciences—held the reins for
taming the savage.[41]

Fosdick's invocations of social control reflected the pervasiveness of this conceptual framework. During the two decades since its introduction by Ross, the idea
had been adopted in diverse academic spheres. The thesis acquired greater scientificity, higher resolution, and more specific technical formulations derived from
engineering, eugenics, physiology, psychology, statistics, sociology, and the mass
media. Social control attained a particularly strong expression in areas of human
engineering, in the new field of behaviorism, and, most significantly, in sociology,
where the emphasis on behavior and the mission of scientism combined to stimulate
highly technocratic formulations of social control.

The term "human engineering" came into general use around 1910. Inspired by
the rising disciplinary and social status of the engineering profession, the ideal of
human engineering initially spread primarily through the works of "welfare secretaries"—social workers affiliated with industries. As an elaboration of Taylorism,
the function of welfare secretaries was to help implement industrial relations programs and work design based on the ideal of cooperation. Midway during the 1910–
1920 decade the term "human engineering" came to denote the application of
scientific principles and technical methods to social and educational processes associated with the maintenance of social order: stable families, work groups, and
rational management of changing sexual and racial relations.[42] In academic circles,
human engineering gained immense scientific legitimacy through research in psychobiology, owing a great deal to the intellectual clout and institutional authority
of Robert M. Yerkes who, with the support of the Rockefeller Foundation, developed models of human engineering based on primatology. These studies gave
primacy to "personality." According to Donna Haraway, "personality" became a
prime object for students of human engineering "because it was central to the two
key levers for psychobiology as a technology of power over labor: the worker and
the family . . . [personality] was an anchor for the control of expanding bodies
through the control of work and sex."[43] The reduction of human engineering to a
strategy of domination of capital over labor tends to underrate the diversity and
complexities of goals and practices behind personality studies. Historical studies
make it clear, however, that the emphasis on personality, especially its more tech-

nical formulations in the study of behavior, was explicitly linked to the science of social control.

By the 1920s behaviorism became an influential trend in academic psychology, its applications ranging from child conditioning to industrial relations. Inspired by the Loebian ideal of biological control, John B. Watson's "Behaviorist Manifesto" (1913) promoted a new psychology whose "theoretical goal is the prediction and control of behavior."[44] Neither social scientists nor social workers could ignore the potent combination of hard experimental science, social utility, and utopian ideology. Many of the problems of "cultural lag"—the slothful adaptation of people and institutions to the technological imperative—were conceptualized during the 1920s as problems of "human behavior." "Behavior" replaced traditional terms such as "conduct," "action," and various mentalistic terms by more aggressive masculine language and empirical-technological currencies. Behavior became during the 1920s the lingua franca of the social sciences.[45]

The Rockefeller philanthropies played a pivotal role in shaping and promoting these trends during the 1920s, through massive funding, vigorous institution building, and the energetic leadership of former Chicago psychologist Beardsley Ruml. Ruml's two-tier strategy for lifting the social sciences up the Comtean scale was intended to prod their philosophical musings from armchair lethargy into the laboratory of social control. He thus intended to mold them after the natural sciences in what Clifford Geertz has critiqued as the quest for a social physics.[46] To be sure, Ruml merely epitomized and empowered the growing trend of scientism during the 1920s, which aimed to endow the human sciences with the power of prediction and control through the quantification of human behavior. As one prominent participant-observer remarked in 1925:

> The primary emphasis in the social sciences now falls on behavior. . . . [These sciences] are seeking both their differentiation from one another and also their cooperation in behavioristic terms. . . . This exaltation of behavior . . . really amounts to nothing less than a rather striking and genuine intellectual revolution . . . [there was an] important shift in scientific interest and emphasis, a shift from understanding to control . . . from knowledge, from the search for truth . . . to management, direction, betterment, greater effectiveness. . . .

The "kingdom of behavior" had arrived.[47]

Between 1922 and 1929, the LSRM under Ruml and the Social Science Research Council allocated about $41 million to American social sciences, social work, and their institutions. The University of Chicago, with its emphasis on empirical methods, became both a model and a target for lavish support. The Rockefeller-supported institutions shaped research agenda in the human sciences by playing an important role in determining not only "what" should be studied but "how" it should be studied.[48] As Dorothy Ross has argued, these developments "stilled most qualms and plunged the profession wholesale into empirical research, from which they hoped a basic science of social control would emerge."[49]

By the mid-1920s social control formed the dominant paradigm in the human sciences. Thus the important text *Means of Social Control* (1925) by F. E. Lumley attempted to synthesize the accumulated knowledge on the subject. Building on the sociological works of Ross, William G. Sumner, John W. Burgess, and Robert E. Park, among others, the text acknowledged the insufficiency of definitions of "social control" but explained that in familiar language, social control meant getting others to do, believe, think, feel, as we wished them to, "using the term 'we' to stand for any authority who can have his way with others."

> Social control has usually meant that kind of life-patterns, which a government, through its officers, imposes upon the citizen. But we have seen that social control means vastly more than that. We might speak of it as the practice of putting forth directive stimuli or wish-patterns, their accurate transmission to, and adoption by, others whether voluntarily or involuntarily. In short, it is effective will-transference. Ideally social control would be in the hands and the interests of the inclusive group whatever it is; practically, however, it is in the hands of, and often in the interests of, some few members who have usurped power and know how to use it. A little reflection will show that all social problems are ultimately problems of social control—capital and labor, prostitution, taxes, crimes, international relations.[50]

This comprehensive elaboration of the concept included techniques of social control that referred not only to symbol organizations or institutional mechanisms but also to innate and acquired human behavior, placing great weight on psychology and biological modes of causation.

The contributions of biology to social control required reassessment and revision, however. Up until the 1920s eugenics was generally viewed as an explanatory framework of many social phenomena. By the 1920s, however, eugenics (both theory and practice) had become problematical, causing considerable unease among its patrons. Fosdick's sentiments and the Rockefeller Foundation's stand on the subject are illuminating for their lack of resolution, reflecting the ambiguous status of eugenics. On the one hand, Fosdick's position revealed a retreat from what only a decade earlier had been a clear panacea; on the other hand, the lure of rationally controlled breeding persisted. Like most educated persons of his time—social commentators, educators, legislators, and scientists—Fosdick strongly believed in the general veracity and applicability of eugenics; in fact he was still on the advisory board of the American Eugenic Association.[51] Invoking the classical Malthusian argument, he appealed throughout the 1920s to rational control based on eugenic principles.

> Can the conscious effort of men in any way steer this biological evolution? Is it possible to adjust the size of population to fit the world's resources so that those who inhabit the earth can do so in seemliness and dignity? Can we shift the emphasis from quantity of human life to quality of human of life? Can the science of eugenics reshape a process that is tumbling with such gigantic forces?[52]

Fosdick did not underestimate the opposing forces, however. He admitted to the somewhat utopian quality of this vision and tempered his expectations with a conservative disclaimer. He had no "illusions as to the speed or ease with which mankind can alter its way of life. There is no royal road to the millennium, no short cut to the Promised Land."[53]

Without a doubt, by the late 1920s the status of eugenics as a bona fide science was on the decline. Many geneticists realized that the efficacy of applying Mendelian genetics to human breeding in order to obtain precise and lasting modifications within a couple of generations was grossly overrated. By that time there was ample evidence that some genes were pleiotropic (simultaneously influencing several traits) and decisive proof that many traits were polygenic (determined by the action of several genes). These findings had seriously challenged the concept of unit characters—that a single gene determined a phenotypic character. Even strict hereditarians realized that complex traits, such as different temperaments and gradations of intelligence, could not be easily outbred or inbred, as they were determined by complex interactions of genes. The gene itself was largely an abstraction. American geneticists had succeeded in mapping the gross structural mechanisms of hereditary transmission, but physical knowledge of the gene and its physiological mode of action was meager. These realizations placed cognitive limits on the extravagant scientific claims of eugenics.[54]

These cognitive constraints, however, did not challenge the fundamental premise of eugenics—the desirability and eventual ability of selectively controlling human reproduction. The paucity of knowledge about physical mechanisms of gene action represented merely a delay in understanding the basis of intervention; lack of knowledge meant a need for fundamental research. The implicit belief in unit characters, even among life scientists, persisted well into the 1940s, along with the intuitive expectations of eugenic intervention.[55] As high school and college textbooks of the 1930s reveal, eugenic doctrines still formed an integral part of biology, social sciences, and education.[56]

Certainly eugenics as a social and political force showed no sign of decline during the 1920s. In fact, the scientific status of eugenics suffered from its political visibility. Eugenics' reputation as a science—as an objective and neutral pursuit—was tarnished during the 1920s by the polemics surrounding eugenic sterilization of the unfit and the debates on race-based immigration restrictions. Geneticists who only a decade earlier participated actively in academic eugenic projects now shied away from associated political entanglements. They largely avoided the congressional hearings leading to the 1924 Johnson Immigration Act and the debates that culminated in 1927 in the notorious *Buck* v. *Bell* ruling in favor of sterilization, though geneticists did not protest these misuses of their science. The polemical style of extremist groups, such as Paul Popenoe's Southern California circle (and later the Human Betterment Foundation in Pasadena) threatened to discredit the reputation of a young aspiring profession, linking it to racial prejudice and political propaganda. For geneticists of a leftist persuasion, among them Hermann J. Muller, John B. S. Haldane, and Julian Huxley, American eugenics symbolized the barbaric forces of racism and capitalism. Their own formulations of selective breeding ostensibly aimed to bypass the evils of class and race biases.[57]

The grip of eugenics was also loosened by the growing authority of the social sciences. Their own framework of social control increasingly offered quantitative explanations of urban problems and social dysfunction (especially immigrant populations) that helped tilt the balance away from racially based determinism of deviance and from viewing slums, poverty, and crime purely as biological categories.

The influence of Franz Boas's school of cultural anthropology also had considerable impact in terms of shifting explanations of human relations from race to culture. In psychology, both Freudianism and behaviorism offered nonherediterian explanations of personality and behavior, though the empirically driven and practically minded behaviorism proved particularly attractive to projects of social engineering. That in the 1920s an absolute cleavage between nature and nurture no longer dichotomized the study of human behavior was evidenced, for example, in the interwar works of Robert M. Yerkes and Edward L. Thorndike. Their eugenically informed studies on individual differences, personality, mental capacities, measurements of intelligence, group psychology, and behavioral norms displayed a power of synthesis that broadened the insurgence of culture into the kingdom of nature, at least to a point of symbiosis.[58]

This diminished stature of eugenics during the mid-1920s—its perceived social promise and political taint, its cognitive limits and intuitive appeal—surfaced in the mixed attitudes of the Rockefeller Foundation. The term "eugenics" became loaded with politics of meaning. The Foundation rejected explicit eugenic projects while at the same time implicitly endorsing eugenic goals and eugenically informed programs. Thus the zealous campaign of Edwin R. Embree (director of the Foundation's Division of Studies) to develop a human biology that integrated eugenics with the social sciences generated unease among the trustees, finally dying an honorable death in 1927 in the front lines of the policy battle.[59] Yet the personal letter to his supporter Ray Lyman Wilbur (Rockefeller trustee and president of Stanford University), in which Embree shared his belated lessons, captured the Foundation's implicit attitude toward eugenics.

> A characteristic of race biology or human biology is that the terms are new. This probably means not necessarily that entirely fresh subjects are to be created but that different groupings of present sciences are involved. . . . The thing which so often discredits new subjects is a sloppy disregard of common accuracy and careful scholarship. This has cursed much of the work in eugenics and mental hygiene and racial studies. So great is the resulting stigma that many scientists refuse to have anything to do with these fields of research. As a matter of fact, there is nothing wrong with the subject; they're more significant than most of the things scientific scholars devote their lives to. The thing that is wrong is the poor work by poor brains that too often has replaced ignorance simply by the confusion of pseudo-science. [60]

As Embree saw it, the works of Raymond Pearl, T. H. Morgan, E. G. Conklin, and F. R. Lillie exemplified the handful of projects that "stand out like peaks above the dismal swamps of general mediocrity and muddle."[61]

Thus by 1930 it was primarily the stigma of American eugenics as pseudoscience and political propaganda, not its potential capabilities, that dissuaded even sympathizers such as Fosdick from seeking social control through large-scale eugenic projects.[62] He seemed to do so with some regret, however, assailing the foes of progress who opposed "the field of eugenics in an attempt to breed a better race."[63] Given its intellectual muddle and political tarnish, Davenport-style eugenics had become an academic liability, its scope further reduced by the massive development of the social sciences. Nevertheless, even in its more limited form, eugenics retained its authority among the educated elite. Throughout the 1930s social control through selective breeding based on scientific principles remained a compelling utopian vision.

Toward a New Science of Man

On January 4, 1929, the *New York Times* announced the largest philanthropic merger in history. "Rockefellers Unite Two Philanthropies—Foundation and Laura Spelman Memorial Are Merged With Wider Program—Assets Are $264,602,447— Sum is Greatest Ever Placed in a Single Endowment Fund of its Kind—Scope of Activities Broad—It Will Include Work in Natural, Social and Medical Sciences and the Humanities."[64] The merger, coinciding with Herbert Hoover's ascent to the White House, projected a social agenda congruent with the plans of the Great Engineer. Hoover intended to plunge the nation into social programs of a scale and urgency comparable to wartime mobilization. As champion of business and a friend of science, he sought to develop these projects of social control by strengthening the cooperation between the two constituencies. His inaugural address, "New Day," declared crime as the nation's greatest danger, promising the growing underclass a future bright with hope.[65] A new era empowered by rational planning seemed to be dawning. The Rockefeller Foundation entered a state of transition, absorbing various agencies that had been founded over the previous two decades. Raymond Fosdick, chairman of the reorganization committee, hailed the merger as a move toward closer programmatic cooperation and greater unity of purpose in addressing social problems.

Indeed, by 1928 the Foundation's civic and academic projects and those of its various boards had overlapped widely with the work of the LSRM. Under the vigorous directorship of Ruml, the LSRM had broadened its range of activities and researches well beyond the original aims of public health and emergency relief to include the social sciences, welfare, family study, education, race relations, and public administration—projects that drew on eugenics and the social sciences. The LSRM promoted and sponsored a great deal of research, but not as an end unto itself; grants were issued for demonstrating concrete applications of newly gained knowledge.[66]

At the time of the merger the Rockefeller Foundation took over the program of the natural sciences from the International Education Board and appointed physicist Max Mason, outgoing president of the University of Chicago, as the interim director of the new Natural Sciences Division and as president of the Foundation in 1929. Though Mason's social background fit closely with the foundation's leadership, the presidential appointment of a scientist (like Hoover's election to the national presidency) set a precedent. Neither the nation nor the Foundation had had a physical scientist for a president. Raised during the 1880s in Madison, Wisconsin in a household inspired more by business acumen and technological invention than by religion, Mason rose to national prominence through the practice and management of physics. His distinctions in mathematics at Göttingen and as a physics professor at the University of Wisconsin and his decisive

Figure 2 Max Mason, undated but probably ca.
1930s. Courtesy of the
Rockefeller Archive Center.

contributions to war projects (notably submarine detection) and to the NRC vaulted
him into the political arena of science.[67]

By the time of his arrival at the Rockefeller Foundation, Mason's concerns had
evolved closer to those of Fosdick; he had developed "a consuming interest in
behavioral research and particularly in the possibility that the physical sciences,
working closely with and through the biological sciences, could shed new and
revealing light on the normal and abnormal behavior of individuals, and ultimately
on the social behavior of groups of men."[68] He spent the first couple of years
orienting himself within the wide terrain of the biological and social sciences,
canvassing potential projects and research institutions, and convening with national
leaders. A visit to the White House at Hoover's invitation (three weeks before the
stock market crash) generated a survey of strategies to address social problems, a
project engaging wide professional networks that linked Hoover's influential Committee on Social Trends, the National Research Fund, and the Rockefeller Foundation to projected scientific research during the 1930s.[69]

During the interregnum period, the Rockefeller Foundation articulated its broad
objective—"the advancement of knowledge"—and proclaimed research as its chief
tool. In October 1930, at their meeting at Princeton, the Foundation trustees
decided to abandon Wickliffe Rose's policy of supporting any "best science" and
reversed his commitment to "pure science" with its emphasis on physics, chemistry,
and mathematics. "The advance of knowledge" would be "the sailing directions
given to the officers by the board," Mason stated, who proceeded to offer a broad
definition of the term, stressing the social utility of research.

> In fundamental facts there must be research in the narrow sense; but advancement
> of knowledge demands also interest in educational processes . . . knowledge is
> gained by applying; and sanity and value brought to research.

After discussion, the trustees unanimously concurred that research devoid of social relevance would become a barren process.[70] They announced that the Foundation would be far more effective if it concentrated its vast resources and targeted them toward research that lent itself to applications in areas of human relations, a field they broadly defined as the "Science of Man."

The agenda for the biological sciences was still uncertain in 1930, reflecting the mixed attitudes of the Foundation toward eugenics and toward its recent incarnations as race and human biology. The case of Frank R. Lillie of the University of Chicago clearly illustrates this ambivalence. Before the founding of Caltech's biology program under geneticist T. H. Morgan, a proposal for an Institute for Racial Biology under Lillie closely matched the Foundation's goals in quality and scope. It was a proposal for developing biological fields relevant to social control. A prolific researcher and adviser to the Foundation, Lillie had recommended in 1924 a broad approach to genetics, beyond the heredity work "so brilliantly developed by T. H. Morgan," including the physiology of reproduction, biology of sex, development, and experimental evolution. Lillie proposed to develop these topics within an Institute for Racial Biology as a means of addressing the social-biological problems of the race; the goals and rationale were formulated through tacit invocations of eugenics.[71]

Language counts. Admittedly, the choice of name (borrowed from the German *Rassenbiologie*) was unfortunate. Both Lillie and Mason agreed in 1931 that "Institute of Genetic Biology" would not be open to misunderstanding. Mirroring the Foundation's own ambivalence, Lillie's implicit eugenic plan explicitly skirted the stigma of eugenics, stressing that his organization "should be kept free from all propaganda concerning eugenics, birth-control, etc.; and in such connections aim merely to furnish the indispensable scientific foundations on which social prophylaxis of the future must depend."[72] Favorably disposed, Mason promised to acquaint the new director for natural science with Lillie's proposal.[73] That during the interim (1929–1933) Lillie's program received lavish support and that Morgan's program at Caltech was initially funded under the same rubric of "human biology and physiological psychology" attest to the perceived unity of purpose between the two programs and to their congruence with the Foundation's broad agenda in the human sciences.[74]

During the fall of 1931 Mason called Warren Weaver, his Wisconsin protégé, to join him in developing the new biology program at the Rockefeller Foundation. By then Weaver's and Mason's history of collaboration reached back more than 15 years, an intellectual lineage enhanced by their common heritage. Like most of his midwestern contemporaries at the turn of the century, Weaver's social life in the small town of Reedsburg, Wisconsin had centered around the church. Growing up as a son of an entrepreneurial druggist, in a home where the virtues of business and "engineering" harmonized with prayer and duty, Weaver's *weltanschauung* matured within a nexus of scientific and Presbyterian rationality. Like Mason, he

Figure 3 Warren Weaver (*center*) as director of the Natural Science Division (undated). Courtesy of the Rockefeller Archive Center.

studied and taught physics and mathematics at the University of Wisconsin. During the years 1910–1920, majoring in civil engineering, young Weaver shone as Mason's favorite pupil, frequently assisting him in teaching and research.[75] Weaver worshipped his mentor, admiring his unbelievable talent for anything to which he turned.

> He was, in my judgment, an absolutely superb teacher. His command of formal mathematical technique was powerful and effortless. He could be exquisitely precise, but he could also accomplish imaginative leaps around and over difficulties. He had a great and lasting influence on a large number of graduate students. . . . I cannot conceal and will not try to conceal the fact that Max was the most brilliant person and at the same time the gayest and most attractive companion I have ever known.[76]

Mason, in turn, groomed Weaver for leadership in science.

Soon after Weaver completed the equivalent of graduate work in mathematics at Wisconsin, Mason steered him toward R. A. Millikan, Mason's friend at Caltech (then Throop College). The years of teaching physics and mathematics at Caltech (1917–1920) were a formative period in Weaver's career and the Institute's ascent; both demonstrated their academic promise. Just out of school and newly wed, the impressionable Weaver basked in Caltech's *gemütlichkeit* and idyllic sunny setting. Frequent trips to the desert and mountains relieved the pressures of research and teaching; the conservative milieu of California's wealthiest families graced academic life with Pacific gentility. The privilege was enhanced by attending Pasadena's luxurious Presbyterian church, where the Scottish brogue of the "truly wonderful" sermons of the charismatic Robert Freeman amplified the intuitive resonance between the work ethic, "the style and temperament of Presbyterianism," "good blood," and material rewards.[77] The Edenic sojourn was truncated in 1920, however, when Mason summoned Weaver back to Wisconsin as his chief collaborator. Millikan, fond of Weaver, reluctantly released him, stipulating that he remain on a permanent leave of absence. Throughout his tenure as officer of the Rockefeller Foundation, Weaver remained officially a faculty member at Caltech.[78]

Of all of Mason's and Weaver's collaborations during the 1920s, *The Electromagnetic Field*, a classic graduate text, testified to Mason's imprint on Weaver's career and to the profound intellectual bond that eventually led to their collective commitment to the new physicochemical biology.[79] Extremely conservative intellectually, Mason applied his mathematical dexterity mainly to problems of classical physics. His intense passion for this subject was matched only by an aversion to the new quantum theory, its concepts of acausality and indeterminacy threatening to undermine a natural order grounded in law and rationality. "As to quantum theory," Weaver recalled,

"his [Mason's] attitude was more than mere avoidance or disregard: he actively disliked the subject, and considered that it was so unpleasantly messy, so full of internal contradiction, and so clearly headed in the wrong direction that he would have nothing to do with it.[80]

Fully sharing his mentor's convictions, Weaver regarded the new quantum physics as a "flash in the pan," reason gone astray.[81] Other collaborations in classical physics followed. A joint study on the sedimentation and diffusion of small particles in a fluid and subsequent papers by Weaver offered significant mathematical tools for the design of the analytical ultracentrifuge of Theodor Svedberg of Uppsala, a visiting professor of colloidal chemistry at the University of Wisconsin in 1923.[82] The Uppsala connection and the analytical ultracentrifuge (a sophisticated device that sorted and sized macromolecules) would be critical to the growth of the new biology into the 1950s. The preservation of classical monuments and the application of deterministic concepts and mechanistic techniques to "backward" provinces of science reflected both the disillusionment and hope behind Weaver's and Mason's collaborations.

Though by no means path-breaking or prolific, Weaver's steady mathematical output ensured his academic future, adding to the stability and contentment of his family life in Madison. During the late fall of 1931, however, his midwestern roots were severed once again after a phone call from New York City. Mason's offer of the directorship of the natural sciences division and a chance to build a new biology on the bedrock of the physical sciences meant giving up an academic career. At the same time, this formidable challenge presented Weaver with a novel identity: that of a scientific manager, a uniquely American professional niche, capitalizing on entrepreneurial skills while tolerant of intellectual limits.[83] As Weaver reflected,

I was both realistic and accurate about my abilities and my limitations. I loved to teach, and knew that I had been successful at it. I had a good capacity for assimilating information, something of a knack for organizing, an ability to work with people, a zest for exposition, and enthusiasm that helped to advance my ideas. But I lacked that strange and wonderful creative spark that makes a good researcher. Thus I realized that there was a definite ceiling on my possibilities as a mathematics professor. Indeed, I think I realized that I was already about as far up in that profession as I was likely to go. So this offer opened whole new possibilities for me.[84]

This remarkably modest self-appraisal, the opportunity to rehabilitate the biological sciences, the call to social service, and the loyalty to Mason guided Weaver's decision to join the newly reorganized Rockefeller Foundation.

The timing of his appointment might have seemed inauspicious. When Weaver assumed the helm of the natural sciences division at the twilight of Hoover's presidency, the Great Depression had already devastated the nation. Since the Rockefeller merger had taken place, 100,000 businesses had failed, corporate profits had fallen 10-fold, and the gross national product had been slashed in half. About one-fourth of the labor force—13 million workers—were idle, while millions clutched onto temporary employment. The soup kitchens and breadlines of the Salvation Army and Red Cross could minister only feeble relief from the spread of malnutrition and disease. Urban jungles sprouted, racial conflict intensified, and bands of homeless roamed the countryside. The erosion of social structures further undermined the efforts to institute and enforce law and order.

We are witnessing a disarmament of the forces of law and order and social control at the very height of the engagement against crime and social deterioration. Large numbers of persons have learned to live by their wits and by their daring in defiance of law. In thousands of cases the incentive to thrift and law abidance seems to have been lost. Hordes of youth and men roam the country without objectives or visible means of support. Our railroad systems are glutted with them, and from the railroad lines they slink off to beg or to steal. Men of advanced years have been reduced to want, and know that they will never again be able to reestablish themselves.

So sounded the alarm bells of Rockefeller's Spelman Fund, calling for intensifying Hoover's war on crime.[85]

The trauma of the Depression catalyzed complex national reactions, unleashing widespread soul-searching and reappraisals of the work ethic, private enterprise, and the equation of prosperity with biological and social fitness. True, blacks and the unskilled lost their jobs first, with whites and managerial personnel being the last to go; joblessness and shiftlessness ranked highest at the lowest echelons. Eventually, though, the Depression claimed a toll among the poor and rich, the slothful and industrious, the godless and the devout. The cloak of benevolence fell from big business, stimulating bolder flirtations with communism and socialism. The stained hand of business tainted its spiritual ally, the Protestant churches. The downward spiral in church attendance paralleled a shift to the Left, and many church leaders and ministers vented their anticapitalist sentiments from the pulpit and in print.[86]

The erosion of the coalition of Protestantism and business damaged their scientific allies. Scientists were trapped in their contradictory claims to purity and utility; as partners of business and industry they were seen by the public as accomplices in avarice. Having lent their expertise to the greed of excessive production, they helped unleash technological unemployment and runaway inflation, thus violating the covenant of prosperity and liberation. Instead of human welfare, their research was regarded as a source of diabolical technologies that spread poverty and misery. Plummeting science budgets and a call for a "science holiday" undermined the prestige of science. As the Depression lingered, a significant number of politicized leaders of science, prodded by leftist British colleagues and inspired by the "Soviet experiment" of planned science, urged a reassessment of the national role of science toward greater social responsibility. The backlash against science, however, struck mainly at the inanimate sciences, whose labors fueled the engines of production. The animate sciences, on the other hand—medicine, biology, and the social sciences—not only escaped the stigma of business and industry but gained favor as custodians of human welfare: nurturing, healing, and socially responsive.[87] The launching of the Rockefeller Foundation's new deal for biology as part of its commitment to the human sciences could not have come at a more auspicious time.

The program was inaugurated in 1933. During April 1933 the trustees and officers of the Rockefeller Foundation held a special meeting at the exclusive Westchester Country Club in Rye, New York to plan future programs. Fosdick's imprint was striking. The Foundation's "Science of Man" agenda, by then fully articulated, echoed Fosdick's 1920s prescriptions for taming the old savage.

> Science has made significant progress in the analysis and control of inanimate forces, but science had not made equal advances in the more delicate, more difficult and more important problem of the analysis and control of animate forces. This indicates the desirability of greatly increasing emphasis on biology and psychology, and upon those special developments in mathematics, physics and chemistry which are themselves fundamental to biology and psychology.[88]

The past 100 years marked the supremacy of physics and chemistry, the trustees observed, but the hope for the future of mankind rested in the development of a new biology and new psychology.

> The challenge of this situation is obvious. Can man gain an intelligent control of his own power? Can we develop so sound and extensive a genetics that we can hope to breed, in the future, superior men? Can we obtain enough knowledge of the physiology and psychobiology of sex so that man can bring this pervasive, highly important, and dangerous aspect of life under rational control? Can we unravel the tangled problem of the endocrine glands, and develop, before it is too late, a therapy for the whole hideous range of mental and physical disorders which result from glandular disturbances? Can we solve the mysteries of the various vitamins so that we can nurture a race sufficiently healthy and resistant? Can we release psychology from its present confusion and ineffectiveness and shape it into a tool which every man can use every day? Can man acquire enough knowledge of his own vital processes so that we can hope to rationalize human behavior? Can we, in short, create a new science of man?

This new science of man would be based on current and future theories and techniques of social control. Many scientists, philosophers, and statesmen shared this conviction and possessed the necessary techniques, the trustees noted; but stimulation, support, and leadership were lacking. The Foundation's goal was to provide the leadership and resources for this coherent and strategic attack on the problems of human behavior.[89]

Mason underscored the wisdom of concentration and the urgency of selecting "fields of critical importance; fields in which the procedures could capitalize on the rich heritage of the Foundation's long experience." In promoting the biological sciences, the officers and members of the board drew on a robust tradition, explaining that "in the Foundation, biology is an interest running back nearly 15 years."[90] The "salients" of the concentration, Mason announced,

> are directed to the general problem of human behavior, with the aim of control through understanding. The Social sciences, for example, will concern themselves with the rationalization of social control; the Medical and Natural sciences propose a closely coordinated study of the sciences which underlie personal understanding and personal control. Many procedures will be explicitly co-operative between divisions. The Medical and Natural Sciences will, through psychiatry and psychobiology, have a strong interest in the problems of mental disease.[91]

Christened "psychobiology," the new cooperative plan originally emerged as a single MS–NS (medical science–natural science) unit, parceled into nine research lots:

1. Psychobiology (psychiatry, neurophysiology)
2. Internal secretions (hormones, enzymes, etc.)
3. Nutrition (vitamins, etc.)
4. Radiation effects
5. Biology of sex
6. Experimental and chemical embryology
7. Genetics
8. General physiology (cell physiology, nerve conduction, etc.)
9. Biophysics and biochemistry

a design bearing a remarkable resemblance to Lillie's proposed "Institute for Genetic Biology."[92]

The Foundation no longer endowed institutes during the 1930s, thus eventually disposing of Lillie's grand blueprint; but there is no ground for assuming that his visions were ignored. His plans, which were referred to as "the spearhead of the [Foundation's] thrust," had the officer's ear for over a decade, and his advisory role in the Foundation shaped the design of the psychobiology program. Although the new biology explicitly excluded the old eugenics, it nevertheless intended to address a similar range of issues with greater breadth and rigor. The fields slated for support would be developed in order to supply biological explanations of the processes underlying human relations, particularly hereditary mechanisms.[93]

Thus in preparation for designing the new cooperative biology program Weaver spent the month of July at the 1934 SSRC Hanover conference on human relations.

These annual conferences, established a decade earlier to provide an exchange forum for creating a social vision, traditionally focused on problems of social control and social engineering.[94] As in the past, the deliberations of practitioners from the social sciences, humanities, education, and the arts (and a few life scientists) centered on the urgent problem of "personality needs" and on "recasting of prevailing ideas and conceptions of human nature and conduct." These conceptions of personality and behavior were constructed within the dominant cultural and intellectual frameworks of the time, tying together genetics, physiology, medicine, psychology, anthropology, sociology, and political science into the service of the managerial sector and its cultural armamentarium. Accordingly, Weaver grouped the salient lessons on human relations into five categories that bore direct relevance to biological research: biological maturation, major features of American culture, development of personality, personal and social problems, and world views. The age-old query "How are things inherited?" led the list of two-score items, spanning physiology, sexuality, intelligence, and mental attributes. The conjunction of heredity and behavior was implicit in Weaver's outline.[95]

The new biology and the new science of man were long-term projects. Even the Foundation's officers did not view the cooperative venture as a union between "consenting adults" but more as an arranged liaison of partners at different stages of maturity. Practitioners in the natural sciences, Weaver stressed repeatedly, had already evolved the proper frame of mind and research ethos: experimentation, quantification, and objectivity. Methodology was in place, and it was just a matter of stimulating and guiding research interests in prescribed directions. The medical sciences, however, were still plagued by second-rate work and lack of proper appreciation for research, especially for genetics and behavior; and of all medical fields, psychiatry was seen to be trapped deepest in the prescientific stratum. Under the leadership of recently appointed director Alan Gregg, the medical sciences could be mobilized toward the rigorous study of human behavior. Gregg's strong eugenic instincts and enthusiasm for psychogenetics were intended to sensitize psychiatry to the hereditary elements of human potential.[96]

Despite their drive toward scientificity, Weaver viewed the social sciences as even more primitive than psychiatry. Descriptive and qualitative, they caused the interdisciplinary partnership considerable embarrassment. "In the social sciences, the institutional program must not only modify the attitude toward research, but must also modify the attitude as to what constitutes research," Weaver complained.[97] The need for rehabilitation was particularly urgent in psychology, "which should receive important emphasis in the program," but presented "something of a dilemma," according to Weaver. Presumably still tied to the apron strings of moral philosophy, psychology would have to be intensively restructured before it could assume its central role in the Rockefeller program. Accordingly, the Division of Social Sciences, under its new director Edmund E. Day, aspired to surpass its precursor LSRM in both style and content. Even beyond Ruml, Day strove to inculcate the social sciences with professionalism, the proper research ethos, and the scientific habit of mind that would accelerate it toward a science of social control. As Day saw it, "the validation of the findings of social science [must be] through effective social control," with human engineering as the endpoint of research.[98]

When discussing the "advancement of knowledge" and formulating the "new science of man," the trustees left no doubt as to the relation between pure and applied research. Research merely led to publication, which had at best restricted possibilities, they believed.

> It is open to question whether the welfare of mankind can more wisely be served by more knowledge or by better dissemination and thorough application of existing knowledge. . . . Indeed we would strongly advocate a shift of emphasis in favor not only of the dissemination of knowledge, but of the practical application of knowledge in fields where human need is great and opportunity is real.

The trustees worried that mere accumulation of facts, untested by practical application, was in danger of becoming a substitute rather than a basis for collective action. As practical men of affairs imbued with social purpose, they believed that under the influence of the scientific method scholarship tended to become obsessed with collecting facts for their own sake while disregarding the lessons implied by the facts.[99]

In 1933 the trustees agreed to return to the point of view of the LSRM, which, they stressed, aptly expressed their own. "The Memorial had no interest in the promotion of scientific research as an end in itself; its motive was not sheer curiosity as to how various human and social phenomena came to be and are; the interest in science was an interest in one means to an end."[100] Their insistence on a tight link between intellectual curiosity and practical goals guided the Foundation's policy not only through the exigencies of the Depression but throughout the 1930s and beyond World War II. Citing that precedent in 1939, Joseph H. Willits, former head of the Wharton School of Business and the newly appointed director of the Social Science Division, warned Fosdick (then president of the Rockefeller Foundation) that it would be a mistake to identify the Foundation's interests entirely with those of the university and research world, where projects tended to be overly theoretical.[101] As director of Natural Sciences, Weaver would constantly need to negotiate the purpose and substance of biological projects with four congruent, though distinct constituencies: the trustees, the medical sciences, the social sciences, and his own. Even during the 1940s, as his confidence rose along with the clout and credential of the molecular biology program, Weaver continued to shuttle between laboratory and the boardroom, repairing and forging links between scientific imagination and political realities.

During the next few years Weaver's program was modified, reflecting changes in intellectual and institutional dynamics. It increased its conceptual and structural autonomy by establishing a greater distance from the medical sciences, shedding along the way the outmoded category "internal secretions" and deemphasizing the rhetorics of sex research. The program's name changed from "Psychobiology," to "Vital Processes," to "Experimental Biology." Finally, in 1938 Weaver unveiled the name "Molecular Biology," defining the term as "subcellular biology" and the "biology of molecules." When analogizing the new biology to modern physics, Weaver compared the divisibility of the cell (the "older" unit of analysis) into subcellular units to the divisibility of atoms to subatomic units. The two new sciences, both predicated on powerful atomic and molecular probes, grounded their

Table 1-1. Rockefeller Foundation Expenditures in Natural Science Fields, 1933–1938

Category		%*
1. Application of physical techniques to biology		
a. Math., physical investigations of tissues, cell, molecules	$175,904	2.5
b. Spectroscopy and biology	168,786	2.4
c. Isotopes and cyclotron	309,604	4.3
d. Physical chemistry and biology	356,137	5.1
e. Organic chemistry and biology	635,794	9.0
f. General biochemistry	369,375	5.2
g. General biophysics	141,290	2.0
Subtotal "vital processes"	$2,156,890	30.5
2. General physiology	651,052	9.3
3. Nutrition	290,153	4.1
4. Genetics	422,365	6.0
5. Embryology	131,026	1.9
6. Endocrinology	326,248	4.6
7. Radiation effects (NRC)	234,340	3.3
8. Sex research (NRC)	530,620	7.5
Subtotal (2–8)	2,585,804	36.7
9. Aid to groups	2,305,485	32.8
Subtotal concentrated program (1–9)	7,048,179	100.0
10. Exceptions to program	245,835	
11. General program	677,500	
12. Old program	1,956,159	
Total (9–12)	$9,927,673	

*Percentages figured on total of items 1–9.

Modifed from R. E. Kohler, "Warren Weaver and the Rockefeller Foundation program in molecular biology." In N. Reingold (ed.), *The Sciences in the American Context*. Washington, DC: Smithsonian Institution Press, 1979. Reproduced with permission.

strength in the study of the "ultimate littleness of things." Their control over nature was to be derived from manipulating miniaturized bits of matter.[102]

How were these miniaturized bits of animate matter to tie to human behavior? How did Weaver's interdisciplinary scheme differ from the research efforts in psychiatry, the medical spearhead of "Science of Man"? "Our objective is broader than psychiatry," Weaver explained in 1935,

> The direction of our activity results from the fact that we are attempting to sponsor "the application of experimental procedures to the study of the organization and reactions of living matter." We have chosen this activity because of a conviction that such studies will in time lay the (only?) sure foundation for the understanding and rationalization of human behavior. In this ultimate goal we merge closely with the ultimate purpose of the MS [medical science] program. These statements

define the *direction* of our activities. The *narrowness* of our aim, within this general direction, is set by the specific definition of the recognized sub-fields.[103]

Thus the thematic narrowness of the new biology program is deceptive when taken out of its broader context of the human sciences. This molecularization of life was to be applied to the study of genetic and epigenetic aspects of human behavior. Based on the faith in the power of upward causation to explain life, Weaver and his colleagues saw the program as the surest foundation for a fundamental understanding of the human soma and psyche—and ultimately as the path to rational social control.

The select subfields which defined the new biology program were targeted for enormous grants (Table 1-1), and a number of American research centers became primary sites for these developments, including the University of Chicago, Caltech, Stanford, Columbia, Harvard, and Wisconsin. When selecting Caltech as a primary site, the officers explained that research in biology (under T. H. Morgan) and in electron diffraction methods (under A. A. Noyes and Linus Pauling) had already progressed to the point where applications commensurate with the Foundation's program could be realized.[104] As Mason saw it in 1933, "C.I.T.-Research in Chem. is at the center of the program of study of vital processes as furnishing aid in the sciences underlying human behavior."[105] Led by world-class researchers in genetics and the physical sciences and well-connected to the Rockefeller hierarchy, Caltech presented a unique setting for the newly designed biology. Freed from the omnipotence of established biological traditions—medical education, agricultural interests, evolutionary biology, the residues of natural history—molecular biology had a clear mandate. Wedded to engineering and the physical sciences, the new biology was implanted in a cognitive-institutional matrix that could foster an understanding of life inspired by visions of human engineering.

Notes

1. Dorothy Ross, *The Origins of American Social Science* (New York: Cambridge University Press, 1991), Ch. 7, quote on p. 230.

2. Ibid., pp. 230–233; and John Kenneth Galbraith, *The Anatomy of Power* (Boston: Houghton Mifflin, 1983), Chs. 2 and 3.

3. D. Ross, *The Origins of American Social Science*, p. 233 (see Note 1).

4. Edward Alsworth Ross, *Social Control: A Survey of the Foundations of Order* (New York: Macmillan, 1918), pp. 439–441. For useful perspectives see Robert N. Bellah, "Creating a New Framework for New Realities: Social Science as Public Philosophy," *Change* (March 1985), pp. 35–39; and Christopher Lasch, *The New Radicalism in America, 1889–1963* (New York: Alfred A. Knopf, 1965), Ch. 5.

5. D. Ross, *The Origins of American Social Science*, pp. 236–237 (see Note 1).

6. On the Progressive Era and reforms see, for example, Robert Wiebe, *The Search for Order, 1877–1920* (New York: Hill & Wang, 1965); and Richard Hofstadter, *Age of Reform: From Bryan to F.D.R.* (New York: Vintage Books, 1955). On the role of the professions see Thomas L. Haskell, ed., *The Authority of Experts: Studies on History and Theory* (Bloomington: Indiana University Press, 1984); Roy Lubove, *The Professional Altruist: The Emergence of Social Work as a Career, 1890–1930* (Cambridge: Harvard University Press, 1965); and Loren Baritz, *The Servants of Power: A History of the Use of Social Science in American Industry* (Middletown, CN: Wesleyan University Press, 1960). On the role of science see Samuel Haber, *Efficiency and*

Uplift: Scientific Management in the Progressive Era, 1890–1920 (Chicago: University of Chicago Press, 1963); Samuel P. Hays, *Conservation and the Gospel of Efficiency* (Cambridge: Harvard University Press, 1959); and Hugh Aitken, *Scientific Management in Action* (Princeton: Princeton University Press, 1960/1985). On the popular images of science see Marcel C. LaFollette, *Making Science Our Own: Public Images of Science, 1910–1955* (Chicago: University of Chicago Press, 1990).

7. D. Ross, *The Origins of American Social Science*, Introduction (see Note 1).

8. "American Empire, 1898–1903," *Pacific Historical Review, 48* (1979), entire issue; Alfred D. Chandler, Jr., *The Visible Hand: The Managing Revolution in American Business* (Cambridge: Harvard University Press, 1977); and Frederick C. Jaher, ed., *The Age of Industrialism in America* (New York: Free Press, 1968).

9. CUB, Frank Vanderlip Papers, Box D-13, Frank A. Vanderlip, "The Americanization of the World," February 22, 1902.

10. E. Digby Baltzell, *The Protestant Establishment: Aristocracy and Caste in America* (New York: Random House, 1964); James B. Gilbert, *Work Without Salvation: America's Intellectuals and Industrial Alienation, 1880–1910* (Baltimore: Johns Hopkins University Press, 1977); Daniel T. Rodgers, *The Work Ethic in Industrial America, 1860–1920* (Chicago: University of Chicago Press, 1974); and T. Jackson Lears, *No Place of Grace: Antimodernism and the Transformation of American Culture, 1880–1920* (New York: Pantheon Books, 1981).

11. On immigration and related social dislocations see Philip Davis, *Immigration and Americanization* (Boston: Ginn & Co., 1920); John Higham, *Strangers in the Land: Patterns in American Nativism, 1860–1925* (New York: Atheneum, 1963); Paul Boyer, *Urban Masses and Moral Order in America, 1820–1920* (Cambridge: Harvard University Press, 1978); and Alan M. Kraut, *The Huddled Masses: The Immigrant in American Society, 1880–1921* (Arlington Heights, IL: Harlan Davidson, 1982).

12. RAC, RG2, Rockefeller Board series, Box 25.256, H. P. Judson, "Suggestions of Policy for the Rockefeller Foundation," 1913.

13. For general histories of the Rockefeller philanthropies see Allan F. Nevins, *Study in Power: John D. Rockefeller, Industrialist and Philanthropist* (New York: Charles Scribner's Sons, 1953); Peter Collier and David Horowitz, *The Rockefellers: An American Dynasty* (New York: Signet Books, 1977); and John Ensor Harr and Peter J. Johnson, *The Rockefeller Century* (New York: Charles Scribner's Sons, 1988). For more specific discussions on the Foundations' policies, see Barry D. Karl, "Philanthropy, Policy, Planning, and the Bureaucratization of the Democratic Ideal," *Daedalus, 105* (1976), pp. 129–149; Barry D. Karl and Stanley M. Katz, "The American Private Foundations and the Public Spheres, 1890–1930," *Minerva, 19* (1981), pp. 236–270; "Foundations and Ruling Class Elites," *Daedalus, 20* (1987), pp. 1–40; and Ellen C. Lagemann, *The Politics of Knowledge: The Carnegie Corporation, Philanthropy, and Public Policy* (Middletown, CN: Wesleyan University Press, 1989). On the support of science see Nathan Reingold, "National Science Policy in Private Foundations: The Carnegie Institution of Washington," in Alexandra Oleson and John Voss, eds., *The Organization of Knowledge in Modern America, 1860–1920* (Baltimore: Johns Hopkins University Press, 1976), pp. 313–342; and Robert E. Kohler, *Partners in Science: Foundations and Natural Scientists, 1900–1945* (Chicago: University of Chicago Press, 1991)

14. Thomas H. Huxley, "On the Physical Basis of Life" (1864) in his *Collected Essays*, Vol. 1 (New York: D. Appleton, 1894); and Gerald L. Geison, "The Protoplasmic Theory of Life and the Vitalist Mechanist Debate," *Isis, 60* (1969), pp. 273–292. For the relation of the theory to genetics and molecular biology see Interlude I, this volume.

15. Edward J. Chamberlin and Sander L. Gilman, eds., *Degeneration: The Dark Side of Progress* (New York: Columbia University Press, 1985), especially chapters by Robert A. Nye "Sociology and Degeneration: the Irony of Progress," pp. 49–72, and Nancy Stepan, "Biological Degeneration: Races and Proper Places," pp. 97–120; Kenneth L. Ludmerer, *Eugenics and American Society: A Historical Survey* (Baltimore: Johns Hopkins University Press, 1972); Mark H. Haller, *Eugenics: Herediterian Attitudes in American Thought* (New Brunswick: Rutgers University Press, 1963); Donald K. Pickens, *Eugenics and the Progressives* (Nashville: Vanderbilt University Press, 1968); Daniel J. Kevles, *In the Name of Eugenics: Genetics and the Uses of Human Heredity*

(New York: Alfred A. Knopf, 1985); Garland E. Allen, "The Eugenics Record Office at Cold Spring Harbor, 1910–1940," *Osiris, 2* (1986), pp. 225–265; and Diane Paul, "The Rockefeller Foundation and the Origins of Behavior Genetics," in Keith R. Benson, Jane Maienschein, and Ronald Rainger, eds., *The Expansion of American Biology* (New Brunswick, NJ: Rutgers University Press, 1991), pp. 262–283.

16. RAC, Special Collections Register, "The Bureau of Social Hygiene," pp. 1–3; RAC, The Laura Spelman Rockefeller Memorial Collection Register, pp. 1–3; Barry A. Mehler, "The Bureau of Social Hygiene Papers," *The Mendel Newsletter, No. 16* (1978), pp. 1–11; D. Kevles, *In the Name of Eugenics,* pp. 208–209; and Garland E. Allen, "Old Wine–New Bottles: From Eugenics to Population Control in the work of Raymond Pearl," in Benson, Maienschein, and Rainger, eds., *The Expansion of American Biology,* pp. 231–261 (see Note 15).

17. Sheila Slaughter and Edward T. Silva, "Looking Backwards: How Foundations Formulated Ideology in the Progressive Period," pp. 55–87; and Donald Fisher, "American Philanthropy and the Social Sciences: The Reproduction of Conservative Ideology," pp. 233–269, in Robert F. Arnove, ed., *Philanthropy and Cultural Imperialism: The Foundations at Home and Abroad* (Bloomington: Indiana University Press, 1982). See also Roger L. Geiger, *To Advance Knowledge: The Growth of American Research Universities, 1900–1940* (New York: Oxford University Press, 1986), Ch. 2.

18. A. F. Nevins, *Study in Power,* pp. 391–392 (see Note 13); B. D. Karl and S. M. Katz, "The American Private Foundations and the Public Spheres," pp. 250–252 (see Note 13); Frank P. Walsh, "Perilous Philanthropy," *The Independent, 83* (August 1915), p. 262.

19. R. E. Kohler, *Partners in Science,* Part I (see Note 13).

20. LCM, Cattell Papers, Box 177, Morris Hillquit file, untitled draft, ca. 1917.

21. LCM, Cattell Papers, Box 157, E. B. Wilson file, Wilson to Cattell, October 23, 1918.

22. Daniel J. Kevles, "Into Hostile Political Camps: The Reorganization of International Science in World War I," *Isis, 62* (1971), pp. 47–61; idem, "George Ellery Hale, the First World War, and the Advancement of Science in America," *Isis, 59* (1968), pp. 427–437; Robert Kargon, ed., *The Maturing of American Science* (Washington, DC: American Association for the Advancement of Science, 1974) Introduction; N. Reingold and I. Reingold, *Science in America* (Chicago: University of Chicago Press, 1981), Ch. 10; Ronald Tobey, *The American Ideology of American Science, 1919–1930* (Pittsburgh: University of Pittsburgh Press, 1971); Alex Roland, "Science and War," *Osiris, 1* (1985), pp. 247–273; David J. Rhees, "The Chemists' Crusade: The Rise of An Industrial Science in Modern America, 1907–1922," Ph.D. dissertation, University of Pennsylvania, 1987; Roger L. Geiger, *To Advance Knowledge,* pp. 94–114 (see Note 17); and Hugh R. Slotten, "Human Chemistry or Scientific Barbarism? American Responses to World War I Poison Gas, 1915–1930," *Journal of American History, 77* (1990), pp. 476–498. On the role of the social sciences see Daniel J. Kevles, "Testing the Army's Intelligence: Psychologists and the Military in World War I," *Journal of American History, 55* (1968), pp. 565–581; Joel Spring, "Psychologists and the War: The Meaning of Intelligence in the Alpha and Beta Tests," *History of Education Quarterly, 12* (1972), pp. 3–15.

23. Stanley Coben, "American Foundations as Patrons of Science: The Commitment to Individual Research," in Reingold, ed., *Science in the American Context: New Perspectives* (Washington DC: Smithsonian Institution Press, 1979), pp. 229–249; Robert E. Kohler, "Science, Foundations, and American Universities in the 1920s," *Osiris, 3* (1987), 135–165; and R. E. Kohler, *Partners in Science,* Part II (see Note 13).

24. George E. Mowry, *The Urban Nation, 1920–1960* (New York: Hill & Wang, 1965); Barry D. Karl, *The Uneasy State: The United States from 1915 to 1945* (Chicago: University of Chicago Press, 1983), chs. 3 and 4; E. D. Baltzell, *The Protestant Establishment,* chs. 8 and 9 (see Note 10); Gabriel Kolko, *The Triumph of Conservatism* (New York: Free Press, 1977); R. G. McCloskey, *American Conservatism in the Age of Enterprise* (Cambridge: Harvard University Press, 1951); and Arthur S. Link, "What Happened to the Progressive Movement in the 1920s?" *American History Review, 64* (1959), pp. 833–851.

25. S. Haber, *Efficiency and Uplift,* Ch. 9 (see Note 6); Robert Moats Miller, *American Protestantism and Social Issues, 1919–1939* (Chapel Hill: University of North Carolina Press, 1958), Chs. 1–3; and Rolf Lunden, *Business and Religion in the American 1920s* (Westport, CT: Green-

wood Press, 1988). A notable example of this genre is Bruce Barton, *The Man Nobody Knows: A Discovery of the Real Jesus* (Indianapolis: Bobbs-Merrill, 1924).

26. R. Lunden, *Business and Religion in the American 1920s*, Chs. 4 and 5 (see Note 25); and Sinclair Lewis, *Babbitt* (New York: Harcourt Brace, 1922). For a useful conceptual framework see Robert N. Bellah, *Beyond Belief: Essays on Religion in a Post-Traditional World* (New York: Harper & Row, 1970).

27. Edwin R. Embree, "Timid Billions: Are the Foundations Doing their Job?" *Harper's Magazine, 198* (March 1949), pp. 28–37; and RAC, RG3, 900, Box 24.180, J. H. Willits to R. B. Fosdick, August 25, 1939.

28. Edward C. Lindeman, *Wealth and Culture* (New York: Harcourt Brace, 1936) p. 46; Joseph C. Kiger, *Operating Principles of the Larger Foundations* (New York: Russell Sage Foundation, 1954), p. 32; and Harold J. Laski, *Dangers of Obedience* (New York: Harper & Brothers, 1930).

29. Lunden, *Business and Religion in the American 1920s*, p. 146 (see Note 25).

30. RAC, RG3.1, 900 Pro-38, Box 23.174, Gregg, "Comments on Personalities . . .", September 15, 1945.

31. RAC, RG 3.2, 900, Box 28.155, Report of the Special Committee on Policy and Program, December 1946, p. 15.

32. Ibid., pp. 14–16. As late as 1948 several trustees recommended the establishment of a "Division of Moral Philosophy." For postwar policy planning, see Interlude II, this volume.

33. Raymond B. Fosdick, *A Chronicle of a Generation: An Autobiography* (New York: Harper & Brothers, 1958); and Robert Moats Miller, *Harry Emerson Fosdick: Preacher, Pastor, Prophet* (New York: Oxford University Press, 1985). See also Harry E. Fosdick, "Science and Religion," *Harpers Magazine, 152* (1926), pp. 296–300.

34. Daryl L. Revoldt, "Raymond B. Fosdick: Reform, Internationalism, and the Rockefeller Foundation," Ph.D. dissertation, University of Akron, 1982.

35. Raymond B. Fosdick, *The Old Savage and the New Civilization* (New York: Doubleday, Doran, 1928), J. A. Hobson's imagery quoted on the cover page.

36. E. D. Baltzell, *The Protestant Establishment*, pp. 197–201 (see Note 10); B. D. Karl, *The Uneasy State*, p. 59 (see Note 24). See also M. C. La Follette, *Making Science Our Own*, passim (see Note 6).

37. R. B. Fosdick, *The Old Savage and the New Civilization*, passim, quote on pp. 80–81 (see Note 35). For a theoretical framework see Karl Mannheim, "The Democratization of Culture" (1933) in his *Essays on the Sociology of Culture* (London: Routledge & Kegan Paul, 1956) pp. 171–246; and Fred E. Baumann, ed., *Democratic Capitalism? Essays in Search of a Concept* (Charlottesville: University Press of Virginia, 1986), especially Chs. 6 and 7.

38. R. B. Fosdick, *The Old Savage and the New Civilization*, pp. 81–85 (see Note 35). On cooperation and the modification of laissez-faire see Herbert Hoover, *American Individualism* (New York: Doubleday, Page, 1922); James Weinstein, *The Corporate Ideal in the Liberal State, 1900–1918* (Boston: Beacon Press, 1968), and Ellis Hawley, "The Discovery and Study of a 'Corporate Liberalism'" *Business History Review, 52* (1978), pp. 309–320.

39. R. B. Fosdick, *The Old Savage and the New Civilization*, pp. 21–32 (see Note 35); for Millikan's response to Fosdick see Robert A. Millikan, "Alleged Sins of Science," *Science and the New Civilization* (New York: Charles Scribner's Sons, 1930), pp. 52–86. In this article, Millikan challenged Fosdick's position as well as the call by the Bishop of Ripon for a 10-year moratorium on science research.

40. William F. Ogburn, *Social Change* (New York: Viking Press, 1924) pp. 200–268; and Wilson D. Wallis, "The Concept of Lag," *Sociology and Social Research, 19* (1935), pp. 403–405. See also D. Ross, *The Origins of American Social Science*, pp. 442, 445 (see Note 1).

41. R. B. Fosdick, *The Old Savage and the New Civilization*, pp. 21–32, quotes on p. 27, and 31, respectively (see Note 35).

42. On the cultural and economic background to human engineering see Stuart D. Brandes, *American Welfare Capitalism, 1880–1940* (Chicago: University of Chicago Press, 1964); Raymond E. Callahan, *Education and the Cult of Efficiency* (Chicago: University of Chicago Press, 1975); Joel H. Spring, *Education and the Corporate State* (Boston: Beacon Press, 1972); and William

Graebner, *The Engineering of Consent: Democracy and Authority in Twentieth-Century America* (Madison: University of Wisconsin Press, 1987). On the relation of the engineering ideal in the human sciences see Philip J. Pauly, *Controlling Life: Jacques Loeb and the Engineering Ideal in Biology* (New York: Oxford University Press, 1987); and Donna Haraway, "A Pilot Plant for Human Engineering: Robert Yerkes and the Yale Laboratories of Primate Biology, 1924–1942," in her *Primate Visions: Gender, Race, and Nature in the World of Modern Science* (New York: Routledge, Chapman & Hall, 1989), pp. 59–83. On the role of the Rockefeller Foundation, see John M. Jordan, *To Engineer Modern America: The Romance of Rational Reform, 1911–1939* (Chapel Hill: University of North Carolina Press, forthcoming, 1992), Chs. 7 and 8.

43. D. Haraway, *Primate Visions*, p. 65 (see Note 42).

44. J. B. Watson, "Psychology as the Behaviorist Views It," *Psychological Reviews, 20* (1913), 158–177. See also P. J. Pauly, *Controlling Life*, pp. 172–177 (see Note 42).

45. Franz Samelson, "Organizing for the Kingdom of Behavior: Academic Battles and Organizational Policies in the Twenties," *Journal of History of the Behavioral Sciences, 21* (1985), 33–47. On the "masculinization" of sociology see D. Ross, *The Origins of American Social Science*, pp. 394–395 (see Note 1). For other (non-mechanistic, non-behavioristic) approaches to human nature during the interwar period see, Sharon E. Kingsland, "Toward a Natural History of the Human Psyche: Charles Manning Child, Judson Herrick, and the Dynamic View of the Individual at the University of Chicago," in Benson, Maienschein, and Rainager, eds., *The Expansion of American Biology*, pp. 195–230 (see Note 15).

46. Martin Bulmer and Joan Bulmer, "Philanthropy and Social Science in the 1920s: Beardsley Ruml and the Laura Spelman Rockefeller Memorial, 1922–1929," *Minerva, 19* (1901), pp. 347–407; D. Ross, *The Origins of American Social Science*, Ch. 10 (see Note 1); J. M. Jordan, *To Engineer Modern America*, Ch. 7 (see Note 42); Martin Bulmer, *The Chicago School of Sociology* (Chicago: University of Chicago Press, 1987); Robert C. Banister, *Sociology and Scientism: The American Quest for Objectivity, 1880–1940* (Chapel Hill: University of North Carolina Press, 1987); and Clifford Geertz, *Local Knowledge* (New York: Basic Books, 1983), especially Ch. 7.

47. F. Samelson, "Organizing for the Kingdom of Behavior," p. 42 (see Note 45).

48. D. Ross, *The Origins of American Social Science*, p. 402 (see Note 1); D. Fisher, "American Philanthropy and the Social Sciences," p. 233 (see Note 17).

49. D. Ross, *The Origins of American Social Science*, p. 429 (see Note 1).

50. Frederick E. Lumley, *Means of Social Control* (New York: Century Co., 1925), pp. 12–14.

51. Gerald Jonas, *The Circuit Riders: Rockefeller Money and the Rise of Modern Science* (New York: W. W. Norton, 1989), p. 170.

52. R. B. Fosdick, *The Old Savage and the New Civilization*, p. 184 (see Note 38).

53. Ibid., p. 185.

54. D. K. Pickens, *Eugenics and the Progressives*, pp. 177–183; J. Ludmerer, *Genetics in American Society*, pp. 121–163; and D. J. Kevles, *In the Name of Eugenics*, Chs. 8–11 (see Note 15 for these references). For a reassessment see Diane B. Paul, "Eugenics and the Left," *Journal of the History of Ideas, 45* (1984), pp. 567–589; and D. B. Paul, " 'Our Load of Mutations' Revisited," *Journal of the History of Biology, 20* (1987), pp. 321–335.

55. On the persistence of the unit character concept in genetics see Norman Horowitz, "Genetics and the Synthesis of Proteins," *Annals of the New York Academy of Sciences, 325* (1979) p. 257.

56. For examples from major textbooks that had gone through several editions since the 1920s see George C. Wood and Harold A. Carpenter, *Our Environment* (New York: Allyn & Bacon, 1938), pp. 893–914; and Perry D. Strausbaugh and Bernal R. Weimer, *General Biology* (New York: John Wiley & Sons, 1947) pp. 352–361 see also Diane B. Paul, "Textbook Treatment of the Genetics of Intelligence," *The Quarterly Review of Biology, 60* (1985), pp. 317–326; D. B. Paul, "The Nine Lives of Discredited Data," *The Sciences* (May 1987), pp. 26–30.

57. K. Ludmerer, *Eugenics and American Society*, Introduction, pp. 121–163 (see Note 15); D. J. Kevles, *In the Name of Eugenics*, Chs. 8–11 (see Note 15); and D. B. Paul, "Eugenics and the Left" (see Note 54). On Popenoe's poor image see APS, Davenport Papers, Morgan to Davenport, January 18, 1915, cited in Garland Allen, *Thomas Hunt Morgan: The Man and His*

Science (Princeton: Princeton University Press, 1979), pp. 228–229; see also Paul A. Lombardo, "Three Generations, No Imbeciles: New Light on Buck v. Bell," *New York Law Review* (April 1985) pp. 30–62.

58. Nicholas Pastore, *The Nature–Nurture Controversy* (New York: King's Crown Press, 1949); Hamilton Cravens, *The Triumph of Evolution: American Scientists and the Heredity-Environment Controversy, 1900–1941* (Philadelphia: University of Pennsylvania Press, 1978), Chs. 2–4. Robert M. Yerkes, *Almost Human* (New York: Century Books, 1925); and Edward L. Thorndike, *Human Nature and the Social Order* (New York: Macmillan, 1940).

59. R. E. Kohler, "Science, Foundations, and American Universities in the 1920s," pp. 157–161 (see Note 23); D. Paul, "The Rockefeller Foundation and the Origins of Behavior Genetics," pp. 262–283 (see Note 15).

60. RAC, RG3, 915, Box 4.33, Embree to Wilbur, July 11, 1927.

61. Ibid. p. 1. Wilbur viewed Embree's project in human biology as "one of the real leads of the future. . . . My only regret is that it is not more extensive." RAC, RG3, 915, Box 4.33, Wilbur to Vincent, May 21, 1926.

62. G. Allen, "Old Wine in New Bottles," and D. Paul, "The Rockefeller Foundation and the Origins of Behavior Genetics," pp. 231–261 and 262–267 respectively (see Note 15).

63. R. B. Fosdick, *The Old Savage and the New Civilization,* p. 45 (see Note 35); his implicit endorsement of eugenics is on p. 184.

64. *New York Times,* January 4, 1929, p. 21, column 1.

65. Ellis W. Hawley, "Herbert Hoover, The Commerce Secretariat, and the Vision of an Associative State," *Journal of American History, 8* (1974), pp. 116–140. Craig Lloyd, *Aggressive Introvert: A Study of Herbert Hoover and Public Relations Management, 1912–1932* (Columbus: Ohio State University Press, 1972); W. Graebner, *The Engineering of Consent,* pp. 65–66 (see Note 42); and Barry D. Karl, "Presidential Planning and Social Science Research: Mr. Hoover's Experts," *Perspectives in American History, 3* (1969), pp. 347–409.

66. RAC, Laura Spelman Rockefeller Memorial Collection Register, pp. 1–7; Kohler, *Partners in Science,* Ch. 9 (see Note 13); M. Bulmer and J. Bulmer, "Philanthropy and Social Science in the 1920s" (see Note 46), and D. Fisher, "American Philanthropy and Social Science," pp. 233–268 (see Note 17).

67. Warren Weaver, "Max Mason, 1877–1961," *Biographical Memoirs of the National Academy of Sciences* (1964), pp. 204–236; Warren Weaver, *Scene of Change: A Life in American Science* (New York: Charles Scribner's Sons, 1970), pp. 28–32.

68. W. Weaver, "Max Mason," p. 224 (see Note 67). Mason's interest in pathological aspects of human behavior was also motivated by tragedies and disappointments in his married life. CUB, Warren Weaver, Transcript of Oral History Memoir, Oral History Office (Record No.343, 3 vols., 1962), p. 333; in Evelyn Fox-Keller, "Physics and the Emergence of Molecular Biology: A History of Cognitive and Political Synergy," *Journal of the History of Biology, 23* (1990), pp. 389–409.

69. RAC, RG12.1, Box 43, Mason Diary, 1928–1930, October 2, 1929, pp. 61–63; Barry D. Karl, "Presidential Planning and Social Science Research: Mr. Hoover's Experts," (see Note 65). See also J. M. Jordan, *To Engineer Modern America,* Ch. 8 (see Note 42).

70. RAC, RG3, 900, Box 24.184, Report of the Committee on Appraisal and Plan, December 11, 1934, pp. 23–24; and Robert E. Kohler, "A Policy for the Advancement of Science: the Rockefeller Foundation, 1924–1929," *Minerva, 16* (1978), pp. 480–515.

71. RAC, RG1.1, 216D, Box 8.104, the resubmittal of Lillie's 1924 proposal followed Mason's visit to Chicago and was attached to his letter to Mason, June 5, 1931.

72. Ibid. p. 2.

73. RAC, RG1.1, 216D, Box 8.104, Mason to Lillie, June 15, 1931.

74. RAC, RG1.1, 216D, Box 8.103; The elements of Lillie's proposal, as Mason put it, were present in the work at Chicago. From 1929 to 1934 biological research at Chicago under the leadership of Lillie received $30,000 a year from the Foundation and $24,000 a year from the NRC Committee for Research in Problems of Sex, which in turn received its funds from the Foundation. In 1933 the Foundation assumed the support of sex research (1934–1938, $50,000 per year) following the Committee's withdrawal of its support. On the early support of Caltech's biology

see RAC, RG12.1, Box 43, Mason Diary, 1928-1930, December 30, 1929, p. 106; and RG1.1, 205D, Box 5.66, Morgan's report, November 8, 1930.

75. W. Weaver, *Scene of Change*, pp. 1-58 (see Note 67); Mina Rees, "Warrren Weaver, July 17, 1894-November 24, 1978," *Biographical Memoirs of the National Academy of Sciences, 57* (1987), pp. 493-530.

76. W. Weaver, "Max Mason," p. 217 (see Note 67).

77. W. Weaver, *Scene of Change*, Ch. 3 (see Note 67); the reference to Freeman is on p. 47. As the remarkably influential pastor of California's largest Presbyterian church, Freeman was one of Pasadena's most effective community leaders. His sermons and writings expressed his conservative Republican ideology and his strong support for negative eugenics. PHS, Robert Freeman biographical file; and PPC Archive. Relevant materials written by Freeman: "Fifty Years of Presbyterianism on the Pacific Coast," *Pasadena Chimes, 5, No. 3* (ca. 1920); "Qualifying for Survival," *Pasadena Chimes, 5, No. 22* (ca. 1920); and his regular column "The Bible Day-by-Day," in *The Continent, 51, No. 2* (1920) p. 50; and *The Continent, 51, No. 15* (1920), p. 474. See also the discussion on Freeman in Chapter 2, this volume.

78. W. Weaver, *Scene of Change* pp. 48, 49 (see Note 75).

79. Max Mason and Warren Weaver, *The Electromagnetic Field* (Chicago: University of Chicago Press, 1929).

80. Weaver, "Max Mason," p. 219 (see Note 67).

81. W. Weaver, *Scene of Change*, pp. 49, 56-60 (see Note 75).

82. Max Mason and Warren Weaver, "The Settling of Small Particles in a Fluid," *Physical Review, 23* (1924), pp. 412-426; and RAC, RG1.1, 800D, Box 8.79, University of Uppsala, "Machines that Sort and Weight Molecules."

83. For a discussion on the emergence of the professional category of a research-entrepreneur and a scientific manager see Charles E. Rosenberg, *No Other Gods: On Science and American Social Thought* (Baltimore: Johns Hopkins University Press, 1976), Part II, Ch. 9; and Robert E. Kohler, "The Management of Science: The Experience of Warren Weaver and the Rockefeller Foundation Programme in Molecular Biology," *Minerva, 14* (1976), pp. 249-293.

84. W. Weaver, *Scene of Change*, p. 62 (see Note 75).

85. RAC, Spelman Fund, Series 4.1, Box 3.205, "Serious Defects In Our Present Methods of Attempting to Cope with the Situation," (1933), p. 14; and Robert S. McElvaine, *The Great Depression: America, 1929-1941* (New York: New York Times Books, 1984).

86. R. M. Miller, *American Protestantism and Social Issues*, Chs. 5-8 (see Note 25); On the cultural impact of the Depression see Warren I. Susman, *Culture as History: The Transformation of American Society in the Twentieth Century* (New York: Pantheon Books, 1984), Ch. 10; and Barton J. Bernstein, "The New Deal: The Conservative Achievements of Liberal Reform," in Barton J. Bernstein, ed., *Towards a New Past: Dissenting Essays in American History* (New York: Vintage, 1968), pp. 263-288.

87. For an excellent account of the politicization of science as a result of the Depression see Peter J. Kuznick, *Beyond the Laboratory: Scientists as Political Activists in 1930s America* (Chicago: University of Chicago Press, 1987), especially Chs. 1 and 2.

88. RAC, RG3, 915, Box 1.1, from NS-Section, Annual Report, April 18, 1933.

89. RAC, RG3, 915, Box 1.7, Weaver's Report, February 14, 1934, pp. 2-3; excerpts in Raymond B. Fosdick, *The Story of the Rockefeller Foundation* (New York: Harper & Brothers, 1952), p. 158.

90. RAC, RG3, 900, Box 21.161, Trustees Meeting, April 11-12, 1933, p. 3; and RAC, RG3, 900, Box 24.184, Report of the Committee on Appraisal and Plan, December 11, 1934, p. 58.

91. RAC, RG3, 900, Box 24.184, Report of the Committee on Appraisal and Plan, December 11, 1934, p. 25.

92. RAC, RG1.1, 216D, Box 8.104, Lillie's proposal of February 20, 1932; included also population problems and ecology. RAC, RG1.1, 216D, Box 8.106, Lillie's proposal of April 7, 1934, included experimental population studies.

93. RAC, RG1.1, 216D, Box 8.106, excerpts from Weaver's Diary, January 18-19, 1934.

94. On the history of the Hanover Conferences see J. M. Jordan, *To Engineer Modern America*, Ch. 8 (see Note 42).

95. RAC, RG12.1, Box 68, Weaver's Diary, 1934, pp. 98–110.

96. RAC, RG3, 915, Box 1.6, "The Proposed Program," (ca. 1933), pp. 12–13. On Alan Gregg see D. Paul, "The Rockefeller Foundation and the Origins of Behavior Genetics," pp. 262–283 (see Note 15); and Jack D. Pressman, "Human Understanding: The Rockefeller Foundation's Attempt to Construct A Scientific Psychiatry in America, 1930–1950," paper presented at the MIT Mellon Workshop "Comparative Perspectives on the History and Social Study of Modern Life Science," April 5, 1991.

97. RAC, RG3, 915, Box 1.6, "The Proposed Program," ca. 1933, p. 13.

98. Quote of E. E. Day "Verbatim Notes on Princeton Conference of Trustees and Officers," RAC, RG3, 900, Program and Policy 1926–1930, October 29, 1930, p. 115; in D. Fisher, "American Philanthropy and the Social Sciences," pp. 236–237 (see Note 17). The objectives of the newly organized division echoed the concerns of the SSRC, noting the ubiquitous techniques of social control grounded in modern methods of communication, advertising, and mass psychology and inviting their objective study on a large scale. *The Social Science Research Council, Annual Report* (1930–1931), p. 29. See also J. M. Jordan, *To Engineer Modern America*, Ch. 8 (see Note 42).

99. The significance and continuity of this position is underscored by the emphasis it received at the time of the launching of the program (RAC, RG3, Box 24.184, Report of the Committee on Appraisal and Plan, December 11, 1934, pp. 42–43) and by its reiteration again during the reappraisal in December 1946; (RAC, RG3.2, 900, Box 28.155, pp. 7–8).

100. RAC, RG3.2, 900, Box 24.184, Report of the Committee on Appraisal and Plan, December 11, 1934; p. 45; Harold J. Laski, *Dangers of Obedience* (New York: Harper & Brothers, 1930).

101. RAC, RG3, 900, Box 24.180, Willits to Fosdick, August 25, 1939.

102. Rockefeller Foundation, *Annual Report* (1938), pp. 34–39.

103. RAC, RG3, 915, Box 1.2, Weaver to Tisdale, February 8, 1935.

104. RAC, RG3, 900, Box 24.184, Report of the Committee on Appraisal and Plan, December 11, 1934, pp. 55–56.

105. RAC, RG1.1, 205D, Box 5.71, Mason to Gunn, December 18, 1933.

CHAPTER 2

Technological Frontier: Southern California and the Emergence of Life Science at Caltech

Machine in the Pacific Garden, 1900–1930

"American social development has been continually beginning over again on the frontier," proclaimed Frederick Jackson Turner in his famous "frontier thesis," unveiled at the 1893 meeting of the American Historical Association during the World's Columbian Exposition in Chicago.

> This perennial rebirth, this fluidity of American life, this expansion westward with its new opportunities, its continuous touch with the simplicity of primitive society, furnish the forces dominating American character. The true point of view in the history of this nation is not the Atlantic coast, it is the Great West.

Escorting Theodore Roosevelt in his imperial ushering of the "Pacific Era," Turner provided Roosevelt's quasimythical epic *The Winning of the West* (1889) with an interpretive framework appropriate to academic discourse. In shaping the national conception of American history, his celebrated "frontier thesis" also inspired and sustained Caltech's self-image as a cultural and scientific pioneer; the institute perceived and presented itself as spearheading the frontier mission.[1]

Speaking for the new "objective history," Turner ventured beyond race ideology and the teleology of Anglo-Saxon destiny. In articulating the significance of the frontier in American history he elaborated a dialectic of closure and opportunity, maturity and rebirth: "the gate of escape from the bondage of the past." Though unwittingly appealing to universal myths of regeneration through eternal return to cosmogony, Turner endowed the timeless acts of creation with temporal, cultural, and political specificity. The "frontier thesis" offered a serviceable historical par-

58

adigm for expansionism and industrialization and supplied a metaphor for Caltech's scientific ambition.[2]

To Turner, as to most Americans at the turn of the century, Southern California symbolized the Western frontier. The north, centering around San Francisco's excellent port, teeming commerce, and rich culture, had by then verified civilization's mastery of the occidental landscape. Southern California and Los Angeles, however—"cow counties," as they were then called—held the intrigue of virgin lands and untapped promise of the Pacific Era. Southern California had experienced minor growth on the coattails of the gold rush, but it was the completion of the transcontinental line to San Francisco in 1869, its extension to Southern California in 1876, and the addition of the Santa Fe line in 1886 that literally put the region on the map.[3]

That Turner's "frontier thesis" was more a trope for economic and social expansion than an authentic historical explanation is evident from its share of ironies and misconceptions. Speaking as he did from the Olympian heights of American academe, Turner articulated the cultural perceptions of the genteel conqueror; his ethnocentric thesis betrayed a blindness to the rich record of non-American frontier experiences in Southern California. The thriving Mexican population of the region had been largely written off as a "vanquished" element, a primitive, picturesque relic. The burgeoning Chinese culture from San Diego to Monterey left a mere trace. This "Cathay in the South" accounted for millions of dollars in export from the fishing and abalone industries, revenue from citrus and produce cultivation, and railroad construction—Yankee ingenuity at work according to local histories. Laboring under the arrows of racial abuse and economic handicaps and following a massacre in one of the worst race riots in American history, most of the Chinese left by 1900. The final shipment of first-generation Chinese—piously packed siftings of dust and bones from 850 Chinese graves in Los Angeles—reached China decades later. By 1900, with the return of the Mexicans and with a large Japanese immigration, the American experience on the "Western frontier" was densely interwoven with that of earlier cultures.[4]

Southern California by then hardly qualified as a frontier. After 1870 the population of the "cow counties" had increased at a phenomenal rate. The enterprising campaigns of railroad baron Collis P. Huntington had lured hundreds of thousands of convalescents, tourists, and land speculators to the "subtropical" paradise and its curative climate, sparking the real estate boom of the 1880s. Between 1887 and 1889 more than 60 new towns with a population of about two million were laid out in Southern California; from 1890 to 1910 the population increased by about 300 percent, due to an enormous migration from the Midwest. Los Angeles and its vicinity had paved streets, sewer systems, hotels, and even a handful of colleges, among them Throop College (1891), the precursor of Caltech. After the discovery of oil deposits during the 1890s, the sprawling city glowed with pride and power— the first city in America (perhaps the world) illuminated by electricity. By 1900 the Los Angeles Chamber of Commerce trumpeted the region's agricultural and commercial potential, envisioning an enormous growth sustained through technological innovations.[5]

Even as Turner spoke, Southern California was already commercially developed. Having absorbed the energy and sweat of Mexicans, Chinese, and Japanese pioneers into their own entrepreneurial projects, the region's business leaders anticipated an industrial revolution, a Pacific reenactment of the Atlantic theme. Their scheme of turning a semiarid province into a lush metropolitan garden hinged on high-level technological developments, aggressive control of key resources (water, timber, fuel, and labor), and management of the complex processes of urbanization. By the decade beginning with 1910, Turner had extrapolated the "frontier thesis" to the "spiritual West"—to the political, social, and academic realms—propounding a conception of national identity grounded in technological mastery and private initiative.[6] This formulation, which resonated with the ideology of Los Angeles' business elite, provided a sense of historical mission for Caltech's participation in the Pacific industrial revolution.

Los Angeles' industrial expansion was predicated on ruthless tactics of water acquisition and distribution, strategies rooted in supply-side thinking that "if you don't get the water you won't need it."[7] Los Angeles had thrived on the abundance of the Los Angeles River watershed, which embraces the San Fernando Valley. However, even with ample water supply for its 1900 population of about 100,000 and its watershed rights secured, Los Angeles soon invented the concept of "water famine" based on the discrepancy between the present supply and anticipated needs. A syndicate of empire builders, men who played a pivotal role in developing Caltech, masterminded a drive to alleviate the so-called water famine: the notorious Owens Valley and Hoover Dam projects.[8]

The syndicate—Harry Chandler, land developer, Caltech trustee, and president of the *Los Angeles Times* (probably the wealthiest and most powerful man in the West), and Henry Huntington (heir to Collis Huntington), transit magnate and Caltech Associate—had acquired most of the San Fernando Valley. After establishing themselves within the city's water board, they seized control of the water supply. Operating within a conspiratorial silence and later benefiting from public misrepresentations in the *Los Angles Times*, the "water ring" engineered the 238-mile city-built aqueduct (1913), which diverted the Owens River to the San Fernando Valley. The ruined community of small growers of Owens Valley, the collapsed real estate, and the lush valley transformed into the first man-made desert was the price of developing hydroelectric power in Southern California. In 1928, after years of bitter litigation ending with the syndicate's victory over Owens Valley, the Boulder Dam bill (Hoover Dam) passed in Congress, securing Los Angeles' rights to tap the Colorado River. By the 1920s, Chandler's "Los Angeles First" philosophy had been expanded to include the entire Southern California region, from the Arizona border on the east to Mexico on the South.[9]

With torrents of water in the aqueducts, the speed of industrialization in the Southwest seemed to surpass the power of the imagination. California's older agricultural industry boomed; "America's natural hothouse" became in the 1920s the nation's supplier of produce. The agribusiness bonanza, capitalizing on technological advances such as shipping, refrigerated cars, and trucking, followed the development of irrigation technologies and hydroelectric power, which harnessed the resources of the Owens Valley and the Colorado River. The rise of the oil industry generated enormous wealth, which in turn stimulated growth in dozens of directions: trucking and automobile industries, ship building, and the nascent aeronautical enterprises. The new motion picture industry created seemingly endless opportunities for fortunes, further contributing to the housing boom, which in turn stimulated the already prosperous lumber industry and further boosted the use of hydroelectric power. The myriad industrial and commercial activities spurred the growth of banking and finance; the decentralization of the Federal Reserve System, which offered money to agricultural regions at low interest rates, also accelerated the region's financial autonomy through the advent of branch banking.[10]

Southern California burst into the national arena during the 1920s sui generis; its indigenous industries (with the exception of the motion picture industry, financed heavily by the Rockefeller and the J. P. Morgan groups) derived their might from the region's own business magnates and were sustained by its new professional elite: law, medicine, and engineering. While big business in the Northeast relied on the old, elite universities for scientific and technological expertise, Caltech, the region's first industrial research liaison, provided an escape from the bondage of the past by cultivating a forward look. Southern California was undergoing a "delayed" industrial revolution; in some sense, the first two decades of the "Pacific century" paralleled the second half of the "Atlantic century." The Southwest's industrial revolution, though, diverged from the experience of the Northeast qualitatively and quantitatively. Southern California not only developed different kinds of industry, they accelerated and compressed the stages of industrialization: vertical integration, investment banking, and urbanization. Even more so than in the East, the swiftness of industrialization in the Southwest contrasted sharply with the sluggishness of political adaptation to a changing social landscape.

As in the Northeast, racial myths informed the projects of expansionism, free enterprise, and industrialization. The influential California booster and visionary Joseph Pomeroy Widney advanced the classic argument that "the Captains of Industry were the truest Captains in the race war."[11] In the Southwest, however, the conjunction of race, ideology, and enterprise acquired a specific significance, grounded in the region's geography and demographics. Charles Fletcher Lummis, Ohio-born editor of the *Los Angeles Times*, promulgated "the new Eden of the Saxon home-seeker," equating the rise of the Anglo-Saxon West with the decline of the foreigner-infested East and the evolutionary improvement of the Nordic stock in Southern California with its loss of vigor in the East.[12] To many American newcomers, Southern California signified a refinement of the notion of racial superiority and manifest destiny.

In his influential two-volume epic, *Race Life of the Aryan People*, the Ohio-born physician Widney, notably a founder of the Methodist-affiliated University of Southern California, articulated the community sentiment by "proving" the superiority of the Engle people (Anglo-Saxons) of Southern California. Educator, polymath, and cultural benefactor, Widney claimed that the Southwest Engles had overcome the Spencerian imperative. Having proved the law of survival of the fittest, they had paved the way for a higher civilization. Los Angeles and Istanbul marked the endpoints of the arch of Aryan civilization; Los Angeles' destiny was the fulfillment of the mission of an Aryan city of the sun.[13]

There were fundamental incompatibilities, however, between the boosters' quest for modernity and their tenacious clinging to tradition; they were soon confronted with the costs of their technological utopia. The business elite that had arrived around the turn of the century had imported the small-town values of the Republican Midwest, a rural culture centered around home owning, church going, and "dry" living. California was not "New Iowa," though, as it was frequently called. The family farm, the traditional unit of stability and continuity, was largely absent in California. Cycles of boom and bust generated perpetual social flux. Constant migration undermined a sense of community. Around 1930 approximately 66 percent of Californians had been born in other states, giving currency to the local expression, "I'm a stranger here myself." The uprootedness and perpetual flux, along with the growing presence of a restless underclass, generated the distinctive social disorders of the urbanizing Southwest.[14]

Perennial confrontations between big business and the region's disenfranchised labor force punctuated the industrialization process. Work opportunities and the promise of wealth attracted not only the enterprising "Nordics" but also the "inferior" social elements and foreign immigrants. As in the Northeast, the foreign labor force—Mexicans, Japanese, Filipinos, and Eastern Europeans—that formed the backbone of the Southwest's industrial and land development consisted mainly of unskilled workers. Unusually ruthless management tactics, coupled with the powerful grip of the "press axis" of the *Los Angeles Times*, squelched most attempts at labor organizing; Los Angeles' notorious "open shop" practices lasted well into the New Deal.[15]

The migratory farm labor that powered California's booming agribusiness endured abysmal treatment. These seasonal foreign workers—mainly Mexicans, but during the Depression also the "Okies" and "Arkies"—were prodded to productivity under inhumane conditions that combined the sweatshop tactics of the North with the stoop labor operations of the southern plantations, practices that earned California farms the notoriety of "factories in the field." The confrontations between labor and capital wrought by these conditions and by the resistance to labor unionizing erupted intermittently in intense racial fury.[16]

Many Protestant churches in the region, as elsewhere, helped to legitimize the social order. At the magnificent Pasadena Presbyterian church, which so captivated young Warren Weaver, Reverend Robert Freeman uplifted his affluent flock, preaching in his Scottish brogue on the virtue and rewards of the work ethic, the Christian mission in California, and the threats to Americanism. He raged against foreign-born college teachers, who "without our wholesome traditions . . . make Bolsheviki out of those American-born children," impressing on his listeners that "We are here to keep up the average morality of the world."[17] He warned Caltech's community of "the threat to our civilization from Mexican immigration now that the Johnson Act restricted European labor" and from the growing presence of the Orientals and their Buddhist temples.[18] One anonymous letter-writer, however, venting moral indignation atypical of the community, cast a shadow on Freeman's sunny alliance with Pasadena's gentry, doubting that the poor of our land, the men and women who earn their bread by the sweat of their brow, had they heard your sermon, would have felt that it was sweet and beautiful to work hard that the capitalistic whales might enjoy themselves. Has it occurred to you, dear reverend sir, that you are not serving the purpose of Christ? Have you ever told yourself, Sir? That to pander to the aristocracy of mammon was an absolute contradiction to the aims of a servant of the Lowly Nazarean?[19]

Signed, A Listener. This devastating criticism contrasted with Freeman's extraordinary popularity with his wealthy, educated congregation.

The denigration of the Mexicans and Chinese during the nineteenth century belonged to a pattern of virulent nativism; but by the 1920s the immense disparity between the poverty of a growing underclass and the luxury of a small ruling elite created conflicts that seemed to confirm Widney's notion of a race war. The grip of conservative Republican politics and its nativist fervor juxtaposed against rising social disorder fostered in Southern California an extreme form of negative eugenics. Under the banner of science and the zealous leadership of Paul Popenoe, Widney's version of the Spencerian doctrine acquired greater legitimacy and social potency. Popenoe argued that race deterioration increased because the immoral and stupid class went unchecked. Because as individuals they were incompetent to practice birth control, only voluntary and compulsory sterilization could preserve the quality of the Anglo-Saxon germ plasm. From the Progressive Era until about 1940, California led the nation in sterilization of the insane, feeble-minded, unfit, and morally degenerate.[20]

Like other Protestant churches, Pasadena's Presbyterian church bestowed on eugenics the power of a moral imperative by sanctifying science and objectifying theology. From his pulpit, Freeman delivered the masterful sermon "Qualifying for Survival," dexterously distinguishing between survival in the animal kingdom and survival in the world of man.

> Theologians are ready to make large concessions to the theories set forth in the Origins of Species, and scientists are less arbitrary and sweeping in their demands for those theories . . . in the world of men these two things are true: there are those who survive despite unfitness, and there are those who, though marked by an initial unfitness, *make* themselves fit to survive. There are the ragweed and the rattler; the mosquito and the despicable housefly in humanity, which although they make no beneficent contribution to life, they only poison and destroy, continue to exist.[21]

Freeman's later participation in California's sterilization program bespoke his social commitments.[22]

The problems of labor unrest, immigration assimilation, and social control in Southern California accelerated with remarkable intensity. The Protestant business establishment began experiencing the dark side of progress less than two decades after it put down roots. As with the technological utopianism underlying the industrialization of the Northeast, the architects of the Pacific civilization had to face the mutually contradictory elements in their plan. To the ruling elite, labor agitations represented irrational behavior that retarded the march of progress and threatened the political order.[23] They did not perceive the fundamental incompatibility between their dependence on a pool of unskilled alien labor and their goal of Aryan hegemony in the sun. Blind to the complexities of human nature and the lessons of history, the Protestant establishment experienced the conflict between the Dionysian will to power and the Apollonian longing for harmony. As with earlier experiences of industrialization in Europe and America, the visionaries mourned the lost image of a *gemeinschaft* in the midst of their frenzied creation of a metropolitan *gesellschaft*. Deifying the machine in their garden, they longed for the simplicity of the very bucolic life they had helped to destroy.[24]

Henry O'Melveny, Southern California's most powerful lawyer, custodian of massive land and industrial enterprises, and Caltech's oldest trustee, unwittingly celebrated these incongruities. "In this land of balmy airs, soft skies and gentle seas," he reminded his Caltech audience posed on the threshold of summer,

> there lived in the old days a people who were indifferent to money, who carried their religion into their daily pleasures and sorrows, were brotherly one towards another, contented, beautiful, joyous. . . . [T]hey, like you have seen and heard the first rains of winter fall gently on hill and mesa . . . the orange light of the poppy kindle the slopes. . . . They knew, as we know, that the August sunshine would plump the fig with sweetness and would be caught and imprisoned by the clustering grapes. . . . From the voices of the past there must come to you the faintly articulate call to remember the people whose dauntless courage discovered this land and who kept it through the years, awaiting your coming.[25]

By this joyous people he of course meant the Mexicans, who even as he spoke stooped in the sun cultivating Pasadena's garden. O'Melveny's paean to the land conveyed to his young technocratic audience a sense of appreciation and loss, entitlement and gain. The benefits of historical hindsight would have illuminated similar dissonances between Edenic myths and civilizing mission. Previous discourse of colonization also wrestled with conflicting tropes: pleasures of temperate weather, bucolic landscape, ripened fruit, and sensuous natives juxtaposed against the threats of naked nature to the culture of mastery and self-denial.[26] O'Melveny's poignant appeal—a tacit expression of Turner's frontier metaphor—created a false sense of continuity between the work ethic and the extinct culture of sensuality and inefficiency, between the lure of naked nature and the technological imperative on the Western frontier.

Cooperative Ideal: Toward a Life Science at Caltech

In 1908 when George Ellery Hale, founder of American astrophysics, began his "big scheme" to establish a premier scientific institution in Southern California, biology did not figure in the plans. He had come to Pasadena in 1903 to head the Mount Wilson observatory under the auspices of the Carnegie Institution, but his goals broadened after becoming a trustee of Pasadena's Throop College. He set out to transform the manual training school into a first-rate college of science and engineering, a project that called for the patronage of the Carnegie and Rockefeller philanthropies and matching contributions by Southern California's private sector. A son of a wealthy Chicago businessman and a cultivated Anglophile, Hale glided gracefully through Los Angeles' circles of wealth, the advisory boards of the large foundations, and the committees of the National Academy of Science. His effectiveness in these interconnected spheres of power—science, industry, business, and philanthropy—stemmed from strategic planning buttressed by an ideological alliance with private enterprise.[27]

In carving a niche for a premier technical institution in Southern California, Hale resounded the themes that animated the region's development projects, the frontier discourse. "In California the conditions and need for technical education are unsurpassed," he told the Throop board of trustees, appealing simultaneously to their sense of destiny and their pragmatic proclivities.[28] His plea echoed their own aspirations and ideology. Henry Robinson, the Ohio-born financial magnate and president of Security First National Bank, invested enormous resources in the new institution. Trained at Cornell as an engineer, he had come to California in 1907 to retire, having by then made a fortune in railroads, newspapers, and (after taking up corporate law) banking, steel, and tin. The lure of Southern California's virgin land turned Robinson's retirement into a big business bonanza. His investment in Throop beginning around 1910 and continuing through the 1920s paralleled his energetic development of the oil, telephone, railroad, metal, and lumber industries, his involvement in banking and investing, and finally his directorship of the Southern California Edison Company.[29]

Throop trustee Norman Bridge contributed financial as well as cultural resources. A former professor at the University of Chicago Medical School and member of Chicago's board of education and election commissioners, Bridge had always led civic projects. After arriving in Pasadena he gave up medicine for bank and oil company directorships, a partnership in a law firm, membership on the board of education, and management of educational and cultural institutions. His contribution of more than $250,000 to establish the Norman Bridge Laboratory of Physics (to be headed by R. A. Millikan), a donation he tripled during the next decade, expressed his dual commitment to Throop College and the Anglo-Saxon mission in Southern California.[30]

Hale centered his plans within the region's political economy. He argued that a first class institute of technology such as Throop would provide the engineering expertise for developing power transmission, oil fields, shipbuilding, and "the extensive hydraulic undertakings, such as the great aqueduct now being built [Owens Valley Project] to supply Los Angeles and the surrounding region with water." The argument empowered the expansionist designs of water czar Harry Chandler; his trusteeship of Throop for a decade and a half overlapped his development projects. With the same persuasiveness with which he solicited Carnegie and Rockefeller money, Hale convinced California's business leaders that their support of Throop College would increase their effectiveness in developing the region and, in turn, further the goals of Western civilization. Not everyone applauded. "You believe that aristocracy and patronage are favorable to science," James McKeen Cattell charged Hale a couple of years later, "I believe that they must be discarded for the cruder but more rigorous ways of pervasive democracy."[31] Cattell's instinctive mistrust of the influence of business on academia contrasted sharply with Hale's championship of private initiative and his rejection of government interference in science.[32]

By 1915, after five years of persistence, Hale had secured a commitment from his close friend at the Massachusetts Institute of Technology, chemist Arthur Amos Noyes, to join him in placing Throop at the center of the academic map. A son of a wealthy and scholarly New England lawyer, Noyes was Hale's academic and social counterpart. Founder of American physical chemistry, lover of poetry, and a genteel negotiator in the academy and the board room, Noyes also shared with Hale what Millikan later extolled as their "Nordic" roots. Hale and Noyes agreed that Robert A. Millikan of the University of Chicago should be the future leader of physics at Throop. In 1917 at the peak of the preparedness period, when the Gates Chemical Laboratory opened (the donation of lumber tycoon C. W. Gates), the two men lured Millikan to Throop through a sizable research fund, hoping that his part-time involvement would turn into a permanent commitment.[33]

In 1921, when he joined the California Institute of Technology (so renamed in 1920) as the director of the Norman Bridge Laboratory and chairman of the Ex-

Figure 4 A. A. Noyes, G. E. Hale, and R. A. Millikan on the steps of Gates Laboratory, ca. 1917. Courtesy of the California Institute of Technology Archives.

ecutive Council, the mature Millikan represented the quintessential American scientist. Raised in Maquoketa, Iowa, the son of a Congregational preacher, Millikan embodied the resonance between small-town traditional values and America's growing scientific cosmopolitanism. A graduate of Oberlin, his mother's alma mater, Millikan completed his doctorate in 1895 under Albert Michelson at the University of Chicago, followed by postgraduate work in Germany's top physics institutes. By 1913 he had attained international renown for his "oil drop experiment," yielding the value of the electron charge, for which he would win the Nobel Prize in 1923. He also gained national recognition for his managerial skills. He was an active member of America's scientific societies and was editor of scientific journals; as director of research for the National Research Council, he directed cooperative war projects for the army and navy. This period was one of close cooperation with physicist Max Mason, who would later become president of the Rockefeller Foundation. Equally important, Millikan emerged as a powerful orator and a popularizer, promoting science as a national resource and moral force on par with Christianity.[34]

Spearheading the campaign for private patronage of science and promoting Caltech's contributions to national power, Millikan trumpeted the lessons of the Great War. Neither soldier, officer, nor industrialist could remain in the lead without the brains of America's researchers. Anointing the scientist as "creator of wealth," Millikan described him as a frontier man, "the explorer who is sent on ahead to discover and open up new leads to nature's gold."[35] Following the rhetorical tradition of Johns Hopkins physicist Henry Rowland, Millikan did not argue so much for the obvious utility of applied science and engineering but for undirected research as the fountainhead of technology. Massive support of pure science, Millikan claimed, was the best long-term investment in economic and political might—not through the paternalism of government but by the private initiative of the captains of business and industry, whose self-reliance had generated America's wealth and power.[36]

Millikan expressed the ethnocentric conception of progress and social control shared by so many of his successful contemporaries: Private initiative and technological development marked an advanced stage of racial evolution. Innately driven, the Nordics continuously thrust forward, whereas Africans, Asians, and Jews (arrested at the stage of the Old Testament) idled below the evolutionary summit. "Why is it that 50 years of Europe is better than a cycle of Cathay?" Millikan posed rhetorically, "Is it not simply because in certain sections of the world, primarily those inhabited by the Nordic race, a certain set of ideas have got started in men's minds, the idea of progress and of responsibility?"[37] Confirmations of the thesis seemed to abound.

Like California's boosters, Millikan endowed the ideology of progress and Caltech's mission with the ideological significance of the Western frontier. "California marks now, as England did three centuries ago, the farthest western outpost of Arian [sic] civilization," Millikan proclaimed, echoing Widney.[38] Caltech stood at the vanguard of the "Pacific Era," he observed, quoting Roosevelt's panegyric of the "third civilization." The new institute captured the essence of the frontier, mirroring the region's "lack of tradition and its spirit of development" and embodying the visions of Caltech's trustees who developed the "Far West." A culturally embedded discursive practice, the frontier trope encompassed science: science as a seed of industrial growth and as a frontier in its own right—an "endless frontier" according to a later extrapolation by Vannevar Bush, director of the Carnegie Institution.[39]

During the 1920s Caltech's powerful triumvirate, Millikan, Noyes, and Hale, guided the Institute to national prominence in engineering and the physical sciences, and they initiated the plans for developing the life sciences. Under Hale's leadership the Mount Wilson Observatory matured into a world-class center of astronomy and astrophysics. His vision of interdisciplinary cooperation, of attracting to Pasadena leading physicists and chemists in conjunction with the astrophysics program, had materialized with remarkable success.[40] The Division of Chemistry and Chemical Engineering under Noyes attained national acclaim for its unique scientific programs. The innovative research in physical chemistry at the Gates Chemical Laboratory offered training in x-ray crystallography and methods of electron diffraction, the only center of its kind in America.[41]

With Millikan's towering presence, the Norman Bridge Laboratory attracted international attention, boasting such luminaries as Albert A. Michelson, Hendrik A. Lorenz, Paul Ehrenfest, and Arnold Sommerfeld as visiting scholars and such physicists as Paul Epstein and Richard Tolman on the permanent faculty. Laboratories equipped with the latest sophisticated apparatus housed the constellation of scientists, and National Research Council fellowships brought aspiring young physicists; Caltech became a training ground for a new generation of scientific leaders.[42] Max Mason not only shaped the career of his protégé Warren Weaver through Caltech, he groomed physicist Lee E. DuBridge for leadership. DuBridge, who spent the years 1926–1928 at Caltech, would succeed Millikan in 1946.[43]

A powerful combination of factors: abundant resources, novel institutional mechanisms, and innovative intellectual strategies sustained Caltech's remarkable growth and influence. As leaders of the National Academy of Sciences and the National Research Council, Hale, Noyes, and Millikan formed the nexus of America's establishment network; their effectiveness in mobilizing vast local resources was matched by their decisive influence on the policies of the large foundations. These policies, in turn, translated into enormous grants that enabled the Institute to construct new facilities and recruit scientific talent. "Whether the Research Council belongs to the National Academy, or the National Academy belongs to the Research Council, or both are satellites of Pasadena is a problem of three bodies that is difficult of solution," quipped Cattell in 1922. "The Carnegie Corporation, the Rockefeller Foundation and the National Research Council are another problem of three bodies," he went on, capturing the essence of political privilege and scientific power.[44]

Within the national network, Caltech's prominence derived from a distinctive organization of knowledge—the interdisciplinary nature of its scientific programs and an emphasis on cooperative research—a feature of principal importance to the implantation of the new biology at the Institute. Hale, an early advocate of interdisciplinary research, had always stressed the view that nature was not parceled into traditional scientific fields (physics, chemistry, or astronomy). By 1912 he had proposed that the National Academy foster "subjects laying between the old-established divisions of science, for example, in physical chemistry, astrophysics, geophysics, etc." Hale's intellectual program, buttressed by the lessons of cooperative war projects, persuaded the Academy during the 1920s to cultivate "borderland sciences."[45]

Unimpeded by intellectual and political residues of prior scientific traditions, Caltech seized the opportunity for erecting a new institutional structure designed to foster interdisciplinary research in borderland sciences. Such cooperative projects, drawing on the combined skills of chemists, physicists, mathematicians, and later biologists, demanded team work and flexibility. The Gates Laboratory was "not the place for the individualist in science or research, but for men who are interested in cooperating with one another," Noyes was fond of saying. The extreme individualist was out of step with the times. Although the lone genius did make important breakthroughs, Noyes admitted, the steady growth of research schools and scientific institutions depended on the cooperative effort.[46]

Noyes captured the spirit of the day. The term "cooperation"—the ideological modification of extreme laissez-faire—reverberated throughout the corporate world during the 1920s and 1930s. "Cooperation not individualism," heralded F. W. Taylor in his *Principles of Scientific Management* at the turn of the century.[47] "Today's business organization is moving strongly toward cooperation. . . . Cooperation in its current economic sense represents the initiative of self-interest blended with a sense of service," wrote Secretary of Commerce Herbert Hoover in *American Individualism* (1922). His ideal of cooperation reflected his notion of "the associative state," a manifold of cooperative institutions (e.g., trade associations, service organizations, professional societies), forming a type of private government.[48] A champion of big business and science, and a friend of Hale, Millikan, and Noyes, Hoover's prose empowered Millikan's own sermons on the blend of private initiative and the spirit of service and altruism that constituted the new scientific ethos.[49]

Caltech's call for cooperation resounded the theme "cooperation in research," which animated the plans of the Carnegie Corporation and guided the scientific policy of the Rockefeller Foundation. Mirroring its industrial and business allies, the new scientific enterprise no longer extolled the virtuosity of the individualist. Just as the multiunit business structure depended on the team player and on the manager who coordinated group projects, the new science relied on cooperative individuals and on the broader interests of scientific managers. Caltech's focus on cooperation, then, was not merely an intellectual strategy, it also was a corporate philosophy; the Institute was at the vanguard of changing social relations of science.

The ethos of scientific cooperation translated into an institutional structure that made Caltech into what John Servos aptly called a "knowledge corporation."[50] The Institute's organization resembled the structure of a large corporation, a multiunit research enterprise coordinated through a hierarchy of top, middle, and lower management. Instead of the traditional university structure consisting of a president, relatively remote from academic departments, and of deans—often inactive academic researchers—Caltech's top management consisted of an executive council, directly involved in overseeing research projects. As is common in corporate management, Millikan, chairman of the Executive Council, also served as the Institute's president, his vote equal in weight to those of the other Council members.

In turn, Millikan, Hale, Noyes, and after 1928 T. H. Morgan functioned as chairmen of their respective division councils; each council was comprised of full professors. In contrast to the traditional chairman, who could single-handedly set departmental policies, Caltech's division chairman had a vote equal to those of other members in the division council, a unit representing middle management. Each council coordinated its own project teams; a full professor would lead three or four research groups. Associate and assistant professors (lower management) directed smaller projects performed mainly by postdoctoral fellows and graduate students. This nested structure kept Caltech's academic hierarchy broadly involved in the Institute's projects. As in business corporations, the knowledge corporation's annual reports encapsulated the Institute's research activities and commodities.

Caltech's corporate structure was strengthened by the direct participation of Southern California's business magnates, who had helped build the Institute and shape its policies. Some of them served for more than a decade on Caltech's Executive Council, a group comprised of four faculty members (Hale, Noyes, Millikan, later T. H. Morgan, and political scientist William B. Munro) and four trustees. The oldest trustees included Harry Chandler, Henry Robinson, multi-millionaire lumberman Arthur Fleming (who by 1930 had donated about $5.2 million to the Institute), and Caltech's attorney, Henry O'Melveny. O'Melveny by the late 1920s represented diverse corporate interests such as Union Pacific Railroads, Goodyear Tire, Bank of America, Shell Oil, Proctor and Gamble, National Biscuit, and Paramount Pictures; his leading cases including real estate, water rights, and public utilities.[51] These eight men, according to Millikan, possessed equal responsibility and authority, and "came to know the Institute from A to Z."[52] Corporate interests, then, directly influenced Institute policies, exerting considerable weight on academic projects.

The trustees' influence was reinforced by the "Institute Associates," a group of 150 men, mainly from Southern California's affluent business circles, who played a crucial role in Caltech's ascendancy. Millikan considered March 23, 1925—the date O'Melveny incorporated the associates—the most significant in Caltech's history; the incorporation constituted a vote of confidence and a pledge of support.[53] Associates poured large sums into new buildings, laboratories, and research funds and kept the Institute afloat throughout the Depression. The key to their effectiveness, according to Millikan, was their constant presence. They constituted an integral part of Caltech's community, visiting the laboratories, participating in Institute functions, and keeping "so close to its workings that they came to understand it—to see with their own eyes that a dollar put into it brought more values to the community, more social returns—than a tax dollar ever does, also more than most philanthropic dollars do." In a single stroke of Republican oratory, promoting voluntarism and downgrading regulation, Millikan underscored that it was the Associates' support that "sold the Institute to the Foundations."[54]

To the local community, a first-rate institution of pure and applied science was a cultural symbol and a practical investment, first in the physical sciences and later in the life sciences. Whereas the Institute's excellence in theoretical work in physics and chemistry generated civic pride, it was applied science that ensured the practical men of affairs tangible returns. Institute projects such as the high voltage engineering research for producing electric power from the Hoover Dam and the program in aerodynamic engineering attested to the industrial utility of research. The same kind of dynamic characterized the effort to develop the biological sciences at Caltech during the early 1920s. The argument for building a first-rate research program in biology was intellectually compelling: Biology would balance the Institute's science curriculum. Life science projects at the Institute also promised to contribute to the welfare of the community—through indirect contributions to the region's agribusiness, for example.[55] First and foremost, though, the rapid industrialization and population growth created in Southern California an acute need for medical facilities: hospitals, clinics, medical laboratories. The region, Noyes stressed, would greatly benefit from a medical research program conducted in cooperation with local physicians and Caltech's biologists, chemists, and physicists.

Plans for developing biological research at Caltech, in cooperation with the divisions of physics and chemistry, were proposed as early as 1922. Three years before Hale invited geneticist T. H. Morgan to lead biological research at Caltech, a local physician, Bernhardt Smith, approached Noyes with a proposal. He offered the income from his medical practice to develop insulin at the Gates Laboratories. Noyes seized the opportunity. In the spirit of Hale's "think big" philosophy, Noyes envisioned this small project as the start of biomedical research at Caltech, a springboard for large-scale developments, supported by the Carnegie and Rockefeller Foundations.[56]

The project proved to be a boon to the chemistry division and to Caltech. Supply of the highly active insulin exceeded local demand, generating in turn an elaborate scheme for a biomedical enterprise at the Institute. "I am writing you briefly, after consulting Evalina [Hale's wife], to acquaint you with dreams which are now floating about the California Institute of Technology," wrote Millikan to Hale in August 1923.

> Rose, President of the General Educational Board, and Prichette [acting president of the Carnegie Corporation] have both been here within the month and coincide in the general view that if anything is done in Southern California in the field of biochemistry, biophysics, and medical education, it must be done in immediate contact with the present work of the Institute. Rose says that Welch's [bacteriologist William Welch of Johns Hopkins Medical School] plant is suffering already from lack of contact with physics and chemistry. He says the Rockefeller Board will not be interested in any medical plan in Southern California which is farther away than across the street at most from the Institute.[57]

The figures tossed around in these schemes were in the order of $10 million. Caltech Associate Henry E. Huntington, in full swing of his own plan to build a hospital in Los Angeles, had to be consulted; O'Melveny cabled Huntington's advisor in Europe to delay the decision until Huntington had discussed his plans with the Caltech group. Millikan had already picked the appropriate site on campus, chosen the architects, and outlined the general features of the new medical complex. "I can see nothing so important as the intensive pursuit of all the sciences, including the biological at the Institute," he wrote to Hale in a celebratory spirit. "This will leave other schools in Los Angeles free to pursue intensively the humanities, social science, and law. This would be the finest thing that could possibly happen to these other institutions."[58]

In October 1923, the Carnegie Corporation awarded the Gates Laboratories $10,000 for insulin research, renewing the grant the following year. John J. Abel, the physiological chemist from Johns Hopkins, noted for his work on adrenaline and pituitrin, came to Caltech in 1924 to coordinate the new research. By 1925 his team had isolated a crystallized form of insulin, and Abel was asked to remain at the Institute to head the plans for medical research.[59] During 1924–1925 Noyes submitted to Millikan several schemes for developing biological research, all as a service to medicine. A proposal for a biology department and an affiliated medical school in Los Angeles called for departments of research and graduate study in organic chemistry, biochemistry, biophysics, and bacteriology and later in evolutionary biology. The undergraduate biology curriculum would conform to Caltech's four-year requirements in physics and chemistry as well as prepare students for entering either a doctoral program in biology or a high-grade medical school.[60]

Noyes' strategy made good intellectual sense. The biochemistry and biophysics laboratories of the new biology division would cooperate with the chemistry and physics divisions, complementing and completing the fundamental science curriculum. The plan was also well situated within the political economy of the region and reflected the wider cultural mission of the institution. Noyes pointed out that biology at the Institute would serve the needs of sanitary engineering and municipal hygiene, and it would add "a highly important cultural study in the all-round training of a broad type of engineer, chemist or other scientific man." First and foremost, however, a biology division would fill Southern California's great lacuna in medical research. Caltech would be "a center for biological research, to which the medical men of this community may look for inspiration and for knowledge as the latest advances in medical science, and to which they may bring their larger problems for research."[61]

In conjunction with the planned biology division, Noyes proposed an Institute of Biological and Medical Research (modeled after the Rockefeller Institute) to be located close to Caltech and closely affiliated with it, either as a separate division or linked through an interlocking Board of Trustees. The staff of the medical center, in cooperation with the Biology group would conduct fundamental research on physiology and special diseases (such as pneumonia or nephritis). The great center of biological and medical research in Pasadena apparently was to start out as the "Metabolic Research Laboratory" under the directorship of Dr. Lorena Breed, assuming that at least $50,000 for building and equipping the laboratory could be secured and that $250,000 for endowment would be obtained. Excess income from the endowment and the income from $100,000 contributed by Caltech would be used "for associating with the proposed research an outstanding research man in the field of physiology or biochemistry (such as Dr. Burrows of St. Louis or Dr. Abel of Baltimore)."[62] The new institute, in cooperation with the medical men of Southern California, would sustain a small research hospital nearby or arrange for a special ward attached to the Pasadena Hospital.

To avoid some of the dangers inherent in this vast plan, Noyes proposed that the main clinical center would be located in Los Angeles. He cautioned of "the very great danger that in the future the main work of the Institute in the basic sciences and in engineering would be dominated by the immense medical development, that its prestige as a *scientific* school would be lost and that it would be difficult to secure the funds needed for its own proper development."[63] His concern (shared by Millikan and Hale) about the dominance of clinical medicine over pure science paled in comparison to objections later raised by T. H. Morgan.

John J. Abel, however, did not accept a permanent position at Caltech. The biochemical research he had initiated came to a halt in 1925 and with it the plans for the proposed Metabolic Research Laboratory. An endowment from Rockefeller's General Education Board had not been secured, and biomedical research at Caltech remained just a grand plan. The institutional machinery had been set in motion, though, and community interest had been stimulated; there were several pledges of support, and the Fleming Trust of about $5 million was set up for the proposed biological laboratories. It was imperative to find a strong leader for the new venture within a short time.[64]

Notes

1. Frederick J. Turner, *The Frontier in American History* (New York: Henry Holt & Co., 1920), pp. 2–3; and Alan Trachtenberg, *The Incorporation of America: Culture and Society in the Guilded Age* (New York: Hill & Wang, 1982), Ch. 1 and his bibliographic essay on pp. 236–238.

2. F. J. Turner, *The Frontier in American History*, p. 38 (see Note 1); and Mircea Eliade, *The Sacred and the Profane* (New York: Harcourt Brace Jovanovich, 1959), Chs. 1 and 2. On Turner and the new history see Peter Novick, *That Noble Dream: The "Objectivity Question" and the American Historical Profession* (Cambridge: Cambridge University Press, 1988), Chs. 1 and 2, passim. On Rooseveltian ideology see Donna Haraway, "Teddy Bear Patriarchy: Taxidermy in the Garden of Eden, New York City, 1908–36," in her *Primate Visions: Gender, Race and Nature in the World of Modern Science* (New York: Routledge, Chapman, & Hall, 1989), pp. 26–59.

3. Carey McWilliams, *Southern California Country* (New York: Duell, Sloan, & Pearce, 1946), Chs. 6 and 7; and Kevin Starr, *Inventing the Dream: California Through the Progressive Era* (New York: Oxford University Press, 1985), Ch. 2.

4. C. McWilliams, *Southern California Country*, Chs. 3, 5, and 15 (see Note 3).

5. C. McWilliams, *Southern California Country*, Ch. 7; Starr, *Inventing the Dream*, p. 49 (see Note 3 for both references).

6. F. J. Turner, *The Frontier in American History*, pp. 309–310, 357–359 (see Note 1).

7. Robert Gottlieb and Irene Wolt, *Thinking Big: The Story of the Los Angeles Times, Its Publishers and Their Influence on Southern California* (New York: G. P. Putnam's Sons, 1977), quote on p. 132. See also Mike Davis, *City of Quartz: Excavating the Future in Los Angeles* (New York: Verso, 1990), Chs. 1 and 2.

8. C. McWilliams, *Southern California Country*, Ch. 10 (see Note 3).

9. R. Gottlieb and I. Wolt, *Thinking Big*, Chs. 7–9 (see Note 7); Frederic Cople Jaher, *The Urban Establishment: Upper Strata in Boston, New York, Charleston, Chicago, and Los Angeles* (Urbana: University of Illinois Press, 1982), pp. 623–625, 662–663; and "Harry Chandler,"*National Cyclopaedia of American Biography, 40* (1955), pp. 498–499.

10. Walton Beam and James J. Rawls, *California: An Interpretive History* (New York: McGraw-Hill, 1983), Chs. 24–26; Royce D. Delmatier, Clarence F. McIntosh, and Earl G. Waters, eds., *The Rumble of California Politics, 1848–1970* (New York: John Wiley & Sons, 1970), Ch. 7.

11. F. C. Jaher, *The Urban Establishment*, pp. 628–629 (see Note 9).

12. K. Starr, *Inventing the Dream*, pp. 89–92 (see Note 3).

13. Ibid.

14. C. McWilliams, *Southern California Country*, pp. 150–164 (see Note 3); and Michael W. Miles, *The Odyssey of the American Right* (New York: Oxford University Press, 1980), Ch. 13.

15. R. Gottlieb and I. Wolt, *Thinking Big*, Chs. 5, 6, and 12 (see Note 9).

16. Carey McWilliams, *Factories in the Field* (Boston: Little, Brown, 1940); and R. D. Delmatier, C. F. McIntosh, and E. G. Waters, *The Rumble of California Politics*, pp. 216–217, 235 (see Note 10).

17. PPC, Robert Freeman, "The Whole Smith Family," *Pasadena Chimes, No. 9* (1921), pp. 6–7.

18. PPC, Robert Freeman, "Sinker and Float," *Pasadena Chimes, No. 18* (1926), pp. 4–6.

19. PPC, 1921 file, July 28, 1921, "A Listener" (Los Angeles) to Freeman. A graduate of the Princeton Seminary and an admirer of H. A. Fosdick, Freeman served as pastor of Pasadena's Presbyterian church for more than three decades (1910–1945?), during which time he also authored two books, poetry collections, and a regular column in *The Continent* (the national Presbyterian publication). A consideration for the pastorate of Riverside Church in 1919 left a paper trail of evaluation letters from scores of church leaders, documenting Freeman's Republican politics, his theology, and, above all, his power of oratory and remarkable popularity. PHS, Freeman file.

20. John Higham, *Strangers in the Land* (New York: Atheneum, 1974), pp. 74, 241, 260; and Donald Pickens, *Eugenics and the Progressives* (Nashville: Vanderbilt University Press, 1968), pp. 96–97.

21. Robert Freeman, "Qualifying for Survival," *Pasadena Chimes, No. 22* (ca. 1920s), p. 3.

22. After its founding in 1928, Freeman joined Pasadena's Human Betterment Foundation, a eugenic organization devoted to mass sterilization of the unfit. See p. 83 in chapter 3.

23. See R. Gottlieb and I. Wolt, *Thinking Big*, Chs. 5 and 6 (see Note 9) for numerous quotes from the *Los Angeles Times* on the destructive and irrational behavior of labor organizers.

24. On the connection between the Nietzschean dichotomies of Appollonian and Dionysian, and Ferdinand Tonnies' differentiation between *gemeinschaft* (community) and *gesellschaft* (society), see Arthur Mitzman, *Sociology and Estrangement: Three Sociologists of Imperial Germany* (New York: Knopf, 1973).

25. YUB, Henry O'Melveny, "Spanish and Mexican Land Grants in Southern California," pp. 32–34, address at the California Institute of Technology, May 9, 1935.

26. See for example, Leo Marx, *The Machine in the Garden: Technology and the Pastoral Ideal in America* (New York: Oxford University Press, 1964), Chs. 3 and 4; and Jean Comaroff, "The Diseased Heart of Africa: Medicine, Colonialism, and the Black Body," in John L. and Jean Comaroff, eds., *Ethnography and the Historical Imagination: Selected Essays*. (Boulder, CO: Westview Press, 1992).

27. Robert H. Kargon, "Temple to Science: Cooperative Research and the Birth of the California Institute of Technology," *Historical Studies in the Physical Sciences, 8* (1977), pp. 3–31; Daniel J. Kevles, *The Physicists* (New York: Vintage Books, 1979), pp. 109–111.

28. F. C. Jaher, *The Urban Establishment*, p. 645 (see Note 9).

29. Ibid, Ch. 6, passim; and "Henry M. Robinson," *The National Cyclopaedia of American Biography, 30* (1943), pp. 127–128.

30. F. C. Jaher, *The Urban Establishment*, Ch. 6, passim but especially p. 636 (see Note 9).

31. Quoted in Nathan Reingold, "National Aspirations and Local Purposes," *Transactions of the Kansas Academy of Science, 71* (1968), pp. 241–42; see also Robert H. Kargon, *The Rise of Robert Millikan* (Ithaca: Cornell University Press, 1982), pp. 119–121.

32. R. H. Kargon, "Temple to Science," p. 17 (see Note 27).

33. R. H. Kargon, "Temple to Science," pp. 10–12 (see Note 27). See also, John W. Servos, *Physical Chemistry from Oswald to Pauling: The Making of Science in America* (Princeton: Princeton University Press, 1990), especially Ch. 6.

34. R. H. Kargon, *The Rise of Robert Millikan*, Chs. 2–4 (see Note 32).

35. Quoted on p. 3 in Robert H. Kargon, ed., *The Maturing of American Science* (Washington, DC: American Association for the Advancement of Science, 1974), Introduction.

36. Robert A. Millikan, "The Practical Value of Pure Science," in *Science and Life* (New York: Books for Libraries Press, 1964), pp. 1–12; and R. A. Millikan, in "The Relation of Science to Industry," *Science and the New Civilization* (New York: Books for Libraries Press, 1958), pp. 32–51.

37. R. A. Millikan, *Science and Life*, p. 11. In his addresses, speeches, and articles from the 1920s through the 1940s, Millikan identified the destiny of the Anglo-Saxon race with its technological superiority, wealth, and power. See, for example, R. A. Millikan, "Science and Modern Life," in *Science and the New Civilization*, pp. 1–31; and Millikan, "Science and Religion," in *Science and Life*, pp. 38–64 (see Note 36 for all references).

38. CIT, Fleming Papers, Box 1.1, R. A. Millikan, "The California Institute of Technology," (ca. 1920s), p. 4.

39. Ibid. For Millikan's explication of Roosevelt's discourse on the "Pacific Era," see *The Autobiography of Robert A. Millikan* (New York: Prentice-Hall, 1950), p. 230. On Vannevar Bush (Washington, DC: National Science Foundation, reprinted 1980) *Science the Endless Frontier* (1945) and later usages of the frontier metaphor see Interlude II and Ch. 8 in this volume.

40. R. H. Kargon, "Temple to Science," pp. 19–31 (see Note 27).

41. John W. Servos, "The Knowledge Corporation: A. A. Noyes and Chemistry at Cal-Tech, 1915–1930," *Ambix, 23* (1976), pp. 176–186.

42. R. H. Kargon, "Temple to Science," pp. 26–27 (see Note 27).

43. For the rise of Lee DuBridge, see Interlude II, this volume.

44. Quoted in R. H. Kargon, *The Maturing of American Science*, p. 8 (see Note 35).

45. Rexmond C. Cochrane, *The National Academy of Sciences* (Washington, DC: National Academy of Sciences, 1978), p. 327; D. J. Kevles, *The Physicists*, Chs. 8 and 9 (see Note 27); D. J. Kevles, "George Ellery Hale, the First World War, and the Advancement of Science in America," *Isis, 59* (1968), pp. 427–437.

46. Quoted in R. H. Kargon, "Temple to Science," p. 21 (see Note 27); see also J. W. Servos, "The Knowledge Corporation," pp. 176–186 (see Note 41).

47. Frederick W. Taylor, *Principles of Scientific Management* (New York: W. W. Norton, 1967, reprint of the 1911 edition), p. 140.

48. Herbert Hoover, *American Individualism* (New York: Doubleday, Page, 1922), pp. 44–45; and Ellis W. Hawley, "Herbert Hoover, the Commerce Secretariat, and the Vision of the Associative State, 1921–1928," *Journal of American History, 61* (1974), pp. 116–140.

49. R. H. Kargon, *The Rise of Robert Millikan*, passim but especially p. 162 (see Note 32).

50. Servos' term, "knowledge corporation," referred to the corporate structure of the Gates Laboratory, a structure later adopted by the Kerckhoff Biology Laboratory. A brief explanation of Caltech's administrative structure appears in Millikan's *Autobiography*, pp. 226–228 (see Note 39).

51. "Henry O'Melveny," *National Cyclopaedia of American Biography, 45* (1962), pp. 290–291.

52. R. A. Millikan, *Autobiography*, pp. 226–227 (see Note 39).

53. Ibid., p. 239.

54. Ibid., p. 250.

55. Ibid., pp. 230–237.

56. J. W. Servos, "The Knowledge Corporation," pp. 178–179 (see Note 41).

57. Millikan to Hale, August 28, 1923. Quoted in Nathan Reingold and Ida Reingold, eds., *Science in America* (Chicago: University of Chicago Press, 1981), p. 317.

58. Ibid., p. 318.

59. J. W. Servos, "The Knowledge Corporation," p. 179 (see Note 41); Robert Kohler, *From Medical Chemistry to Biochemistry* (Cambridge: Cambridge University Press, 1982), pp. 319–320.

60. CIT, Millikan Papers, Box 18.7; Biology 1924–1925, Noyes to Millikan.

61. Ibid., "Plans for the Development of Biology at the California Institute of Technology," Noyes to Millikan, March 20 (1924–1925 file), p. 1.

62. Ibid., "Plans for the Development of Biology at the California Institute and Its Relation to a Medical School in Los Angeles," Noyes to Millikan, February 4 (1924–1925 file), pp. 1–3.

63. Ibid., "Outline of a Possible Plan for the Cooperation of the California Institute in Developing the Proposed Metabolic Research Laboratory," Noyes to Millikan (1924–1925 file), pp. 1–3.

64. Ibid., "Plans for the Development of Biology at the California Institute and Its Relation to a Medical School in Los Angeles," Noyes to Millikan, February 4 (1924–1925 file), pp. 1–3.

CHAPTER 3

Visions and Realities: Biology Division During the Morgan Era

Morgan and the New Biology: Problem of Service Role

When Hale approached Thomas Hunt Morgan during the spring of 1925, Morgan's presence towered over American biology: innovator, champion of fundamental research, and a statesman of science. Raised during the 1870s in an upper crust milieu in Lexington, Kentucky, Morgan was no stranger to capitalist fortunes and political power. The Hunt enterprises (his grandfather) had reached unprecedented regional wealth, and the Morgan name evoked memories of Confederate pride and international diplomacy; T. H. Morgan belonged to a "cavalier stock," as it was called in the South. A Ph.D. degree from the newly founded Johns Hopkins University in 1891 placed T. H. Morgan in the first generation of American-trained doctorates—symbol of American academic self-sufficiency. From his early research in embryology at Bryn Mawr to his leadership of genetics at Columbia, Morgan's trajectory mirrored the maturing of academic biology. Like Caltech's triumvirate, Morgan was a member of the nation's science aristocracy, sharing its "Nordic roots," social privilege, and professional visions.[1]

Yet Morgan's candidacy posed a paradoxical choice for Caltech's projected enterprise. Although he stood as a great leader of biology, he was surely least likely to promote biology as a branch of medicine. Beyond charting new cognitive pathways, Morgan's research program signified the disciplinary clout of a nonmedical academic biology. *Drosophila* genetics epitomized the autonomy of graduate-level research extricated from service roles to agriculture and medicine and, by the 1920s, disavowing most of its earlier ties to eugenics. Within the national ecology of knowledge, *Drosophila* genetics inhabited the prestigious gray zone of pure science.[2]

77

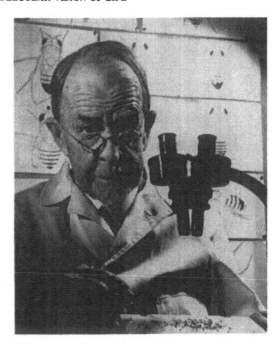

Figure 5 T. H. Morgan.
Courtesy of the California
Institute of Technology
Archives.

Hale, Millikan, and Noyes had implicit confidence in Morgan's managerial style, activities that had placed American genetics on the international map of science. A prolific writer, Morgan by 1925 had published several books and nearly a hundred articles on genetics and its relation to cytology, embryology, and evolution. Spanning a broad spectrum of specialized and popular journals, Morgan's exposition commanded a wide audience. The *Drosophila* group as a whole (Morgan, Hermann, J. Muller, Alfred H. Sturtevant, and Calvin B. Bridges) altered the undergraduate biology curriculum by publishing a laboratory manual on *Drosophila* experiments for college courses in genetics.[3] Concentric circles of the *Drosophila* influence spread from Morgan's own center as well as through his disciples in the United States and Europe: Norway, Sweden, Denmark, and the Soviet Union (and to a lesser extent Germany, England, and France). His administrative activities buttressed his influence: service on editorial boards of biological journals and professional societies, and the presidency of the Society of Experimental Biology and Medicine for the years 1910–1912.[4]

In one respect, though, Morgan differed radically from Hale, Noyes, and particularly Millikan. In his own reserved manner, Morgan relished unconventionality and nonconformity; he was a perpetual devil's advocate, an imp. Cultivating an unorthodox manner, he delighted in shocking people—as he would when flaunting his militant atheism at Millikan's evangelical torrents—and hid his softer, more conventional side under a protective cover of eccentricity. He dressed himself so poorly that there was at least one occasion when the laboratory janitor was taken to be Professor Morgan—and Professor Morgan to be the janitor.[5] This extreme form of individualism, however, was more a display of style than substance. His cultivated eccentricity blended smoothly with the instincts and social skills of the well-born, a modus operandi that also exemplified his science. He merged strong personal initiative with group effort. The success of his *Drosophila* program, coupled with the founding of a research school, attested to the ethos and effectiveness of cooperative individualism.

Hale and Millikan had only minor disciplinary and intellectual ties to life science, but they knew Morgan well through his activities in the American Association for the Advancement of Sciences, the National Research Council, and the National Academy of Sciences. In fact, they voted for Morgan in his 1927 election to the presidency of the Academy, and Morgan together with Millikan served as vice-chairmen of the Academy's executive board. As friend of Abraham and Simon Flexner (friendships dating back to their boyhood days in Lexington), Morgan moved gracefully through the corridors of the biomedical establishment and was frequently invited to serve on boards of medical and biological institutions.[6]

Like his Caltech counterparts, Morgan's institutional network centered around the large foundations: the Carnegie Corporation and the Rockefeller Foundation. The continuous aid of the Carnegie Corporation had by 1928 supported 20 doctoral theses in genetics, Morgan acknowledged. "These students are now scattered over the world and many have become centres [*sic*] of *Drosophila* enlightenment," he boasted to John C. Merriam, president of the Carnegie Institution of Washington. "I am writing this, not to boast of our accomplishments but to set down what we have been doing and add that there is every reason to think that we can continue to carry on in the same way as in the past."[7] His academic enterprise benefited from the patronage of the Rockefeller Foundation. American postdoctoral fellows trained under Morgan through fellowships of the Foundation's sponsored National Research Council, in which Morgan was an active member. The International Education Board supported the work of Rockefeller Fellows; by 1928 they included Norwegian geneticist Otto L. Mohr, Theodosius Dobzhansky from Russia, and Curt Stern from Germany.[8]

Like Hale, Noyes, and Millikan, Morgan cultivated a grand vision. Even before the opportunity to build a new program at Caltech emerged, Morgan had been self-consciously molding the future of biology, pinning the advance of the field on an intensified cooperation with physics and chemistry. He admired the work of Jacques Loeb and reciprocated Loeb's accolades. The grip of the mechanistic view of life had already during the 1910 and 1920 decades turned Morgan into an enthusiastic advocate of a new cooperative physicochemical biology. His efforts to promote the Loebian ideal in biology led him to a collaborative educational project with Loeb and general physiologist W. J. V. Osterhout (Loeb's successor as head of physiology at the Rockefeller Institute in 1924). Together, the three edited the noted Lippincott series, publications devoted primarily to promoting the mechanistic approach to life.[9]

Thus in most respects, Morgan's acceptance of the Institute's offer in 1927 attested to a close match between his own visions for biology and those of Caltech's academic leaders. The intellectual and institutional advantages for implanting the new biology at Caltech were compelling. The institutional mechanisms for inter-disciplinary cooperation were already in place; the requisite ethos of teamwork had animated the Institute's projects since its founding. As the hub of America's scientific establishment, Caltech attracted massive resources through its links to industry, business, and the foundations. Funds for biological research abounded, especially as the Rockefeller Foundation was just then embarking on large-scale support of the human sciences, with clear emphasis on the problems of heredity. Moreover, the splendid track record of community support of Caltech now prom-ised to extend to the life sciences. Local support was doubly crucial, as the phi-lanthropic foundations operated on the basis of matching funds.

What of the medical promise and the local relevance of the new biology? How would the new division validate its purported contributions to the region's devel-opment and welfare? The business community, of course, took pride in the ad-vancement of fundamental research, but it was utility that ultimately measured scientific progress. Fund-raising efforts by Hale, Noyes, and Millikan repeatedly underscored both explicit and implicit returns: hydroelectric power, chemical pro-cessing, aeronautics, medical facilities, and racial progress. Although Morgan's appointment hardly required intellectual justification, the social legitimation of the new enterprise—its local utility and political economy—did pose difficulty. Mor-gan's leadership meant resistance to the touted medical goals of the new biology; it also entailed activating pledges while skirting community expectations.

When Morgan moved to Pasadena in 1928, the Kerckhoff Laboratory was al-ready under construction. The endowment by Caltech trustee William Kerckhoff—lumber magnate and one of the principal developers of hydroelectric power in Southern California—represented a response to earlier calls of developing medical research in the region. Morgan's plans for the new division made no mention of biomedical groups, however. His scheme included the development of five de-partments—genetics, embryology, general physiology, biophysics, and biochem-istry—which in cooperation with the divisions of physics and chemistry would focus on the study of fundamental, rather than clinical, vital processes. Given the option of assembling either a purely research division or a division that would accept undergraduate and graduate students and knit itself into the general plan of the Institute, Morgan selected the latter. He thus set a precedent for biological edu-cation: The biology division admitted only those undergraduates and graduate students able to meet the Institute's unmodified requirements in mathematics, physics, and chemistry. With this policy, Morgan reasoned, the division would stress the physicochemical side of biology right from the start. At the same time his choice fulfilled Caltech's goal of rounding the Institute's science curriculum.[10]

The issue of service role remained shrouded in ambiguity. Morgan fully agreed with Millikan on the marginal ties between Caltech's biology and the region's agriculture. A plant physiology group would be of service to California's agribu-siness and attract support from corporations and public agencies. However, with several agricultural colleges in California already flourishing, the uniqueness of Caltech's biology division would diminish if it were to follow a similar path.[11] The connection to medicine was far more complex. As champion of fundamental re-search, Morgan, like his colleagues earlier (C. O. Whitman, E. B. Wilson, J. Loeb,

Figure 6 William G. Kerckhoff Laboratories of the Biological Sciences. Courtesy of the California Institute of Technology Archives.

and F. R. Lillie), had steadfastly resisted the development of biology as a service to medicine. Their mission had been to foster for biology an independent disciplinary identity. Institutional histories testified to the inverse relation between the health of academic biology and the growth of the medical curriculum. Although his earlier research in embryology bordered on the fringes of physiology, traditionally a medical field, Morgan himself had no medical training or interest in medicine. Genetics distanced him intellectually from medicine even further. Despite his administrative and social ties to the biomedical community, Morgan wished to see biological research freed from the influence of the hospital.[12]

Morgan rejected Noyes' plan for building a medical school and hospital facilities at Caltech and for offering medical education. Caltech presented a unique opportunity for creating a new institutional niche, an insurance policy that biology would not degenerate into a handmaiden to medicine. In the 1928 *Bulletin of the California Institute of Technology*, when first introducing the new biological curriculum, Morgan advertised this point as a strength: "Most physiological laboratories had in the past, for practical reasons been associated with medical schools; and few of them have been in intimate contact with research staffs, and had the use of research facilities, of laboratories which are primarily devoted to fundamental investigations in the physical sciences."[13] By contrast, at Caltech, researchers in genetics, embryology, physiology (plant and mammalian), biochemistry, and biophysics would collaborate with physicists and chemists. By 1929 he appointed department heads in all areas except animal physiology. The marginalization of medical interests was explicit.

What therefore would be the service role of the new biology, its social legitimation? Still hovering in the background was the promise and peril of biological race betterment. Even more so than for the Rockefeller Foundation, from a cultural standpoint, the interest in Morgan's work and in the new biology was informed by the eugenic sentiments of the affluent circles of the region. A plea for pure science could hardly untangle the Gordian knot of genetics and eugenics, especially in Pasadena's conservative milieu during the late 1920s. The shared ideological commitments of the community and the Institute, their Anglo-Saxon self-conception, and their Aryan civilizing mission was not likely to be deflected simply through an operational cleavage between genetics and eugenics. Morgan's enterprise signaled the promise of biological improvements of the race. It did so despite the absence of a programmatic conjunction of the new biology and local eugenic interests that contributed financially to Caltech's project.

The eugenic movement in Southern California, of course, predated Morgan's arrival. Paul Popenoe had by then attained national recognition, and Ezra S. Gosney, Pasadena's upright community leader and Caltech Associate, had been active in eugenic and genetic organizations for nearly two decades. A native of Kentucky and a former corporate lawyer, Gosney typified the wealthy transplanted midwesterner; having moved to Pasadena in 1910 he quickly established himself as director of several large corporations and civic projects. His Spencerian devotion to the "problem of preparing young people for the struggle of life" guided his leadership in the Boy Scouts of America and the YMCA. The same ideology inspired his participation in the American Eugenics Society, American Social Hygiene Society, Eugenic Research Association, American Association for the Study of the Feeble-Minded, and American Genetics Association; abroad he participated in the Eugenics Society of London, the Institute International d'Anthropologie (Paris), Deutsches Bund Volksaulartung Erbkunde (Berlin), and the International Conference for Social Work. The founding of a biology department in Pasadena around the world's leading geneticist undoubtedly promised new opportunities for Gosney's eugenics crusade.[14]

In 1928, the same year Caltech's biology division was founded, Gosney consolidated his diverse eugenic commitments to become president of the Human Betterment Foundation. A 25-member elite organization committed to mass sterilization of the unfit, the Foundation boasted a distinguished roster of men of letters and men of affairs. During its 14-year life-span the names of Stanford's David Starr Jordan and Lewis M. Terman graced the Foundation's stationary. Its membership included officer and trustee R. A. Millikan, who by then had identified the escape from Malthusianism as the world's greatest scientific task; Vice-President William B. Munro, former Harvard scholar, advisor to the Rockefeller Foundation, and after 1930 Caltech professor of political science; and trustees Henry M. Robinson and Harry Chandler. Members of the clergy—Rev. Merle N. Smith, pastor of Pasadena's First Methodist Church, and Rev. Robert Freeman—supplied Gosney's organization with moral authority.[15]

The purpose of the Human Betterment Foundation, Gosney stressed, was not original research but coordination of new knowledge coming out of leading biological research laboratories and then the spread of that information to social workers, educators, and physicians nationwide. "Strong, intelligent, useful families are becoming smaller and smaller," the Foundation's brochure documented.

> Irresponsible, diseased, defective parents on the other hand, do not limit their families correspondingly. There can be but one result. That result is race degeneration. The law of self-preservation is as necessary for a nation as for an individual. When families that send a child to an institution for the feeble-minded average twice as large as families that send a child to the university, it is time for society to act.[16]

During the years 1929–1930 the Foundation conducted an exhaustive study of 6000 sterilizations of the eugenically unfit, followed eight years later by a similar survey of 10,000 cases. Reams of statistics on "social degeneration," sterilization and follow-up data on "pleased patients," and the benefits passed on to society buttressed the campaign; the book *Sterilization for Human Betterment*, pamphlets, and articles recorded the merits of Gosney's enterprise.[17]

Caltech's eugenic connection must have placed Morgan in an awkward position of both disapproval and tolerance. Like other leading geneticists, Morgan had promoted the aims of eugenics during its early years, but by 1915 he had severed his ties with the eugenic movement. He objected to the reckless uses of genetics for political ends, especially the propaganda tactics of Popenoe. Yet Morgan's opposition surfaced only in a few private communications and addressed mainly style rather than aims.[18] Even during the mid-1920s, at the height of the debates over immigration restriction and eugenic sterilization, Morgan did not publicly oppose eugenic programs per se. On both cognitive and social grounds, he might have opposed extreme versions of genetic determinism and negative eugenics, but he did not object to broader interventions in social evolution.

If Morgan's 1924 Mellon Lecture "Human Inheritance," may be used to judge his views on eugenics, it shows him to be uncommited. Extrapolating from *Drosophila* work, he stressed that complications due to genetic linkages and the polygenic and pleiotropic aspects of heredity were likely to greatly complicate the simplistic schemes for controlling human breeding. Evolutionary considerations introduced other difficulties: Is it more advantageous to breed for variability or uniformity? Furthermore, Morgan argued, social and economic inheritance modified the effects of biology. A person "may not inherit the bodily or mental characters that his parents have acquired through training, but in another way he inherits the results of their experience." Memory, speech, writing, education, and economic resources contributed to man's evolutionary fate. Though heredity was the primary determinant in human evolution, according to Morgan, exceptionally creative individuals—artists and ministers, for instance—may shape social evolution more favorably through their labors than through their quota of offspring.[19]

Morgan underscored Galton's conclusion "that nature (heredity) plays a far more important role than nurture (environment)," but genes were not the sole determinants of human evolution. He therefore believed that the geneticist alone could not hope to solve a problem as complex as human progress. Competent specialists—psychologists, physiologists, pathologists, anthropologists, economists, and statisticians—should join the geneticists in investigating the problem of social evolution.[20] Morgan did not oppose sociobiological engineering; he certainly did not publically condemn the Nordic sentiments animating the eugenic movement. Had this been the case, he would have objected to its racial discourse two decades earlier before lending his name to the eugenic crusade.

Although he disapproved of Davenport-style eugenics, Morgan in fact shared some of the tacit nativist beliefs of America's ruling elite during the 1920s. "Morgan had definitely an ambivalent social attitude," recalled the Russian geneticist Theodosius Dobzhansky:

> He was distinctly *against* social claims and pretentions [*sic*]. He was outspokenly against eugenics, which at that time was race and class eugenics. Rationally, he would—I don't remember him specifically speaking about it, but I am sure that rationally he would take the standpoint that everybody is born equal. At the same time, he had the curious subconscious division of people into a kind of superior and inferior class. . . . He had a curious classification which was important for us. Russians to him were of two varieties, the "black" and the "white." The black and the white were not referring either to political parties or, as far as I know, not even to pigmentation. But somehow or other, the old man had an idea that there is a kind of dimorphism there. Thank God, my wife and myself were classified among the "white"![21]

Morgan, however, did separate his cultural instincts from the rigors of the laboratory bench; his private criticisms of eugenics were directed primarily against the superficiality and unprofessionalism of the extremists, against the zealots and propagandists. "I believe," he concluded his Mellon Lecture, "that they [men] will not much longer leave their problems in the hands of amateurs and alarmists, whose stock in trade is to gain notoriety by an appeal to human fears and prejudices— an appeal to the worst and not to the best sides of our natures."[22] There is little doubt that in Morgan's mind Gosney and Popenoe counted among these alarmists, all the more so because of their proximity.

During the years when Morgan's new division cultivated its scholarly reputation and intellectual autonomy, the Human Betterment Foundation mailed free of charge tens of thousands of brochures and articles each year to teachers, social workers, physicians, universities, publishers, and numerous organizations. The literature dramatized "America's Burden"—billions of tax dollars lost each year for caring for the 18 million defectives or potential defectives. Inspired by the efficient and rational approach of German eugenics, Gosney urged readers of *Scientific American* to embrace the views of the noted Viennese surgeon Adolf Lorenz, predicting that the sterilization of the unfit would eventually reach "all civilized countries as [the] means of getting rid of the scum of humanity."[23] By 1940, a couple of years before its dissolution and the transference of its assets to Caltech's biology division, the Foundation would report more than 30,000 sterilizations, nearly 50 percent performed in California.[24]

Morgan may have privately opposed the modus operandi of the Human Betterment Foundation. He surely must have disliked the publicity it received while he was building up the biology division during the 1930s. Throughout this period, especially after the rise of the Nazi party, Gosney's organization would attract considerable attention. Although his project was hailed in some circles—a leading high school textbook approved of the Foundation's sterilization campaign, as did Watson Davis, director of *Science Service*—others condemned it.[25] In a 1934 interview in *Spectator* magazine, Columbia geneticist L. C. Dunn, speaking for other leading geneticists, criticized eugenic sterilization as a form of Hitlerism devoid of scientific validity. Psychologist Bronson Price and his colleagues characterized the Foundation and its leaders as practitioners of a "sort of social demonology."[26] The Foundation defended its practices by referring to R. A. Fisher's calculations of the short-term benefits of sterilization and by upholding the scientific merit of German eugenics. Though by no means front-page news, the debates over Gosney's sterilization campaign found their way into the *New York Times* throughout the 1930s.[27] Given Gosney's support of Caltech and the membership of the Institute's leaders in his organization, however, a formal opposition from Morgan's division might have been politically inexpedient. The institutional posture of the new biology and its service role straddled the makeshift fence between eugenics on one hand and medicine on the other in hopes of procuring funds. Local support was particularly important as matching funds for Rockefeller grants, and these funds were endangered by Morgan's rejection of medical research.

Naturally, Morgan needed to respond to the expectations in Pasadena's community regarding a medically oriented biology. Millions of support dollars were at stake, and Morgan had to articulate his goals broadly in order not to jeopardize the division's economic future. In his 1929 letter to Arthur Fleming, president of Caltech's Board of Trustees, Morgan agreed that a research laboratory and a hospital in connection with biology were matters of tremendous importance. He acknowledged that "in the short time I have been here I have already been approached by a number of medical men of the highest standing." Morgan gracefully reassured Fleming of the possibility of future work in experimental medicine but warned that "it would be premature now to attempt even to outline a program for such a center of medical research." His tentative list included human physiology, cellular physiology, bacteriology, parasitology, organic chemistry, and public hygiene, "including not only hygiene in the narrower sense, but the movement of populations, expropagation of the race and the heredity of the human race." He also outlined research possibilities in cancer and immunology, fields that he had little inclination to pursue.[28]

A somewhat different version went to the Rockefeller Foundation. Morgan's 1930 proposal to the Foundation did not mention medical topics as potential fields of development. Instead Morgan hoped to develop "physiological psychology and bacteriology from the biological and genetic point of view," areas of primary interest to the Foundation's new program. Millikan's report assured Mason of the bright fiscal prospects of the biological sciences at Caltech. "We are also hoping that in view of Mr. Kerckhoff's establishment of Kerckhoff Biological Laboratories, Mrs. Kerckhoff will be willing to contribute new endowment funds for the support of the biological sciences," he projected in 1930.[29]

In 1930 Max Mason, then President of the Rockefeller Foundation, agreed to commit $5 million for developing the natural sciences at Caltech, with $1,140,000 earmarked for Morgan's biology program and $1,900,000 for the eventual development of organic chemistry in cooperation with the new biology division. "The Institute has shown a remarkable and sound growth," commended the statement attached to the first installment of $500,000 in December 1930. "It has carried forward its development conservatively and has shown great ability in its organization of community interest in support of its work."[30] As Millikan had noted, it was the support of the Pasadena business community that "sold" the Institute to the Foundations.

By 1933 the fiscal forecast changed. The Fleming Trust had been wiped out by the Depression, and the embryonic biology division was hit hard. Assuming the availability of the Kerckhoff endowment, the division would now lean more heavily on the Rockefeller Foundation, which had just launched its new biology program under the aegis of the new "Science of Man." Mason and the new program director Warren Weaver (still technically on Caltech's faculty), slated Caltech as a primary site for developing the new biology. The Institute was to be a recipient of enormous grants.

In May 1933 Morgan requested from the Rockefeller Foundation $100,000 for biological work, stressing the relation of his own cooperative program in physico-chemical biology to the Foundation's project in biology and human behavior.

> Genetics is broadening its scope; the gross structural features of inheritance are today fairly well-known, and the workers are turning to the physiological aspects of heredity. Here we are endeavoring to bring our genetic group into closer contact with the physiological group. . . . In order to round our work in physiology we need a man, or a group of men, interested in the physiological side of biochemistry. We have in mind a man versed primarily in the physiology of the nervous system in order to bring our more specialized work into closer contact with the broader fields of human relations.[31]

Whether rhetoric or substance, Morgan's projected program legitimated the Foundation's agenda. Morgan's application, part of a joint grant with the chemistry division, had excellent chances of approval, provided of course that the Institute kept up its part of the agreement to raise matching funds.

That summer, however, panic struck the Institute. A crisis that threatened the future of the new division erupted out of the chronic tension between biology and medicine. Mrs. Kerckhoff, one of the most powerful of Pasadena's wealthy widows, was now inclined to divert the rest of the Kerckhoff endowment toward establishing other memorials to her husband, such as an institute of medical research similar to the Rockefeller Institute, independent of Caltech and most likely to be affiliated with the medical community of Los Angeles. Through her chief legal counselor, Henry O'Melveny, Mrs. Kerckhoff had been in communication with Rockefeller Institute's director Simon Flexner concerning the new plans. The legal resolution of the crisis rested in O'Melveny hands and in his power of persuasion.

On August 8, just before leaving for a month's vacation in Canada, Millikan dispatched a lengthy letter to Flexner expressing his "disappointment and concern" with Mrs. Kerckhoff's "present state of mind." He described the situation and urged Flexner to help.

> I hope that if you agree with our point of view (as I think you do, and that very emphatically), you will be able to find a way to express to her your opinions sufficiently strongly to induce her to abandon this wavering state of mind. . . . The point which needs to be gotten into her consciousness is that if she is going to leave a lasting and influential memorial to her husband . . . she can do it enormously more effectively through the world-wide influence which adequately supported Kerckhoff Laboratories can have.[32]

Millikan complained of Mrs. Kerckhoff's failure to understand that fundamental work such as that of Morgan on genetics was vastly more important, for both the perpetuation of the name of her husband and the progress of the race, than the palliative work represented by hospital activities. With Flexner's medical authority buttressing O'Melveny's legal influence, Millikan hoped to quickly secure $2 million to $3 million of the Kerckhoff endowment.[33]

Flexner's appeal to Mrs. Kerckhoff via O'Melveny communicated neither thorough knowledge nor strong conviction; in any case he did not succeed in convincing Mrs. Kerckhoff to give her money to Caltech. Two weeks later Hale, Flexner's old-time friend and war buddy, impressed upon him the urgency of the matter. Hale had implicit confidence in O'Melveny, "He is one of the oldest and wisest of trustees of the California Institute. . . . He is 'one of us'; and he will try to convince her it is best to establish her memorial in connection with the California Institute."[34] O'Melveny was now in his seventies, though, and his health was failing rapidly. Time was of the essence. To add to Flexner's persuasiveness, Hale attached copies of memos prepared for O'Melveny by Morgan (now away at Woods Hole) and by Caltech's biochemist Henry Borsook on the relation between physics, chemistry, physiology, and medicine.

Morgan and Borsook argued that advances in modern medicine, as evidenced by Nobel Prizes during the preceding 20 years, were no longer the domain of physicians or surgeons. Instead, recipients had been bacteriologists, immunologists, biochemists, and biophysicists, whose researches supplied the rational principles underlying the treatment of diseases. In fact, Morgan pointed out, Pasteur was a chemist, and x-rays and radium were discovered by physicists. A roster of diseases and medical syndromes that had been explained or alleviated by researchers, rather than clinicians, followed. A historian of science's delight for its elasticity and selectivity, the list documented, among other things, the exclusive role of biochemists in proving the existence and need of vitamins, the contributions of bacteriologists to the work on scarlet fever, and protein chemists' advancement of allergy research. It is interesting, Borsook mused, that in the list—vitamins, scarlet fever, pernicious anemia, diabetes, kidney disease, acidosis, toxemias of pregnancy, surgical shock, and allergies—only one important advance had come from a physician.[35]

Noyes, anxious to see his dreams for biomedical research at Caltech realized, dispatched a six-point plan to Hale:

1. Build *at once* an extension to the existing Kerckhoff laboratories with a fine library (with a place for Mr. Kerckhoff's portrait).
2. Appoint quickly a man in "biochemical neurology."
3. Lay the whole personal Mrs. Kerckhoff situation confidentially before Mason and Weaver; Morgan could do this at Woods Hole.
4. Write Morgan *at once* the whole story and urge him to return to Pasadena as soon as he can.
5. Prepare an additional memo on the work of the biology division.
6. If Mrs. Kerckhoff insists on medical research, perhaps she could endow a laboratory as a major department within the biology division.

Realizing that these plans conflicted with those of Morgan, Noyes wrote to Hale: "I do not know whether Morgan would approve of this but perhaps he would, if thereby many millions of dollars could be secured for research, and if it were understood that laboratory researches fundamental to medicine (rather than clinical studies), such as the work in biochemistry, biophysics, and neurology, would be carried on."[36]

For Noyes the affiliation of the biology division with medical research touched on a deeper issue, its significance transcending the Kerckhoff affair. Aside from Mrs. Kerckhoff's involvement in the matter, he wrote to Morgan at Woods Hole, "your Division would through the years be more successful in securing able research students in biology itself, and in getting financial support from wealthy donors if you could make it obvious that the Division had some relation to and interest in 'medicine.' " This proposal was of course Noyes' plan from its inception a decade earlier. He urged Morgan to see Mason before returning to Pasadena, because if Mason understood Mrs. Kerckhoff's attitude it would increase his likelihood of helping.[37]

The negotiations intensified through the summer, lubricated by the camaraderie of privilege and aquatic leisure. "I am here on the yacht 'Day after Tomorrow,' " wrote Noyes to Morgan on August 30 from aboard the "Pasado Manāna" off Catalina Island,

> with the owner, Mr. Lee Phillippe, of the Pacific Mutual Co., Herbert Hoover, Robinson, and Munro [Millikan joined the following day]. I have had a chance to talk with Robinson. He had a good talk with O'Melveny, and he asks me to tell you confidentially the following about Mrs. Kerckhoff. In her will as it *now* stands she leaves *ultimately* most of her property (which may amount to four millions) to trustees with authority to give it to the Institute; but it would not come to us immediately.

O'Melveny was trying to get the delay eliminated. Noyes emphasized the importance of not losing this opportunity and to begin as soon as possible the construction of a new Kerckhoff Laboratory and secure "the new man in neurology (whose direct relation to medicine could be emphasized with Mrs. Kerckhoff), and in general to make any possible developments or gestures in the direction of medical research."[38]

Morgan had a long emergency meeting with Weaver at Woods Hole early in September, going over the entire background of biology at Caltech. Morgan recounted that upon coming to Caltech he had been promised a budget of $100,000 per year; the unexpended portions would be held for the division's future use. Morgan, however, had released these reserves to the Institute during the Depression years, with the result that in 1933 the division's annual budget was $61,000. The Rockefeller Foundation grant for $1,140,000, awarded in 1930, went into building up the work in biochemistry, biophysics, and plant physiology (genetics was then still supported mainly by the Carnegie Foundation); even these areas were now suffering from lack of technical assistance and research fellows. There were no funds for new appointments, and it was urgent that the division have a group in physiology.[39]

On September 9, 1933, Morgan wrote to Millikan, who had just returned from Canada, about his meetings with Weaver and with F. B. Hanson (officer in the Foundation's natural sciences division). Caltech's position was reasonably secure, he thought. From what Weaver said, if Mrs. Kerckhoff had definitely made up her mind to support a medical research institute, there was a good chance of working toward that end through the medical branch of the Rockefeller Foundation. "In any case," Morgan noted, "nothing is going to happen until Weaver comes to Pasadena about the Middle of October. By that time I hope the Kerckhoff situation will be clearer than it is to me at present."[40]

Morgan, for good reasons, exuded serenity. The Rockefeller Foundation's new program explicitly emphasized the cooperative effort between the Foundation's divisions of natural, medical, and social sciences. Owing to the anticipated coordination of research efforts between the Foundation's divisions, there appeared to be no special difficulty with a Kerckhoff endowment that ultimately aimed at medical applications. There was sufficient interchangeability between biological and medical projects within the new cooperative model and enough flexibility of definition to accommodate biomedical research.[41] In November 1933, after Weaver's October visit to Caltech, Morgan resubmitted his grant application to Weaver. This time he added: "There is, as you know, in the background another consideration that must be taken into account in reaching a decision. There is the possibility of a future development—either as a part of the work of the Kerckhoff laboratories, or in intimate relation with it—of other research laboratories devoted more nearly to the study of the fundamental aspects of medical sciences, i.e., physical and chemical, and biochemical."[42]

Morgan's resubmitted grant proposal was also more heavily weighted toward neurophysiology and hormonal development, central elements in the Foundation's psychobiology program. The theory of the humoral mechanisms of nerve activity, he wrote, suggested ways of detecting nerve current transmitted over the synapses in the central nervous system and means for studying chemical action of nerves in organs, muscles, and glands. Such researches, in turn, would be linked with studies of neurally stimulated secretions of hormones from the endocrine glands, affording, for the first time, a deeper insight into the coordination of the organism as a whole. These proposals, Morgan hastened to explain, were made in collaboration with trustees Hale and Robinson and with Millikan and Noyes.[43] Neurophysiology presented a perfect solution: a cognitive link between soma and psyche and an institutional bridge between biomedicine and behavior.

At the height of the Kerckhoff negotiations, Morgan was awarded the Nobel Prize for Physiology or Medicine, the first biologist with no medical training to win the Nobel Prize in this category. Announced late in October 1933, the prize could not have been better timed. The grant application to the Rockefeller Foundation had just been resubmitted, and Mrs. Kerckhoff, owing largely to the efforts of O'Melveny and Flexner, was a bit more inclined toward Caltech, though yet undecided. A renewed attempt by "the Germans" to get the balance of Mrs. Kerckhoff's funds for a medical institute in Germany and similar pressures from the University of California (where Mrs. Kerckhoff's intimate friend was a trustee) stood between the endowment and Morgan's plans.[44]

Morgan's Nobel Prize created a unique opportunity for focusing community attention on Caltech's biology program and on the international recognition of genetics as an important area in physiology and medicine. In celebration of the event, a lavish dinner was planned for mid-December at Caltech's Athenaeum, and an inspired Mrs. Kerckhoff requested that she be permitted to give the dinner. Flexner arrived as a guest of honor, briefed well in advance by Hale: "You will thus understand how much store we set on the dinner and your participation in it." Hale urged Flexner to lend his authority and oratory to an address devoted to the relation of physics, chemistry, and mathematics to medicine, avoiding of course any reference to the question of endowment. "By doing so," Hale projected, "I fully believe you would turn the balance and bring down the beam safely and surely on the right side."[45]

The second Kerckhoff Laboratory was dedicated in June 1938. Designed to foster close cooperation between bioorganic chemists, biochemists, and physiologists, the new Kerckhoff Laboratory adjoined the new Crellin Laboratory for bioorganic chemistry, dedicated a month earlier. The social goals of the new scientific venture were articulated broadly enough to encompass both medical and eugenic benefits. As the *New York Times* described it, the current Rockefeller gift of $1 million was awarded for research "for biological improvement of the race through an intensive attack on organic chemistry as related to life."[46] Caltech's biology was thus constrained and liberated by its ambiguous service role, navigating between the Scylla of medicine and the Charybdis of eugenics until World War II. By avoiding a firm commitment to eugenics and medicine, the Division was somewhat encumbered in its local fund-raising strategies, though it managed to garner sufficient resources by holding out vague promises to both enterprises; this openendedness created an intellectual space for developing a distinctive biological identity. Under the auspices of the Rockefeller Foundation, the new biology could focus on long-range visions rather than immediate returns.

Contradictory Elements

Articulating a service role for research was not the only difficulty in Morgan's biology division. Underneath the incongruities of institutional policy lay intra-divisional conflicts of interest. Several issues of substance and style remained unresolved in Morgan's own mind throughout his tenure, reflected in hiring practices that impeded the development of the division. Although Morgan was a successful leader in physiological genetics, he did not provide effective leadership in biochemistry, biophysics, or physiology. Poor judgment, ethnic biases, and cross-purposes compounded by financial losses during the Depression contributed to his failure in these areas. As Theodosius Dobzhansky observed, Morgan was by nature a contradictory person and on several counts did not practice what he preached. Ultimately, Morgan succeeded best in the areas closest to his own interests and expertise.

Morgan brought with him to Caltech in 1928 his Columbia associates Alfred H. Sturtevant and Calvin B. Bridges, his and E. B. Wilson's protégé Jack Schultz, and the young Russian geneticist Dobzhansky. One of the few geneticists in America to combine research interests in evolutionary biology and genetics, Dobzhansky led several expeditions along the Pacific Coast from Alaska to British Columbia to study variations in new *Drosophila* strains. Members of the genetics group were still deeply involved in classical genetics: ironing out minute details of linkage, crossing-over, recombination, and mechanisms of sex determination in several *Drosophila* strains and deliberately perpetuating the *Drosophila* group's tradition of intellectual criticism, cooperative spirit, and regular seminars.[47]

Plant physiologist James Bonner, then a graduate student at Caltech, recalled: "In the genetics laboratory, Sturtevant and Dobzhansky had tried to recreate the famous fly-room at Columbia. They sat at two ends of the long table and worked at their flies. The students sat in between and listened to the wise conversation and contributed to it when they could."[48] To be sure, Sturtevant, Dobzhansky, and Bridges were enormously productive in their intricate genetic analyses, especially after the 1933 discovery by Theophilus Painter of the giant chromosomes of the salivary glands in *Drosophila*. Bridges was generating new chromosomal maps, and Dobzhansky disentangled the mechanisms of sex determination and the "position effect." These researches were now "normal science."[49]

Classical genetics during the early 1930s was perceived to be past its prime and would not be the principal focus at Caltech. Most younger scientists were no longer intrigued by pure Mendelian analyses of *Drosophila* or maize. They got their training at Caltech's "fly room" or from plant geneticist Ernest G. Anderson at the Institute's Arcadia farm, an hour's bicycle ride from the Kerckhoff building; but they soon branched out to newer problems of physiological genetics. Graduate students were now attracted to the mechanisms that linked Mendelian transmission to the cellular and biochemical processes leading to phenotypic expression in animals, plants, and fungi. Thus rather than building up a research empire, Sturtevant and Dobzhansky generally assumed more of a pedagogical role, providing rigorous genetic training and imparting a research ethos. Sturtevant's influence in the division was particularly strong, second only to that of Morgan. For Sturtevant, Schultz reminded George Beadle years later:

> [Students] were not commodities for one's aggrandizement, but people who were interesting to have around. There was no need to be aggressive about them. All Sturt ever did was be himself in his lab. Dodik [Dobzhansky] was the one who really wanted students, and you may recall how the bright undergraduates (Boche, Bonner) started with him and went elsewhere. . . . Later the situation changed, particularly as Dodik became more prominent.[50]

Dobzhansky, dynamic and broadly educated, was an inspiring and provocative teacher. However, his interests in evolutionary biology were peripheral to Caltech's focus on experimental biology and the Rockefeller Foundation's new biology program.[51]

Morgan brought with him his Columbia graduate student, embryologist Albert Tyler, in 1928 the first biologist to receive a doctorate from Caltech. A marine biological station was set up at Corona Del Mar, Noyes's former beach house, where students received training in experimental embryology and histology. The Corona Del Mar experience was intended to capture the orientation and atmosphere of the Marine Biological Laboratory at Woods Hole, Morgan's favorite scientific retreat. Even beyond the traditional developmental mechanics, Morgan hoped to develop a mechanistic physicochemical embryology linked to general physiology and physiological genetics. As he stated in his 1933 Nobel Prize address, the biochemical and cellular processes that linked gene action with embryonic development remained unexplored, a wide gulf separating the two fields. Yet his own work, the 1934 monograph *Embryology and Genetics*, which contributed little to a synthesis of the two fields, only reinforced the cognitive and disciplinary divide.[52]

To promote a mechanistic conception of life and to probe reproduction, growth, development, and related physiological processes on a fundamental level common to all organisms, the new biology program would depend exclusively on the causal explanations of physics and chemistry. Amplifying and formalizing the dominant trend of modern science, animate nature within the new program increasingly retreated from the field into the confines of the laboratory. Rather than being viewed as a historical process, life would be fixed within eternal spatial structures and timeless laws of nature. On introducing the new curriculum in the 1928 *Bulletin of the California Institute of Technology*, Morgan underscored that "It is with a desire to lay emphasis on the fundamental principles underlying the life processes in animals and plants that an effort will be made to bring together, in a single group, men whose common interests are the discovery of the *unity* of the phenomena of living organisms rather than in the investigation of their manifold *diversities* [my emphasis]." Inasmuch as the curriculum description was a programmatic statement of the new biology, it also represented a revolt from the tradition of natural history and an implicit negation of the relevance of ecology and evolutionary biology.[53]

Morgan of course acknowledged the importance of interactions of organisms with their environments and appreciated the immense diversities of form and function in the biological universe, more than he cared to admit, according to Dobzhansky.

" 'Naturalist' was a word almost of contempt with him, the antonym of 'scientist,' " Dobzhansky recalled. "Yet Morgan himself was an excellent naturalist, not only knowing animals and plants but aesthetically enjoying the observing of them."[54] Under the Loebian spell, Morgan now pronounced that enough insight had already been gained to demonstrate that the diversity was mostly due to permutations and combinations of relatively few fundamental and common principles. Mendelian principles of heredity, for example, applied equally well to plants and humans, Morgan argued, as did the responses of organisms to light and such cellular activities as anabolism, catabolism, and respiration. Once elucidated by the laws of physics and chemistry, Morgan insisted, the complexities of vital phenomena and the diversity of life forms dissolved into interactions of fundamental unifying principles.[55]

Morgan trumpeted the virtues of the new physicochemical biology with the zeal of a new convert, though he had known Loeb since the 1890s and championed experimental biology for two decades. Trends in biological research were changing, he argued. In England, Germany, Russia, Scandinavia, and France the specialized institutes in diverse biological fields focused primarily on the application of mathematical, physical, and chemical methods to biological problems; these research centers had been growing steadily since the early 1920s, he observed. American embryologists, botanists, zoologists, and geneticists should no longer be content with traditional topics, working on narrowly circumscribed problems with exclusively biological methods. Instead, they should be routinely borrowing analytical tools from their colleagues in biochemistry, biophysics, and physiology departments.[56] Even the term "physiology" or "general physiology" had by the 1920s assumed a broad meaning, epitomized during the 1920s by the researches of Jacques Loeb at the Rockefeller Institute. In contrast to the earlier medical physiology, which concentrated on the human body and dealt with functions of specific organs, the new physiology, in a sense a precursor of the new biology, focused on basic questions regarding vital processes common to all organisms from protozoa to man. As Morgan explained in his course descriptions:

> General physiology differs in its aims from the traditional physiology, which relates more particularly to man and the higher vertebrates, in so far as it encompasses the whole field of living things, selecting those for investigation that are particularly suited to solve specific problems.

Organisms became mere probes. To accomplish this convergence on fundamental unifying principles, Caltech's physiology group would bring together biochemists, biophysicists, and "others whose interests lie in the reactions taking place in the sense organs and central nervous system (physiological psychologists)."[57] "The terms 'physiology' and 'general physiology' have come to be used with wider and wider meaning in biology," echoed the 1930s reports of the Rockefeller Foundation:

> General physiology, as exemplified by Jacques Loeb, is especially concerned with the treatment of basic problems by chemical and physical techniques. . . . Furthermore, modern physiology is often concerned with cells, single nerve fibres [sic], and tissues, rather than with whole organs. . . . The program at California Institute of Technology is primarily concerned with studies designed to bridge

the gap between the gene-chromosome theory of genetics and the developed characteristics of the mature organism.[58]

Whereas in Europe the application of physical quantitative methods did not lead to the miniaturization of life, Morgan projected an unambiguous vision. The growing dependence on physicochemical techniques and on interdisciplinary efforts would correspond to parceling life into ever smaller units.

To accomplish these objectives, and in conformity with Caltech's cooperative ideal and the Rockefeller Foundation's policy of cooperation, the new biology was presented not only as an intradivisional enterprise but as a cooperative venture with the physics and chemistry divisions. The Foundation had agreed in 1930 to Noyes's plan of expanding and strengthening Caltech's chemistry division in order to develop properly Morgan's biological work. They had figured it would cost a minimum of $50,000 per year to develop organic chemistry "without which Dr. Morgan's own program cannot be carried out, and we shall have to provide a new wing to the Chemical Laboratory [the Crellin Laboratory, built in 1938] costing, with equipment about $400,000."[59] Accordingly, when planning the physical layout of the departments, Morgan insisted that the biology wing should be contiguous on one side with the planned organic chemistry laboratories, rather than with geology, as Hale had originally proposed. Clearly, Morgan hoped that the proximity of the two buildings would encourage close contact between the two departments.[60]

The effectiveness of the cooperative enterprise was predicated on the strength of its departments of biochemistry, biophysics, and physiology and on their own productive cooperation. Here Morgan was on a relatively unfamiliar territory. In 1927 Morgan confessed to Hale that "the time is not far off when individual names will have to be considered. In the genetic field, where I know my ground, this will not be difficult; but when it comes to the physiologists, I shall have to go more slowly, and perhaps hold up the situation until I go abroad next spring."[61] He might have added "biochemistry and biophysics." Indeed, it was in the crucial areas of physiology, biochemistry, and biophysics that Morgan's weaknesses and contradictions manifested most clearly.

To begin with, Morgan had little direct knowledge of the physicochemical principles of biology. Although exuding enthusiasm over Loeb's physiological researches and for quantitative mechanistic biology, Morgan was uneasy with physicochemical biology.[62] Having been trained in embryology during an era when biology was largely descriptive, Morgan's mathematical tools were limited to ordinary algebra; he knew little physics and even less chemistry. According to Henry Borsook, the head of the biochemistry department since 1929, Morgan was ignorant of the most rudimentary chemical techniques. He was impressed even by a simple set of pH standards (prepared by Borsook) that corrected the unexplained inconsistencies in the data of Morgan's embryological experiments. Morgan developed enormous respect, bordering on worship, for biochemistry, biophysics, and the new physiology but did not have the necessary background for following the work being done in these areas, except in broad outline.[63]

As is often the case when critical understanding is absent, Morgan enveloped all things physicochemical in an intellectual mystique. With the zeal of a naive convert, Morgan insisted that biology had to be explained in terms of physics and chemistry. According to Dobzhansky, "Morgan himself knew little chemistry, but the less he knew the more he was fascinated by the powers he believed chemistry to possess. There was no surer way to impress him than talk about biological phenomena in ostensibly chemical terms."[64] Consequently, Morgan himself was a poor judge of researchers' abilities in these fields, relying instead on the assessments of others.

That Morgan never fully internalized the gospel of physics and chemistry in biology was clearly manifested during his 1931 meeting with Albert Einstein at Caltech. According to Borsook, when Einstein asked Morgan why he had moved to Caltech, he received the standard reply that the future of biology rested in the application of the methods and ideas of physics, chemistry, and mathematics. Einstein, expressing skepticism, gently pointed out that even physicists could handle only the very simplest substances—hydrogen, helium, and a few other inorganic molecules—and were unable to analyze complex organic molecules. Challenging Morgan, Einstein asked if biologists could ever explain in terms of physics and chemistry so important a biological phenomenon as first love. Although Morgan struggled to explain something about the connections of sense organs to brain and hormones, he later admitted that he himself did not quite believe the expedient response he had offered.[65] If his research at Caltech serves as an indicator, Morgan's own interests lay closer to the morphological and developmental biology he helped marginalize. Morgan abandoned even *Drosophila* genetics by the early 1930s. Until shortly before his death in 1945, he pursued the invertebrate embryology with which he had launched his career. Although a herald of the future, Morgan was basically a scientific icon of the past.

A similar discrepancy between intent and practice muddled his relation to American and European science; his actions betrayed a conflict of loyalties. Like others of his generation, Morgan shared the uneasiness of American researchers toward European science during the interwar period. Morgan took pride in American biology as an emblem of a matured national science and was committed to its growth. Like most American leaders in science, he had generally weaned himself from the European, especially the German, influence. At the same time, like his contemporaries in academe who had spent several years at European research institutes, Morgan had retained a romanticized perception of European research.[66]

Again and again, during the summers while he was still on the East Coast, Morgan returned to the revered Marine Biological Station at Naples (and other European institutes) in order to keep up with the latest developments in embryology and physiology and to bring them to America. True, by the 1920s Morgan had become disillusioned with German biology, finding it too descriptive and speculative, even metaphysical and vitalistic. Nevertheless, he did believe that, with few exceptions (e.g., genetics), the best researchers in biology were Europeans. The luster of European science blinded his vision, eventually impairing his judgment when choosing candidates for departmental leadership.[67]

The overestimation of European science was accompanied by an anti-Semitic bias, muddling his plans with nonrational elements that stunted the growth of physiology and biochemistry at Caltech. A pervasive sentiment during the interwar period, anti-Semitism was buttressed by restrictive quotas in elite academic institutions. The fact that some of Morgan's friends and associates were Jews did not preclude an entrenchment of his nativist proclivities. "Time and again," Dobzhansky recalled, "he [Morgan] would make, especially when irritated, anti-Semitic remarks of the most crude sort."[68] This prejudice influenced his hiring decisions.

In 1928, a year before Henry Borsook was appointed head of the biochemistry group, Morgan attempted to lure to Caltech a number of prominent researchers, all working at the intersection of biochemistry, biophysics, and general physiology: William J. Crozier from Harvard, Selig Hecht from Columbia, and John Northrop from the Rockefeller Institute. Although these attempts failed, he did have an opportunity to bring in the noted Rockefeller Institute enzymologist Leonor Michaelis. He wrote to Millikan:

> Indirectly it has been conveyed to me that Doctor Michaelis, biophysicist, now at J. H. U., who is recognized as one of the leading men in this field, will be free at the end of this year, and probably would accept a call. I know him fairly well, since he comes to Woods Hole, and I know he has the esteem of the best men. On the other hand, he is not young, and already has collected about himself a few young Jews. He himself is markedly Semitic. I have my doubts whether we should want to start under these conditions, and shall make no moves. Possibly next year we might invite him for a year, but this, too is dubious.[69]

A cultivated German émigré, Michaelis had raised quantitative biochemistry to new heights with his seminal studies on reaction kinetics beginning around 1910. Focusing on the physicochemical nature of proteins, especially enzymes, Michaelis was an ideal choice for cooperative projects in biochemistry, biophysics, and physiology.

Instead, Morgan appointed in 1929 Henry Borsook, an English-born medically oriented biochemist from the University of Toronto, the only M.D. on Caltech's faculty. To be sure, Borsook's traditional work was competent; and his recent investigations on the relation of thermodynamics to biochemical reactions in plants and animals were commendable. However, his true interests lay in traditional metabolic research and clinical nutrition, a subject far removed from the interests of Caltech's physicists and chemists. Borsook's contributions as a teacher were valuable indeed, but he possessed neither the intellectual imagination nor the managerial style necessary for building a department. An extreme individualist, Borsook was ill-suited for the large cooperative enterprise at Caltech.

Borsook admitted as much. He regarded himself a maverick, a scientific loner whose amateurish research style led to eclecticism. He disliked conformity, or perhaps competition and scrutiny, and did not appreciate fashionable branches in science, recalling that "if anybody began to work in something I was working at I would drop it and turn to something else. . . . In a way it was an amateur's way of looking at science rather than a professional's but that's the way I was, you see."[70] Perhaps not surprisingly, Borsook disliked the aggressive managerial style of Linus Pauling, the head of the chemistry division; and although there was considerable overlap in their research interests—the relation of amino acids to peptides and the chemistry of vitamins—minimal cooperation developed between their departments. Borsook preferred intellectual isolation, did not like keeping up with the literature, and had few professional contacts.[71] His provincial academic style did not necessarily conflict with his own research projects, but it was hardly suitable for leadership in a cooperative interdisciplinary program.

Dominated by the influence of classical genetics and medical biochemistry, Morgan's division during the early 1930s appeared traditional in outlook. Caltech's biophysicist Robert Emerson, a specialist in photosynthesis, complained in 1931 of the orientation and practices in Morgan's division, all "milk-bottle-molasses and beef-hash-muscle in outlook. . . . The biochemistry section is highly medical in outlook and seems to me very narrow."[72] The medical angle might have been useful for fund raising, but it impeded the growth of alternative research paths. Within a few years, however, after the arrival of George W. Beadle in 1931 and with frequent visits by leading biologists, Caltech distinguished itself as an international center in physiological genetics. Physiology, however, remained somewhat of an embarrassment.

The plant physiology group, comprised largely of Utrecht scientists led by Fritz Went, progressed well—perhaps too well for a division where agricultural research was not intended to be a primary focus. By 1933, Morgan complained to the Rockefeller Foundation that there was still no general physiologist or neurophysiologist at Caltech. Given the Foundation's emphasis on psychobiology and following the award of a long-term Rockefeller grant, Morgan was now in a position to make two appointments in neurophysiology and to build up the physiology and psychobiology groups at Caltech.

In April 1934, soon after the announcement of the grant, Morgan informed Weaver of his plans to go to Europe to recruit physiologists, never even considering the possibility of hiring American candidates. Here too anti-Semitism played a role. He intended to go to London to meet with possible appointees and to discuss "the general problem with men like Dale, Hill, and Haldane." From there Morgan would extend his search to Sweden, Denmark, Holland, and perhaps Belgium.[73] Rockefeller Foundation officer W. E. Tisdale, who met up with Morgan in May 1934 at a Royal Society soirée noted with some embarrassment that:

> He [Morgan] has announced to all who will listen that the Rockefeller Foundation has given him money to secure the services of a physiologist. He is combing England and the Scandinavian countries to find one who is not Jewish, if possible. From the English reception of this announcement, I am inclined to believe that he will have difficulty in finding a first-rate Englishman who will be willing to go to Pasadena.[74]

Indeed, Morgan was unsuccessful in his efforts. Although he did not return empty-handed, he ended up hurriedly selecting two solid but rather ordinary neurophysiologists from Utrecht just four days before boarding the ship to America. Cornelis Wiersma and his assistant Anthonie von Harriveld were competent enough; certainly their researches on electrical conductivity of nerves matched the interests of the Rockefeller Foundation, but they distinguished themselves by their unremark-ability. Neither possessed outstanding imagination or leadership qualities. As Rockefeller Foundation officer H. M. Miller later observed: "Both are probably quite competent, but HMM is inclined to wonder whether *Morgan* might not have secured equally good or even superior young Americans, if he had made half the effort here that he did in Europe."[75] He might have also enlarged his talent pool had he not been limited by his ethnic preferences.

Excellence aside, the presence of neurophysiologists at Caltech had significant implications. By focusing on physiological genetics and psychobiology, the biology program fit squarely within the Foundation's "Science of Man" agenda in terms of both rhetoric and practice. In 1938, two years after Raymond Fosdick's as-sumption of the presidency of the Rockefeller Foundation and Max Mason's move to Caltech, the Hixon Fund was established at the Institute to support research on human behavior. Administered by a committee consisting of Mason, Sturtevant, Borsook, and Pauling, the fund supported the studies of R. Laurente de No from the Rockefeller Institute who, in consultation with Caltech's staff in physics and mathematics, worked on nerve action currents for half a year. The fund also aided a cooperative project between members of Caltech's staff under the leadership of Wiersma, van Harreveld, and representatives of the "state department of institu-tions," the agency for the feeble-minded that had cooperated with the Human Betterment sterilization campaign. The team investigated the effects of electroshock as a means of psychotherapy; and later, in cooperation with the Department of Psychiatry at Los Angeles County Hospital, clinical experience was obtained on electronarcosis as a treatment for mental disorder. These projects would later place Linus Pauling in an advisory role to the Ford Foundation when it launched its program in behavioral science during the early 1950s.[76]

At the end of the 1930s, the realities at Morgan's division contrasted markedly with the visions of strong cooperative physicochemical biology at Caltech. Plant physiology and physiological genetics were thriving, but the other groups lagged far behind. In fact, the weaknesses of the physiology, biochemistry, and biophysics departments contributed to a lack of cooperation between biology and the chemistry and physics divisions. The sympathetic attitude toward cooperation and Caltech's social cohesion did foster some casual scientific exchange among the Institute staff, but no formal joint projects developed—only a couple of collaborative publications between the biology and chemistry divisions during the 1930s.

Viable interdisciplinary cooperation demanded more than rhetoric and good intentions. Despite the purported cognitive overlap between biological phenomena and the laws of physics and chemistry, there was in reality little disciplinary overlap between the three disciplines; they represented vastly different intellectual tradi-tions and social contexts. Perched at the apex of the Comtean ladder, physicists and chemists generally had little appreciation or respect for the descriptive re-searches of biologists, works they tended to view as unrigorous. Gazing up the rungs, biologists generally suffered from an inferiority complex rooted in their ignorance of mathematics and physics, as well as from their lower academic status. This dynamic initially played a role in inhibiting collaborations between these constituencies at Caltech.[77]

Throughout most of the Morgan era the biology division remained isolated from the rest of the Institute. Bonner remembered the first Kerckhoff Laboratory, built on the extreme northwest corner of the campus, as "completely isolated from all the other buildings—from the administration building, Throop Hall and from the chemistry building, Gates Laboratory, and from the physics buildings across the quadrangle. . . . So we were not only intellectually isolated from the rest of the campus pretty well, but also physically isolated. This didn't change until 1938, with the great building spree of 1938."[78]

The building spree of 1938 created favorable conditions for cooperation by joining the new Crellin Laboratory of bioorganic chemistry with the new Kerckhoff building, but the physical improvements did not solve the structural problems of the biology division. Distanced from medicine and disinterested in eugenics, the ambiguous service role of the biology division continued to place it in a precarious position with respect to the envisioned needs of the region. The mediocre appointments, which retarded the growth of the division, created long-standing internal strife. In the absence of strong leadership in physiology, biochemistry, and biophysics, the biological enterprise trailed far behind Pauling's thriving chemistry division, hardly creating the atmosphere for a viable interdisciplinary partnership. During the 1930s the division's productivity and acclaim derived primarily from the researches in physiological genetics of Jack Schultz, George Beadle, and Max Delbrück.

Figure 7 Biology staff in 1931. *Seated (left to right)*: Henry Borsook, Herman E. Dolk, Henry S. Sims, A. H. Sturtevant, Sterling Emerson, Hugh M. Huffman, Thomas Hunt Morgan. *Standing*: Hermann F. Schott, Charles R. Burnham, Walter E. Lammerts, Kaj Linderstrom-Lang, Emory L. Ellis, Geoffrey Keighley, James Bonner, Albert Tyler, George W. Beadle, Jack Schultz.

Notes

1. Garland Allen, *Thomas Hunt Morgan: The Man and His Science* (Princeton: Princeton University Press, 1981), Chs. 1 and 2. On Morgan's graduate training see Keith R. Benson, "American Morphology in the Late 19th Century: The Biology Department at Johns Hopkins University," *Journal of the History of Biology, 18* (1985), pp. 163–205.

2. Philip L. Pauly, "The Appearance of Academic Biology in Late Nineteenth-Century America," *Journal of the History of Biology, 17,* (1984), pp. 369–397; and Charles E. Rosenberg, "Toward an Ecology of Knowledge: On Discipline, Context and History," in Alexandra Oleson and John Voss, eds., *The Organization of Knowledge in Modern America, 1860–1920* (Baltimore: Johns Hopkins University Press, 1979), pp. 440–455.

3. G. Allen, *Thomas Hunt Morgan*, pp. 257–262 (see Note 1).

4. Ibid. See also Jan Sapp, "The Struggle of Authority in the Field of Heredity, 1900–1932: New Perspectives on the Rise of Genetics," *Journal of the History of Biology, 16* (1938), pp. 311–342; Jonathan Harwood, "National Styles in Science: Genetics in Germany and the United States Between the World Wars," *Isis, 78* (1987), pp. 390–414; and Richard Burian, Jean Gayon, and Doris Zallen, "The Singular Fate of Genetics in the History of French Biology, 1900–1940," *Journal of the History of Biology, 21* (1988), pp. 357–402.

5. APS, Dobzhansky Papers, B:D65, Oral History, pp. 250–251.

6. G. Allen, *Thomas Hunt Morgan*, pp. 339–342 (see Note 1); Rexmond C. Cochrane, *The National Academy of Sciences* (Washington, DC: National Academy of Sciences, 1978), pp. 294–303; and CIT, Oral History, Henry Borsook, p. 10.

7. CIT, Morgan Papers, Box 1.10, Morgan to Merriam, December 14, 1928.

8. G. Allen, *Thomas Hunt Morgan*, pp. 257–262 (see Note 1); and Bently Glass, "Curt Stern, 1902–1981," *Biographical Memoirs of the American Philosophical Society, Yearbook* (1982), pp. 514–520.

9. G. Allen, *Thomas Hunt Morgan*, pp. 330–332 (see Note 1).

10. Ibid., pp. 339–342; RAC, RG1.1, 205D, Box 5.71, Weaver's interview with Morgan, September 8–10, 1933.

11. Robert A. Millikan, *The Autobiography of Robert A. Millikan* (New York: Prentice Hall, 1950), pp. 230–237; G. Allen, *Thomas Hunt Morgan*, p. 338 (see Note 1).

12. P. L. Pauly, "The Appearance of Academic Biology," pp. 369–397 (see Note 2); Robert E. Kohler, *From Medical Chemistry to Biochemistry: The Making of Biochemical Discipline* (Cambridge: Cambridge University Press, 1982), pp. 318–323; Nathan Reingold and Ida Reingold, eds., *Science in America* (Chicago: University of Chicago Press, 1981), pp. 287–288, 344–345.

13. Thomas H. Morgan, "Study and Research in Biology," *Bulletin of the California Institute of Technology, 36* (1928), p. 87; the undergraduate curriculum did offer a premedical option.

14. "Ezra S. Gosney," *National Cyclopaedia of American Biography, 31* (1944), pp. 504–505; The Gosney/Human Betterment Foundation Papers are deposited at the California Institute of Technology Archives. Because these papers are closed to researchers, it is impossible to establish the precise nature of the relationship between Gosney's organization and the biology division. It is clear from the administrative record, however, that some connection did exist. CIT, Biology Division Records, Box 10.11, "Gosney Research: (Formerly Human Betterment Foundation)." See also additional discussion on the Gosney Research Fund, Interlude II, this volume.

15. APS, Human Betterment Foundation (HBF) file, MS. Collection 16; the roster of members appears in several pamphlets and official stationary. For Millikan's concern about Malthusianism see his *Science and the New Civilization* (New York: Charles Scribner's Sons, 1930), p. 27, followed by praise of the 1927 World Population Conference in Geneva (a forum addressing a range of eugenic issues). On the conference see Meriley Borell, "Biologists and Birth Control Research," *Journal of the History of Biology, 20* (1987), pp. 51–87.

16. APS, HBF file, "Human Sterilization Today," p. 1.

17. APS, HBF file, passim; Donald K. Pickens, *Eugenics and the Progressives* (Nashville: Vanderbilt University Press, 1958), pp. 94–101. George W. Beadle, "The Gosney Research Fund," *Engineering and Science* (May 1947), p. 26.

18. G. Allen, *Thomas Hunt Morgan*, pp. 227–234 (see Note 1).

19. Thomas H. Morgan, "Human Inheritance," *American Naturalist, 58* (1924), pp. 385–409, quote on p. 406.

20. Ibid., p. 409; quote on p. 400.

21. APS, Dobzhansky Papers, B:D65, Oral History, p. 254.

22. T. H. Morgan, "Human Inheritance," p. 409 (see Note 19).

23. APS, HBF file, passim; E. S. Gosney, "Eugenic Sterilization," *Scientific American, 151, No.1* (1934), pp. 18–22.

24. APS, HBF file, "Human Sterilization Today," p. 8.

25. G. C. Wood and H. A. Carpenter, *Our Environment* (New York: Allyn & Bacon, 1938), pp. 900–904; Watson Davis, *The Advance of Science* (New York: Doubleday, Doran, 1934); Ch. 24, "Race Betterment," singled out the efforts of Gosney's foundation. For a discussion on eugenics during the 1930s see Chapter 1, this volume.

26. APS, HBF file, Popenoe to Dunn, January 22, 1934, pp. 1–3; Dunn's interview in the *Spectator*, January 11, 1934; and F. C. Reid (Asst. Secy, HBF) to Price, November 20, 1940.

27. *New York Times*, July 21, 1932, 14:1; July 22, 20:2; July 26, VIII, 4:2; January 14, 1934, IV, 5:4; March 1, 1936, X, 6:3; and March 30, 7:2.

28. CIT, Millikan Papers, Box 18.11, Biology—Morgan to Fleming, January 19, 1929, pp. 1–4.

29. RAC, RG1.1, 205D, Box 5.66, Morgan's report, November 8, 1930; Millikan's report, March 7, 1930.

30. Ibid., grant resolution, December 10, 1930.

31. RAC, RG 1.1, Box 5.71, Morgan to Mason, May 15, 1933, pp. 2–3.

32. CIT, Millikan Papers, Box 18.14, Biology 1933, Millikan to Flexner, August 8, 1933, pp. 1–3; quote on p. 1.

33. Ibid., p. 2.

34. CIT, Noyes Papers, Box 84.9, Correspondence File; Hale to Flexner, August 21, 1933, p. 1.

35. Ibid., "Memo sent to Mr. O'Melveny by Prof. Morgan," and "A Summary of Some of More Important Medical Advances Since the War."

36. Ibid., Noyes to Hale (undated, but ca. mid-August 1933). From the content it is evident that Morgan had already left for Woods Hole.

37. CIT, Millikan Papers, Box 18.14, Noyes to Morgan, August 25, 1933, p. 3.

38. CIT, Noyes Papers, Box 84.9, Correspondence File; Noyes to Morgan, August 30, 1933, pp. 1–2.

39. RAC, RG1.1, Box 5.71, Weaver's interview with Morgan, September 8–10, 1933.

40. CIT, Millikan Papers, Box 18.14, Biology 1933, Morgan to Millikan, September 9, 1933.

41. A couple of illuminating discussions explain the interchangeability of projects between divisions; the support of R. A. Fisher's eugenic research was just such a borderline case. RAC, RG3, 915, Box 1.2, R. A. Fisher—Galton Laboratory, March 27, 1935; and "Relations Between the Medical Sciences and the Natural Science," R. B. Fosdick to Weaver and Gregg, December 10, 1937.

42. RAC, RG1.1, Box 5.71; Morgan to Weaver, November 9, 1933, p. 1.

43. Ibid., pp. 2–3.

44. CIT, Millikan Papers, Box 18.14, Biology 1933, Hale to Flexner, November 15, 1933.

45. Ibid.

46. "New Rockefeller Gift," *New York Times*, June 12, 1938, II 4:5; the sum included support of the chemistry division.

47. On Dobzhansky's contributions see Ernst Mayr and William B. Provine, eds., *The Evolutionary Synthesis*, (Cambridge: Harvard University Press, 1980).

48. CIT, Oral History, James Bonner, p. 10.

49. "William G. Kerckhoff Biological Laboratories"—publications. *Bulletin of the California Institute of Technology, 42, No. 140* (1933), pp. 12–15.

50. APS, Schultz Papers, MS Collection No. 27, Beadle file; Schultz to Beadle, July 31, 1970, pp. 1–2.

51. Although Dobzhansky was a pivotal figure in the development of evolutionary biology, he was peripheral to Caltech's molecular biology program. Aside from a minor grant application to the Rockefeller Foundation in 1934, mainly for travel expenses, there is almost no mention of him in the context of the new program. Interestingly, evolutionary biology could have been an intellectual ally to eugenic supporters, but the paucity of documentation prevents a reasonable examination of the perceived service role of Dobzhansky's research.

52. Thomas H. Morgan, "The Relation of Genetics to Physiology and Medicine," *Nobel Lectures in Molecular Biology* (New York: Elsevier North-Holland, 1977), pp. 5–13; G. Allen, *Thomas Hunt Morgan*, pp. 298–301 (see Note 1); Scott F. Gilbert, "Cellular Politics: Ernest Everett Just, Richard B. Goldschmidt, and the Attempt to Reconcile Embryology and Genetics," in Ronald Rainger, Keith R. Benson, and Jane Maienschein, eds., *The American Development of Biology* (Philadelphia: University of Pennsylvania Press, 1988), pp. 311–346; and Boris Ephrussi, "The Cytoplasm and Somatic Cell Variation," *Journal of Cellular Comparative Physiology, 52 (suppl.)* (1958), pp. 35–54.

53. Thomas H. Morgan, "Study and Research in Biology," *Bulletin of the California Institute of Technology, 36, No. 117* (1927), pp. 86–88, quote on p. 87.

54. Theodosius Dobzhansky, "Morgan and His School in the 1930s," *The Evolutionary Synthesis*, pp. 445–452, quote on p. 446 (see Note 47).

55. T. H. Morgan, "Study and Research in Biology," p. 87 (see Note 53).

56. Ibid., pp. 86–88.

57. T. H. Morgan, "Research and Study in Biology," *Bulletin of the California Institute of Technology, 37, No. 121* (1928), p. 103.

58. RAC, RG1.1, 205D, 5.70, "General Statement on Experimental Biology," April 17, 1935, pp. 2–3. See also, Philip Pauly, "General Physiology and the Discipline of Physiology, 1890–1935," in Gerald Geison, ed. *Physiology in the American Court, 1850–1940* (Washington, DC: American Physiological Society, 1987), pp. 195–207.

59. RAC. RG1.1, 205D, Box 5.66, Millikan to Mason, March 7, 1930.

60. G. Allen, *Thomas Hunt Morgan*, p. 357 (see Note 1).

61. CIT, Millikan Papers, Box 18.9, Morgan to Hale, August 15, 1927; quoted in R. E. Kohler, *From Medical Chemistry to Biochemistry*, p. 321 (see Note 12).

62. On Morgan's promotion of mechanistic biology see G. Allen, *Thomas Hunt Morgan*, pp. 330–331 (see Note 1); and Reingold and Reingold, *Science in America*, pp. 284–288 (see Note 12).

63. CIT, Oral History, Henry Borsook, pp. 6–7, 15–17.

64. Dobzhansky, "Morgan and His School in the 1930s," p. 447 (see Note 54).

65. CIT, Oral History, Henry Borsook, pp. 4–5.

66. Kargon, *The Maturing of American Science* (Washington, DC: American Association for the Advancement of Science, 1974), Introduction, pp. 2–10; Reingold and Reingold, *Science in America*, pp. 1–16, 216–222 (see Note 12). On the tensions between European and American life scientists see Kenneth R. Manning, *Black Apollo in Science* (New York: Oxford University Press, 1983), passim.

67. CIT, Oral History, Borsook, pp. 7–8; G. Allen, *Thomas Hunt Morgan*, p. 331 (see Note 1). On American development of genetics see Charles E. Rosenberg, "The Social Environment of Scientific Innovation: Factors in the Development of Genetics in the United States," in *No Other Gods* (Baltimore: Johns Hopkins University Press, 1978), pp. 196–209.

68. APS, Dobzhansky Papers, Oral History, p. 254.

69. CIT, Millikan Papers, Box 18.10, Biology 1928, Morgan to Millikan, May 28, 1928.

70. CIT, Oral History, Henry Borsook, p. 19.

71. Ibid., pp. 23–24.

72. Emerson to Crozier, March 24, 1931; quoted in R. E. Kohler, *From Medical Chemistry to Biochemistry*, p. 322 (see Note 12).

73. RAC, RG1.1, 205D, Box 5.72, Morgan to Weaver, April 4, 1934.

74. Ibid., W. E. Tisdale Diary, May 9, 1934.

75. Ibid., Box 6.74; H. M. Miller's report on his visit to Caltech, September 25–27, 1935.

76. Max Mason, "The Hixon Fund," *Engineering and Science* (May 1947), p. 26. For Pauling's research in "psychobiology" see Epilogue, this volume.

77. Some of these thoughts were expressed by Morgan himself at the dedication of the new Kerckhoff building. CIT, Morgan Papers, Box 2, "Opening Exercise," June 10, 1938, pp. 1–5.

78. CIT, Oral History, James Bonner, pp. 10–11.

Protein Paradigm

The interest in physiological genetics during the 1930s stimulated the formation of linkages between studies of genetic function and analyses of the composition and structure of genes and chromosomes. These new cognitive strategies, backed by financial and institutional resources of the Rockefeller Foundation, brought a number of American geneticists into interdisciplinary cooperation with biochemists and biophysicists. These convergences were based on the protein paradigm, on the dominant explanatory framework that endowed proteins with determinative powers over heredity and related vital phenomena. Most life scientists, whether explicitly or tacitly, subscribed to the protein view of life; and a coalescence of cognitive interests and institutional dynamics sustained the authority of the protein paradigm until the early 1950s.

This intellectual-institutional nexus of the protein paradigm had a profound effect on the funding strategies of the Rockefeller Foundation and on the rise of molecular biology at Caltech. The architects of molecular biology placed protein chemistry at the core of the program and premised many biological studies on the knowledge gained from the various projects of protein research. The leadership of Linus Pauling in protein research epitomized this trend. Although by the late 1940s several groups had directed their efforts toward research on nucleic acids, Caltech remained the bastion of the protein paradigm until 1953.

This Interlude traces the origins of this programmatic commitment to the early part of the twentieth century and follows the interactions of intellectual patterns and institutional forces that led to the conceptualization of molecular biology in terms of the protein paradigm, placing Caltech at its vanguard.

Heredity and the Protein View of Life

When Morgan moved to Caltech in 1928 with Alfred H. Sturtevant and Calvin B. Bridges, classical genetics was already perceived to be past its prime. The conceptual bases of heredity—the gross structural mechanisms of transmission and mutation—were by then well understood. The *"Drosophila* group" of Columbia's famed "fly room" had identified nearly 100 natural mutant types in the fruit fly, relating each visible mutation to a specific location in the chromosomes. Guided by these mutations as chromosomal markers in Mendelian crosses and informed by cytological knowledge, Morgan's group had generated reams of data on relative positions of genes in the chromosomes, mechanisms of linkage, crossing-over, genetic inversions, and translocations. Bypassing physicochemical manipulations, they pried out information with logical inference and algebraic tools; and the knowledge was displayed in intricate genetic maps that charted the course of character transmission and genetic change. By the 1920s, Morgan's work and its counterparts in other research centers had firmly established that genes, in plants or animals, were arranged in the chromosomes in linear order, like beads on a string. Jacques Loeb's praise of Morgan's genetics—"the most exact and rationalistic part of biology, where facts cannot only be predicted qualitatively but also quantitatively"—expressed the view shared by many biologists during the 1920s.[1]

The discipline-building phase too had been completed; the *Drosophila* trail linked East Coast laboratories, Midwest agriculture schools, and Pacific colleges. Recognizing that cognitive advances depended on networks of knowledge, Morgan and his associates introduced a new disciplinary feature: easy access to standardized stocks of mutant flies at Cold Spring Harbor and Caltech and the newsletter *Drosophila Information Service*. Materials and methods circulated among geneticists around the world, stimulating an ethos of cooperative competition. By the late 1920s *Drosophila* genetics represented the dominant approach to heredity research in America, much of Europe, and even Russia.[2]

Yet Morganian genetics attracted diverse critics. Because genes were too minute for cytological methods and escaped even the best of microscopes, because Morgan's work did not deal with "real" physicochemical entities, and because genetic mechanisms in *Drosophila* did not explain development or speciation, skeptics continued to challenge the cognitive scope and disciplinary power of genetics. The medical sciences found *Drosophila* genetics of little clinical relevance. Biochemists and physiologists—notably German physiological geneticist Richard Goldschmidt—tended to doubt the physical existence of the theoretical entities altogether.[3] Students of cytology, embryology, and zoology in the United States and Europe questioned Morgan's verdict that the "cytoplasm may be ignored genetically." Even in his accolade of Morgan's genetic rigor Loeb did not mean to imply nuclear determinism.[4]

To the extreme doubters, Morgan responded with a well-articulated 1933 rebuttal. "What is the nature of the elements of heredity that Mendel postulated as purely theoretical units? What are genes? Now that we locate them in the chromosomes, are we justified in regarding them as material units; as chemical bodies of a higher order than molecules?"

> Frankly, these are questions with which the working geneticist has not much concern himself, except now and then to speculate as to the nature of the postulated elements. There is no consensus of opinion amongst geneticists as to what

the genes are—whether the gene is a hypothetical unit, or whether the gene is a material particle. In either case the unit is associated with a specific chromosome, and can be localized there by purely genetical analysis. Hence, if the gene is a material unit, it is a piece of a chromosome; if it is a fictitious unit, it must be referred to a definite location in the chromosome—the same place as on the other hypothesis. Therefore, it makes no difference in the actual work in genetics which point of view is taken.[5]

His was a two-pronged argument for logical sufficiency and disciplinary autonomy.

Even *Drosophila* geneticists acknowledged the cognitive peculiarity of their program. As a quantitative science, *Drosophila* genetics was somewhat of a paradox. Although arguing persuasively for the existence of concrete units of heredity, Morgan's experimental methods did not deal directly with physical quantities. As Sturtevant and George W. Beadle pointed out:

Physics, chemistry, astronomy, and physiology all deal with atoms, molecules, electrons, centimeters, seconds, grams—their measuring systems are all reducible to these common units. Genetics has none of these as a recognizable component in its fundamental units, yet it is a mathematically formulated subject that is logically complete and self-contained.[6]

As a reaction to his predecessors, who had arbitrarily postulated physiological entities and mechanisms of heredity, Morgan had adopted what Nathan and Ida Reingold called "a strategy of almost geometric chasteness." The gene was simply a location on a line.[7] But what was a gene? Of what was it made, and how did it work physically?

Having isolated the gene conceptually and having demonstrated the general validity of the chromosome theory of heredity, Morgan now expanded the scope of his program. He turned during the 1930s to questions of physiological genetics, to topics Frank R. Lillie had been advocating for a decade: the physicochemical nature of replication and mutation in relation to cellular growth and organismic development. In 1932, in the presidential address to the Sixth International Genetics Congress in Ithaca, Morgan outlined the new priorities as follows:

First then, the physical and physiological process involved in the growth of genes and their duplication (or as we say "division") are phenomena on which the whole process rests. Second: an interpretation in physical terms of the changes that take place during and after conjugation of the chromosomes. Third: the relation of genes to characters. Fourth: the nature of the mutation process—perhaps I may say the chemicophysical changes involved when a gene changes to a new one. Fifth: the application of genetics to horticulture and to animal husbandry.[8]

These priorities (with the partial exception of the last item), predicated on the hybridization of genetics, physiology, biochemistry, and the physical sciences, formed the backbone of Morgan's agenda at Caltech. These very topics, which were to constitute the fundamental approach to development, maturation, and inherited behavior, represented the salient features of the newly minted Rockefeller Foundation's program in psychobiology.

The disciplinary and cognitive tangles confronting the amalgamation of physiology, biochemistry, and genetics stemmed from divergent definitions of the functional units of heredity and from contested views of genetic action. The one point on which Morgan, his supporters, and his critics could generally agree was the protein nature of the hereditary material. During the first four decades of the twentieth century, physicochemical explanations of the gene and its mode of action—constancy, change, and regulation—were constructed mainly within the explanatory framework of the protein view of life. According to that premise, proteins formed the physical bases of life and were the principal determinants of reproduction, growth, and regulation. The belief in the primacy of proteins did not represent a detour from the path to the double helix or a mere episode in the century of DNA.[9] The protein view of life was not a divergence but a dominant concept that defined the mainstream of heredity research, forming a coherent and internally consistent scientific theory and laboratory practice. For a quarter of a century the protein paradigm explained, albeit inadequately, the cellular and subcellular processes subsumed under the term molecular biology, and it guided the entire research program at Caltech.

The conceptualization of the material basis of life in terms of proteins reached back to the nineteenth century, of course, deriving conviction and imagery from the enormous influence of T. H. Huxley's protoplasmic view of life (1864). His theory bestowed upon the protoplasm—simple or nucleated—all the physical and mental attributes of life, enshrining the gelatinous substance as the source of biological diversity and the locus of material and cognitive control. August Weismann's work focused attention on the nucleus as the site of hereditary transmission but shed little light on the chemical nature of chromosomes. Cytologists and physiological chemists generally agreed that the nuclear material, chromatin, was composed of two principal components: the part Friedrich Miescher had named "nuclein" in 1869 (and believed it to be the hereditary material) and the albuminoid substances that characterized the cell's protoplasm. Unequivocal evidence demonstrated that the amount of nuclein varied with the stages of mitosis, but researchers suspected that the quantity of nuclein in the chromosomes was too minute in comparison with protoplasmic matter to transmit hereditary traits. The intellectual momentum of the protoplasmic view of heredity thrust it into the next century.[10]

Various scientists and popularizers during the first three decades of the twentieth century strengthened the grip of the protoplasmic view of life on scientific thought and culture. Eugenicists in particular infused potency into the linkage of hereditary fitness and protoplasmic endowment by deploying images of stewardship over the national protoplasm.[11] Two figures, D'Arcy W. Thompson and Jacques Loeb, stand out as visionaries who shaped scientists' concepts of the animate. With the protein view of life as a point of departure, both scholars articulated the essence of life in physical terms (with only passing attention to genetics). They diverged in a fundamental way, however, mirroring the Aristotelian categories of matter and form. In a sense, the differences between the intellectual agendas of Thompson and Loeb may be viewed as conceptual prototypes, representing, respectively, the structural and biochemical approaches to animate phenomena. Both these lineages have manifested their imprints on the molecular biology program. The ideas of D'Arcy Thompson struck a resonant chord with the crystallographers, who would lead the principal programs in protein research.

A British aristocrat of learning and a staunch anti-Darwinian, Thompson inspired the international scientific community with his classic *On Growth and Form* (1917). His exposition drew on the age-old idea of living crystals—crystals as animate structures and crystallization as a biological process.[12] But Thompson empowered the old metaphor with the armamentarium of molecular mechanics, encouraging scientists to visualize growth and form in geometric and mathematical terms. Whereas his work was limited by his deliberate trivialization of chemistry, his argument emanated from the protoplasmic view of life. Thompson's idea that growth was an intricate but quantifiable process of protoplasmic packing and its form an array of space lattices reflecting precise molecular arrangements appealed to physicists, mathematicians, and crystallographers. Complexity was no obstacle to logic, Thompson argued, "That Nature keeps some of her secrets longer than others— that she tells the secret of the rainbow and hides that of the northern lights—is a lesson taught me when I was a boy." His confidence that physicomathematical explanations would ultimately demystify the uniqueness of protoplasmic attributes inspired scientists in Europe and America.[13]

Jacques Loeb, on the other hand, concentrated on matter. An arch-determinist and a skeptic of Darwinian explanations of organismic adaptation, Loeb strove to persuade students of the life and human sciences of the equivalence of life and matter: both were a product of the blind forces of necessity. Often in the public eye for his iconoclasm, Loeb designed his sensational studies of tropism and artificial parthenogenesis in order to prove a point. Fundamental attributes of life, such as fertilization and instinct, were reducible to the laws of physics and chemistry and thus amenable to human control. His conviction that the solution of the riddle of heredity would be written in physicochemical language stemmed from the perceived primacy of proteins. According to Loeb, all living matter, liquid or solid, was principally constituted of proteins, and their physicochemical properties governed all animate processes.[14]

By the first decade of the twentieth century, studies of the amorphous protoplasm tended to differentiate along the cleavage lines of competing approaches to protein chemistry; and chemical explanations of reproduction varied accordingly.[15] Of these explanations, enzyme action and the concept of autocatalysis carried the greatest weight. Fortified by the chemical triumph of Buchner's zymase over the living cell (1896), biochemists rushed through the ruins of the protoplasmic monarchy to establish the republic of ferments.[16] Despite technical hurdles—harsh extraction procedures, hand-powered bench centrifuges, almost no electrical equipment, and no refrigeration—scores of enzymes were isolated in crude form.[17] Some were autocatalytic. Such an enzyme displayed the astonishing property of self-duplication, generating more of itself by accelerating the reaction in which the enzyme itself was an end-product. By the 1920s, autocatalysis, often analogized with crystal growth in the mother liquor, had become a popular catch-all term for a range of vital processes involved in cellular reproduction and organismic growth.

Theoretically, both organismic and chemical growth could be described as exponential functions; the mathematical-graphical representation of the autocatalytic process was the sigmoidal or S-shaped curve, plotting matter against time. It appeared that whether one dealt with the chemical constituents of an enzyme, food for the organism, or crystal solution, one began with a "seed" particle that eventually regenerated out of its raw materials until reaching steady state and leveling off. Plotted against elapsed time, such growth (in terms of quantity of matter) yielded an S-shaped curve. Whether the curve was flattened or steep depended on the rate of the reaction, which in turn was determined by other supposedly measurable variables—the quantity of catalyst, for instance. In more complex systems such as organisms, life scientists envisioned myriad autocatalytic reactions regulated by a "master reaction" and adding up to the final S-shaped curve.[18]

Propagated by the colloid chemist Wolfgang Ostwald and the school of *entwicklungsmechanik*, the autocatalytic theory of growth and reproduction gained support in America beginning around 1910 through the joint influence of A. L. Hagedoorn (Loeb's student), T. B. Robertson, and especially Jacques Loeb. Reasoning on the basis of analogy, the advocates of the autocatalytic theory argued for the ontological unity of diverse phenomena—enzymatic autocatalysis, chromosome duplication, protoplasmic growth, and organismic development—as all of these processes could be described as S-shaped curves. Apart from the lure of theoretical unity, the "autocatalycists" appealed to material reality. Colonizing the conceptual vacancy created by Morgan's inferential nonmaterial approach, they transported enzymes onto the center stage of heredity research.[19]

Most classical geneticists in America, preoccupied with crosses and maps, found the autocatalytic theory of reproduction peripheral to their epistemic universe. The noted German biologist Richard Goldschmidt, however, a vociferous opponent of Morganian genetics, promoted the autocatalytic theory of heredity in American journals and at Woods Hole. His search for a unified theory of development and reproduction converged on autocatalysis as an unequivocal solution of heredity. If genes were real, he argued, they were a definite quantity of something; that something, according to convergent evidence, was an enzyme. Genes were autocatalysts that carried out certain reactions at rates determined, at least in part, by their quantity. The only alternative, Goldschmidt concluded, was agnosticism.[20]

Goldschmidt's frustrations and the need for a physicochemical theory of heredity was especially well articulated by Harvard's flamboyant psychophysicist Leonard T. Troland. In his important article, "Biological Enigmas and the Theory of Enzyme Action," Troland lamented in 1917 that despite the fact that several Mendelians had hinted that "unit characters" were enzymes, no worker in genetics, with the exception of Goldschmidt, had seen the light. He urged geneticists to adopt the premise that Mendelian factors were autocatalytic enzymes, as such an approach offered a single synthetic solution to all biological enigmas: the mysteries of the origins of living matter, viruses, the course of variations, the mechanisms of heredity and ontogeny, and general regulation.[21]

As an outsider to biological research, Troland drew on an impressive array of studies, interweaving the findings of Irving Langmuir, Ostwald, Loeb, Robertson, and Goldschmidt with his own intuitive sense of gene action. Troland intended to convince biologists that the apparent complexity of organismic growth was merely the additive effect of a large number of simple growths, each governed by its specific autocatalytic mechanism and regulated by an overarching "master reaction." From an evolutionary standpoint, the "first enzyme," which later mutated, gave rise to the multitudes of complex life forms; and according to Troland, strong evidence pointed to free autocatalytic enzymes still existing in the biological universe in the primitive form of "filterable viruses."[22]

Morgan had several objections. He explicitly rejected the claim that genes were enzymes and considered the whole approach much too speculative. In a 1926 lecture in Woods Hole he analyzed the problems of physiological genetics, drawing on European and American studies, and outlined the weaknesses in Loeb's and Robertson's arguments. Morgan thought that, even if the putative autocatalytic reactions were experimentally valid, they were no proof of gene action. These reactions could be in fact many stages removed from the gene itself. He refuted the notion of a multiplicity of regulated autocatalytic reactions, showing that the concept of "master reaction" was completely arbitrary. Challenging Goldschmidt's fait accompli, Morgan argued that the "quantitative theory" contained no measured quantities of postulated enzyme, only some assigned arbitrary values. Morgan advocated caution. He concluded that "we may not be warranted in speaking of the genes as enzymes, the genes may be protein bodies, one of whose activities is to produce enzymes which, being set free, act in each cell, and take part in catalytic reactions in the cytoplasm."[23]

Morgan's views carried considerable weight but did not represent a consensus. Sturtevant, whose ideas guided the younger Caltech geneticists, framed the problematic relation between genes and enzymes in terms nearly identical to those of Morgan. The *Drosophila* geneticist Hermann J. Muller, on the other hand, was captivated by the autocatalytic concept of reproduction, by Troland's prose, and by Loeb's engineering ideal in biology. A leftist eugenicist, Muller was motivated by the twin passions for a material genetics and for a technological control of heredity. His Nobel-winning feat of inducing artificial mutations in *Drosophila* with x-rays (1927) was intended to prove the existence of a physical gene and to demonstrate a capability for its manipulation.[24] He had been advocating a physicochemical approach to genetics and promoting "naked genes" (filterable viruses) since the early 1920s. Viruses, and especially bacteriophage, simulate genes in their autocatalytic action, Muller argued—they were the key to the riddle of life.[25]

This world picture greatly stymied the curiosity about the chemical structure and physiological role of nucleic acids, despite their known unequivocal connection to replication. By comparison to the robust research programs on proteins, American research on nucleic acids during the first four decades of this century was negligible. Ironically, the most extensive and important work on nucleic acids discouraged further interest in the topic. The studies (1909–1929) conducted at the Rockefeller Institute by the Russian-born maverick Phoebus A. T. Levene led to identification of the components of nucleic acids, distinguished ribonucleic acid (RNA) from deoxyribose nucleic acid (DNA), determined their acidic character, and linked these substances to regulatory chemical functions in the nucleus.[26] This work served to trivialize the physiological function of nucleic acids, however.

Based on these studies Levene constructed the "tetranucleotide hypothesis" during the early 1920s. The hypothesis, literally a theoretical construction, postulated that the four nitrogenous bases, derived from the nucleotides, were present in nucleic acids in equal proportions and combined in a fixed manner. The putative repetitive sequences suggested that nucleic acids had little biological specificity. By 1930 biochemists had come to regard nucleic acids as simple, uninteresting substances, incompatible with the complexities of genetic functions: replication, mutation, cellular regulation, and organismic development.[27] The short-lived interest in nucleic acids was eclipsed by the intellectual promise, institutional resources, and technological innovations of protein research.

Just around the time when genetics was broadening its scope toward biochemistry, protein chemistry and the autocatalytic theory of reproduction received an additional boost. In 1930 Loeb's protégé at the Rockefeller Institute, John H. Northrop, crystallized the proteolytic enzymes pepsin and trypsin, and he soon after demonstrated their autocatalytic properties. Northrop had also worked on bacteriophage and viruses and, like many of his predecessors and contemporaries, was convinced of the identity of viruses and enzymes, a functional reproductive unit that formed a link between the chemical and physiological realms. His Nobel Prize-winning work not only reinforced the conceptual foundations of the autocatalytic theory of life but introduced the study of enzymes into the new domain of x-ray crystallography.[28]

The most sensational "proof" of autocatalytic reproduction of proteins came in 1935, when Wendell M. Stanley of the Rockefeller Institute crystallized the tobacco mosaic virus (TMV). Adopting the techniques and interpretative framework of his senior colleague Northrop, Stanley successfully isolated the virus in apparently pure crystalline state. This technical feat, as Stanley pointed out in *Science*, demonstrated that the "tobacco-mosaic virus is regarded as an autocatalytic protein which, for the present may be assumed to require the presence of living cells for multiplication."[29] His studies were confirmed and elaborated by the most advanced protein technologies at the laboratories in Uppsala. Stanley's work seemed to supply concrete evidence for the compositional and functional equivalence of viruses, enzymes, and genes, linking them to the age-old concept of crystallization. The objections to the autocatalytic theory and to Stanley's disregard of the viral nucleic acids were ignored with the euphoria of the protein victory.[30]

Life scientists and the popular media alike credited Stanley with finding the key to the riddle of life. Muller touted Stanley's work as an epochal discovery, spreading the lessons of protein crystals among geneticists. George Beadle, Max Delbrück, and the officers of the Rockefeller Foundation singled out Stanley's work as the most important breakthrough in understanding the molecular basis of the gene; the discovery has been described as the symbolic beginning of molecular biology.[31] A 1938 article on heredity in *Scientific American* marveled at the molecular unraveling:

> Here is an almost unbelievable, a wholly novel, ability of a molecule: to create its like out of the lesser molecules of a suitable surrounding medium. Only in the gigantic virus protein have we discovered such a remarkable property. . . . For an approach to this problem of self-creation, or autosynthesis, we must consider the enzyme, also believed to be a formidable protein, though not as accomplished a one as the gene. We have to speculate that the gene, as a super-enzyme, causes a bafflingly complex chain of chemical processes in the protoplasm in which the chromosomes swim.[32]

The fate of physiological genetics was bound with the course of protein research. The Rockefeller Foundation targeted massive funds to develop the field; the diversity of proteins—enzymes, amino acids, colloids, globulins, and crystals—provided abundant research projects. "All that association of phenomena which we term life is manifested only by matter made up to a very large extent of proteins," argued Warren Weaver in justifying the centrality of protein research in the Foundation's new program:

> [Proteins] enter into nearly every vital process. They are the principal component of the chromosomes which govern our heredity; they are the basic building stuff for the protoplasm of each cell of every living thing. Our immunity to many diseases depends upon the mysterious ability of serum globulin. . . . Several of the hormones, including insulin, are protein in nature. . . . The invasion of certain huge protein molecules, otherwise known as viruses, give us common cold, influenza. . . . Enzymes, those strange chemical controllers of so many of the detailed processes of the body, those perfect executives which stimulate and organize all sorts of activities without using up any of their own substances or energy— these enzymes are now believed to be protein in nature. Indeed many diverse scientists, each with his own special enthusiasm, would be willing to agree that these proteins deserve their names of "first substances."[33]

During the 1930s and into the early 1950s, protein chemistry marked the vanguard of diverse researches subsumed under the term "molecular biology."

Chemistry of Proteins During the 1930s: Theories and Technologies

Protein chemistry during the 1930s was in a state of flux. Research programs competed to explain protein composition and structure; some diverged even on the basic definition of a molecule. More than any area of life science, protein research depended on instrumentation, particularly on ultracentrifugation, electrophoresis, and x-ray diffraction studies. Beyond new measurements, these technologies shaped the organization of biological research, literally creating instrument-centered "borderland" subspecialties. Acquiring costly apparatus demanded financial resources, and in turn access to the sophisticated apparatus endowed scientists with cognitive resources and prestige. At the same time, novel technologies produced scientific artifacts and transdisciplinary evidence that complicated the process of adjudication between competing theories. The prestigious Uppsala–Rockefeller network, which dominated protein research, epitomized these interdisciplinary currents of knowledge and power; these developments, in turn, influenced the research program at Caltech.

Theodor Svedberg, professor of colloid chemistry at the University of Uppsala, was awarded the 1926 Nobel Prize in Chemistry as soon as his experiments on proteins were published. His ultracentrifuge studies, which determined for the first time the molecular weights of hemoglobin, ovalbumin, phycocynin, and phycoerythrin, received instant recognition owing to their multiple levels of significance. After a year's collaboration with physical chemist John W. Williams at the University of Wisconsin (1924), and drawing on Mason's and Weaver's analysis of sedimentation, Svedberg constructed the analytical ultracentrifuge. The giant machine, resembling a jet plane's cockpit, sorted protein molecules according to size by subjecting them to a centrifugal force of 400,000 times gravity. A rotating optical device recorded the rate of sedimentation (a run took more than 24 hours), and the data were then interpreted mathematically.[34] The primary significance of this technological feat lay in its adjudicatory potential. The ultracentrifuge tested the validity of colloid chemistry, a field then being challenged on several fronts.

Founded and propounded by Wolfgang Ostwald during the decade beginning with 1910, colloid biochemistry focused on the aggregate nature of protoplasmic substances—gelatin, albumin, and cellulose, for example—emphasizing their micellar properties. The new science sought to reveal the "world of neglected dimensions," the submicroscopic region between molecules and cells (10^{-5} to 10^{-7} cm). Those biochemists and biophysicists who adhered to the Van't Hoff-Arrhenius theory of solutions demonstrated that solution chemistry sufficed to explain the properties of colloids; Loeb, Northrop, Winthrop J. V. Osterhout, and Leonor Michaelis were among these scientists (all were at the Rockefeller Institute). Ostwald's concept appealed to many serious researchers, especially to physically minded life scientists who had found Emil Fischer's chain theory of proteins counterintuitive. Fischer, formulating his theory at the University of Berlin at a time when many of the 20 amino acids had already been identified (1899–1908), demonstrated that proteins consisted of linear arrangements of amino acids held together by peptide bonds. However, Fischer also argued that molecular weights of polypeptides could not exceed 5000 daltons. How could such tiny snippets account for the obvious unity and cohesion of protoplasmic matter?[35]

Hermann Staudinger supplied an answer that challenged Fischer's conservative verdict. His work on biopolymers (e.g., rubber) at the University of Freiburg indicated that colloid-like substances consisted of giant molecules—"macromolecules," as he christened them in 1924. Predictable and original objections followed. Colloidalists protected the integrity of their fundamental units and the autonomy of their specialty; organic chemists urged Staudinger to purify his contaminated samples; the x-ray crystallographers made a forceful argument: A unit cell of silk, or cellulose, is composed of molecules. Therefore how could a molecule be larger than the unit cell?[36]

On one level Svedberg's ultracentrifuge supplied a decisive answer. Centrifugal forces sheared the colloids into regular fragments, demonstrating that substances such as hemoglobins and albumins had well-defined molecular weights. "Macromolecule" was indeed an appropriate term for these heavy weights, ranging from four to five orders of magnitude. Aided by large Rockefeller grants, Svedberg's group calculated during the following decade the molecular weights and shapes of more than 30 large proteins. As sole owners of the analytical ultracentrifuge (1926–1937), they dominated the molecular study of proteins. Did these molecular substances, however, conform to the traditional definition of a molecule—the smallest particle existing in a free state? Opinions diverged on what exactly Svedberg was measuring.[37] In fact, Svedberg himself noted that small changes in hydrogen ion concentration (pH) sufficed to dissociate the sedimented proteins into units of lower molecular weight, suggesting that the forces holding the aggregates together were electrical. This observation did not help to define the molecularity of proteins, but it sparked the development of yet another powerful technology, further strengthening Svedberg's protein establishment.[38]

Svedberg assigned the difficult task of building an apparatus for the electrical separation of proteins to his graduate student Arne Tiselius (1930). In 1937, following a year's visit with the protein chemists at the Rockefeller Institute, Tiselius unveiled the new electrophoresis apparatus: an enormous piece of equipment, spanning 20 feet in length and about 5 feet in height. For the first time, complex protein mixtures were separated reliably; and shadow photography recorded the colorless boundaries. As in the case of the ultracentrifuge, the costly electrophoresis apparatus became an organizing focus for cooperative interdisciplinary research, an indispensable tool for exploring the science of proteins. It also became a source of scientific authority. It defined the research on antibodies, confirmed the homogeneity of Stanley's viral proteins, and promised to test Mendelian inheritance by distinguishing chemically between pure lines and hybrid proteins. The "Tiselius," as it was called (winning its engineer the Nobel Prize), would become crucial for Caltech's molecular biology program.[39]

But beyond techniques, Svedberg aspired to elucidate the mystery of protein structure. Discerning regularities in the voluminous ultracentrifuge data, around 1930 he postulated that all proteins, regardless of apparent size, represented aggregates of subunits of molecular weight 17,500—half a Svedberg unit of 35,000.[40] Yet Svedberg's results could not be checked independently outside Uppsala: Technological monopoly thus generated intellectual property. Those who objected to Svedberg's arbitrary interpretations also mistrusted his technological authority. "Fair is foul and foul is fair/Hover through the fog and murky air," quoted the noted Wisconsin biochemist Karl P. Link, describing Svedberg's enterprise:

> Have the U.C. [ultracentrifuge] boys, in contrast to the organic chemist, defined the "molecules" they are measuring in such terms that fellow scientists and border scientists, whom they presumably are attempting to help, know what they are talking about? Have the U.C. boys not appropriated for their use a term (molecule) which has a definite, precise, and workable connotation in the chemical laboratories of the world at large, and are they not using it in a manner that leads to absurdity and confusion rather than clarity? Are the U.C. boys not making up new rules of their game as they go along. . . ? And have they not called in the advertising [sic] departments and the prestige of prizes and honorary degrees to tell the world of their great mechanical and engineering skill?

Marshaling the evidence of other Svedberg critics, Link reminded his audience of Paracelsus' writing: "The power to recognize and follow the truth cannot be conferred by academical degrees."[41]

Link's closest friend, the German émigré Max Bergmann at the Rockefeller Institute, generally approved of the assault. "I have been skeptical of several of Svedberg's findings and interpretations." he responded. "However, I have been in doubt as to the extent to which my criticism was justified. Therefore I decided to investigate the composition of several of the proteins by our chemical methods." A former colloid chemist, Bergmann adopted Fischer's chain theory and during the 1930s had been studying the proteolytic digestions of proteins in order to determine their amino acid compositions.[42] Between 1936 and 1938, with the help of Link's former graduate student Carl Niemann, the Bergmann–Niemann team generated some intriguing results. They found that the total number of amino acid residues per protein "molecule" obeyed a series of multiples of 288, corresponding to Svedberg's multiple rule; ironically, their own analysis seemed to "independently" corroborate Svedberg's theory. The Bergmann-Niemann law did not just denote the amino acid content. The numerical convolutions also traced a periodicity within proteins, patterns that purported to account for genetic specificity. This concept greatly appealed to Goldschmidt, who found in it confirmation of his theory as well as suggested mechanisms of gene action.[43]

The cumbersome formulation enjoyed a wide hearing, firming up Niemann's reputation and winning him a faculty position at Caltech in 1938. Levene was so impressed with the work he was moved to write to Bergmann: "I have looked over your reprints and cannot find words enough to tell you how much I admire your work, for the things you have now accomplished seemed unattainable only a short time ago."[44] Having effectively terminated his interest in nucleic acids, Levene was now working with Bergmann and other biochemists to form a separate division of protein chemistry within the American Chemical Society and to establish a "central laboratory for protein chemistry."[45] Link, recovering from his slow comprehension of Bergmann's theory, confessed: "Lieber Max: . . . I thought I was the only dumbbell in the world, but when I read the reviewer's words that the Bergmann-Niemann Law was tough going I felt much better."[46]

Other biochemists, however, among them A. C. Chibnall, N. W. Pirie, and R. D. Hotchkiss, were not as easily assured. They objected to the arbitrary nature of the numerical interpretations. If one picked a random region in the bond listings in the *New York Times*, Hotchkiss demonstrated, and calculated their molecular weights (as he did), the fit of the protein data was within an error of 2.3, less than half the 4.9 error of Bergmann's amino acid data.[47] His and Pirie's judgments coincided with Chibnall's verdict that these quantitative theories "demonstrated nothing more than the hypnotic power of numerology."[48] It was not a clear-cut case. The theories exhibited the requisites of the hypotheticodeductive model and the robustness of a sound experimental science. The theoretical construction of protein composition was based on convergent empirical evidence drawn from several supposedly independent lines of research.

The Bergmann-Niemann theory became more difficult to resist when x-ray studies marshaled corroborative evidence. X-ray crystallography of proteins was still an embryonic specialty during the mid-1930s. When the Rockefeller Foundation began supporting the physicochemical approach to vital processes, materials and methods lagged behind D'Arcy Thompson's visions of an animate geometry. A. L. Patterson of Philadelphia's Johnson Foundation had just developed a technique of applying Fourier series to x-ray spectra, enabling researchers to calculate interatomic distances in crystals. John D. Bernal in Cambridge recorded in 1934 the first x-ray diffraction pattern of crystallized pepsin in its mother liquor; the lattice constants agreed with ultracentrifuge determinations of the molecular weight of pepsin. Based on his studies of the x-ray diffraction patterns of keratin, William

Astbury at Leeds inferred that the fibrous protein had hexagonal folds, suggesting a ring structure for the protein.[49]

Due partly to Fischer's own suggestions, several researchers during the 1930s postulated various ring structures for proteins involving covalent bonds. The Australian-born Oxford crystallographer Dorothy M. Wrinch, however, fortified the ring postulate with mathematical rigor and geometric flair. Deeply inspired by D'Arcy Thompson's ideas on form, Wrinch capitalized on topological considerations. She proposed during the mid-1930s a honeycomb-like cage structure, a cyclol, for native globular proteins. That the cyclol consisted of 288 amino acid residues— and thus supposedly offered yet another independent source of evidence for the Svedberg and Bergmann-Niemann units—only served to enhance the "hypnotic power of numerology."[50] Like her Rockefeller and Uppsala colleagues, Wrinch anticipated that her protein structure would illuminate the molecular mechanisms of heredity. "W. [Wrinch] is very anxious to get some person, Chambers presumably being the obvious candidate, to attempt the removal of a large number of chromosomes of the salivary gland and arrange these in a rodlike bundle which would then be x-rayed for structure," Weaver recorded, after a visit with Wrinch, "W. says that this is the first item on the 10-year plan she and Muller have recently agreed on for genetics."[51] Wrinch's theory possessed mathematical elegance, aesthetic appeal, and genetic promise.

The cyclol theory received wide exposure, generating both fanfare and controversy. Wrinch, "the woman Einstein," as she was portrayed at the peak of her fame, presented her beautiful geometric models in print and in numerous lectures in England and America, and she received the enthusiastic endorsement of Nobel laureate physicist Irving Langmuir.[52] Yet both Bergmann and Niemann objected to her cyclol theory and resented her acclaim. "[I]t seems that she is taken much more seriously in America than England," Niemann complained to Bergmann (from London). "In fact she is looked upon as the No. 1 menace and people go out of their way to avoid her. Around the University College her name is an object of horror."[53] The popularity of her theory was short-lived. Within the next few years, following Niemann's move to Caltech, the Pauling–Niemann team would orchestrate the demise of Wrinch's cyclol and of her scientific career.

Thus as the interest in physiological genetics began to converge on the structure and composition of proteins, and as Linus Pauling plunged into the turbulence of protein research, the field churned with technological, cognitive, and disciplinary cross currents. The opposition to Svedberg's arbitrary approach prolonged the gasp of colloid chemistry. Advocates of the chain theory confronted the proponents of the ring structure of proteins. Even those who did grant that denatured or "dead" proteins were linear polypeptides believed that biologically active "live" proteins were nonlinear. Many biochemists thought that the distinctive properties of biologically active proteins resulted from covalent linkages, manifested in globular structures. By the early 1940s Pauling would spur scientists toward a consensus, grounding molecular biology in the most recent models of protein generated in his laboratory. Throughout the 1930s, though, representations of gene function—replication, mutation, and phenotypic expression—diverged along the different contours of the protein researchscape.

Notes

1. Quoted in Jacques Loeb's *The Mechanistic Conception of Life* (Cambridge: Harvard University Press, 1964), p. 25. The conceptual foundation of genetics were by then published by Thomas H. Morgan, *The Theory of the Gene* (New Haven: Yale University Press, 1928). See also Elof A. Carlson, *The Gene: A Critical History* (Philadelphia: W. B. Saunders, 1966).

2. On the growth of the *Drosophila* genetics network see Garland E. Allen, *Thomas Hunt Morgan: The Man and His Science* (Princeton: Princeton University Press, 1979).

3. An informative contemporary analysis of the gene problem is given by Milislav Demerec, "Eighteen Years of Research on the Gene," *Cooperation in Science, 501* (Washington, DC: Carnegie Institution of Washington Publication, 1938), pp. 295–314. On the criticisms of Morgan's work and on alternative approaches see Garland E. Allen, "Opposition to the Mendelian-Chromosome Theory: The Physiological and Developmental Genetics of Richard Goldschmidt," *Journal of the History of Biology, 7* (1975), pp. 49–92; Jan Sapp, "The Struggle for Authority in the Field of Heredity, 1900–1932: New Perspectives on the Rise of Genetics," *Journal of the History of Biology, 16* (1983), pp. 311–342; Jan Sapp, *Beyond the Gene: Cytoplasmic Inheritance and the Struggle for Authority in Genetics* (New York: Oxford University Press, 1987); Jonathan Harwood, "National Styles in Science: Genetics in Germany and the United States between the World Wars," *Isis, 78* (1987), pp. 390–414; Richard Burian, Jean Gayon, and Doris Zallen, "The Singular Fate of Genetics in the History of French Biology, 1900–1940," *Journal of the History of Biology, 21* (1988), pp. 357–402; and Peter J. Bowler, *The Mendelian Revolution: The Emergence of Hereditarian Concepts in Modern Science and Society* (Baltimore: Johns Hopkins University Press, 1989).

4. Morgan's 1926 quote is in Sapp's *Beyond the Gene*, p. xiii (see Note 3 for reference.)

5. T. H. Morgan, "The Relation of Genetics to Physiology and Medicine," *Nobel Lectures in Molecular Biology* (New York: Elsevier North-Holland, 1977), p. 5.

6. Quoted in N. H. Horowitz, "Genetics and the Synthesis of Proteins," *Annals of the New York Academy of Sciences, 325* (1979), p. 256. From G. W. Beadle and A. H. Sturtevant, *Introduction to Genetics* (Philadelphia: W. B. Saunders, 1939).

7. Nathan Reingold and Ida H. Reingold, *Science in America: A Documentary History, 1900–1939* (Chicago: University of Chicago Press, 1981), p. 147.

8. T. H. Morgan, "The Rise of Genetics," *Proceedings of the Symposium of the International Congress in Genetics, 1* (1932), pp. 101–103.

9. The standard sources on the early history of molecular biology assume various apologetic postures with respect to the protein "embarrassment." See Franklin H. Portugal and Jack S. Cohen, *A Century of DNA* (Cambridge: MIT Press, 1977); Robert C. Olby, *The Path to the Double Helix* (Seattle: University of Washington Press, 1974); and Horace F. Judson, *The Eighth Day of Creation* (New York: Simon & Schuster, 1979). Because there was little historical continuity in nucleic acid research before 1945, even Olby's effort to avoid writing a linear history often succumbed to bouts of whigism. See also Robert C. Olby, "The Protein Version of the Central Dogma," *Genetics, 79* (1975), pp. 3–27.

10. Thomas Huxley, "On the Physical Basis of Life" (1864), in his *Collected Essays*, Vol. 1 (New York: D. Appleton, 1894); and Gerald L. Geison, "The Protoplasmic Theory of Life and the Vitalist-Mechanist Debate," *Isis, 60* (1969), pp. 273–292. On Weismann's contributions see William Coleman, "Cell, Nucleus, and Inheritance: An Historical Study," *Proceedings of the American Philosophical Society, 109* (1965), pp. 124–158; Frederick B. Churchill, "August Weismann and a Break from Tradition," *Journal of the History of Biology, 1* (1968), pp. 91–112; and Churchill, "From Heredity Theory to Vererbung: the Transmission Problem, 1850–1915," *Isis, 76* (1987), pp. 337–364. For a conceptual overview see John Farley, *Gametes and Spores: Ideas about Sexual Reproduction, 1710–1914* (Baltimore: Johns Hopkins University Press, 1982), Chs. 6–10.

11. References to the protoplasmic nature of life and heredity have been ubiquitous in scientific publications from around 1910 until the early 1950s. They may be found, for example, throughout issues of the *Journal of Heredity*, *Science*, and *Scientific American*. The primacy of the protoplasm is particularly well articulated in "Vorwort," *Protoplasm, 1* (1926–1927), pp. 13–15, the intro-

ductory issue of the international journal. Contributors during the 1930s included B. Ephrussi, W. T. Astbury, J. D. Bernal, J. B. S. Haldane, A. V. Hill, and A. Hollaender. Of greater significance is the paradigmatic presentation of the protoplasm in textbooks for high schools and colleges, for example, George C. Wood and Harold A. Carpenter, *Our Environment* (New York: Allyn & Bacon, 1938), pp. 893–914; Douglas Marsland, *Principles of Modern Biology* (New York: Holt Rinehart, 1947, 1951), passim; and William C. Beaver, *General Biology* (St. Louis: C. V. Mosby, 1939, 1942, 1946), passim. For an in-depth analysis of the pedagogic displacement of the protein paradigm after the "DNA revolution," see Barak Gaster, "Assimilation of Scientific Change: The Introduction of Molecular Genetics into Biology Textbooks," *Social Studies of Science* 20, (1990), pp. 431–454. On the connection between eugenics and protoplasm see Chapter 1, Notes 14 and 15, this volume.

12. D'Arcy Thompson, *On Growth and Form* (Cambridge: Cambridge University Press, reprinted 1988), pp.132–172. Crystallization as an animate phenomenon was discussed already by Francis Bacon in *Sylva Sylvarum* (1627) and continued to influence life scientists, especially after the analysis by Theodor Schwann in the final section of his work, *Microscopical Researches* (1839). On the significance of crystallization in the life sciences see Donna Haraway, *Crystals, Fabrics, and Fields* (New Haven: Yale University Press, 1976), pp. 41–54.

13. Quoted in D'Arcy Thompson, *On Growth and Form*, pp. 170–171 (see Note 12). For a discussion on the influence of his work see G. E. Hutchinson, "In Memorium, D'Arcy Wentworth Thompson," *American Scientist, 36* (1948), p. 577; Peter B. Medawar, *The Art of the Soluable* (Harmondsworth: Penguin Books, 1967); and Pierre Laszlo, *A History of Biochemistry: Molecular Corrolates of Biological Concepts* (New York: Elsevier, 1986), Chs. 13 and 14.

14. Jacques Locb, *Proteins and the Theory of Colloidal Behavior* (New York: McGraw-Hill, 1922). On the significance of the work see also Joseph Fruton, *Molecules and Life* (New York: John Wiley & Sons, 1972), pp. 131–144; Francois Jacob, *The Logic of Life: A History of Heredity* (New York: Pantheon Books, 1982), pp. 246–248; and Philip J. Pauly, *Controlling Life: Jacques Loeb and the Engineering Ideal in Biology* (New York: Oxford University Press, 1987), pp. 151–153.

15. See next section of this chapter.

16. For Buchner's discovery and its influence see Robert E. Kohler, "The Background to Eduard Buchner's Discovery of Cell-Free Fermentation," *Journal of the History of Biology, 4* (1971), pp. 35–61; R. E. Kohler, "The Reception of the Eduard Buchner's Discovery of Cell-Free Fermentation," *Journal of the History of Biology, 5* (1972), pp. 327–353; R. E. Kohler, "The Enzyme Theory and the Origins of Biochemistry," *Isis, 64* (1973), pp. 181–196.

17. John T. Edsall, "The Development of the Physical Chemistry of Proteins, 1898–1940," *Annals of the New York Academy of Sciences, 325* (1979), pp. 53–77.

18. For a discussion on the autocatalytic theory of life see A. W. Ravin, "The Gene as a Catalyst; The Gene as an Organism," *Studies in History of Biology, 1* (1977), pp. 1–45; R. C. Olby, *Path to the Double Helix*, pp. 149–152 (see Note 9); D. Haraway, *Crystals, Fabrics, and Fields*, pp. 41–54 (see Note 12); and Lily E. Kay, "W. M. Stanley's Crystallization of the Tobacco Mosaic Virus, 1930–1940," *Isis, 77* (1986), pp. 450–472.

19. See for example T. B. Robertson, "On the Normal Rate of Growth of an Individual and Biochemical Significance," *Archiv fur Entwicklungsmechanick, 25* (1908), pp. 581–615; T. B. Roberstson, "On the Nature of the Autocatalyst of Growth," *Archiv fur Entwicklungsmechanick, 37* (1913), pp. 497–508; Wolfgang Ostwald, "Über die zeitlichen Eigenschaften der Entwicklungsvorgange," *Vortrage und Aufsatze über Entwicklunsmechanik des Organismus, 5* (1908), pp. 1–18; A. L. Hagedoorn, "Autokatalytical Substances," *Vortage und Aufsatze uber Entwicklungsmechanik des Organismus, 7* (1911), pp. 26–35, a discussion that includes filterable viruses; and Jacques Loeb and M. M. Chamberlain, "An Attempt at Physico-Chemical Explanation of Certain Groups of Fluctuation Variation," *Journal of Experimental Zoology, 14* (1915), pp. 559–568. Some aspects of Robertson's autocatalytic theory came under attack during the 1920s: see Sharon Kingsland, "The Refractory Model: The Logistic Curve and the History of Population Ecology," *Quarterly Review of Biology, 57* (1982), pp. 29–53.

20. Richard Goldschmidt, "The Determination of Sex," *Nature, 107* (1921), pp. 780–784. Quoted in R. Goldschmidt, "The Gene," *Quarterly Review of Biology, 3* (1928), p. 321; and R. Goldschmidt, *Physiological Genetics* (New York: McGraw-Hill, 1938), p. 282.

21. Leonard T. Troland, "Biological Enigmas and the Theory of Enzyme Action," *American Naturalist, 31* (1917), pp. 321–350. Troland died in his early thirties, and his obituary in *Science, 76* (1932), p. 26 provides brief biographical information.

22. L. T. Troland, "Biological Enigmas," p. 347 (see Note 21).

23. T. H. Morgan, "Genetics and the Physiology of Development," *American Naturalist, 56* (1926), pp. 32–50; quote on p. 50.

24. Hermann J. Muller, "Artificial Transmutations of the Gene," *Science, 66* (1927), pp. 84–87. For an uncritical account of Muller's program see Elof A. Carlson, *Genes, Radiation, and Society: The Life and Work of H. J. Muller* (Ithaca: Cornell University Press, 1981); for a penetrating analysis see P. J. Pauly, *Controlling Life*, pp. 177–183 (see Note 14).

25. Hermann J. Muller, "Resume—Symposium on Genes and Chromosomes," *Cold Spring Harbor Symposia in Quantitative Biology, 9* (1941), pp. 290–308; H. J. Muller, "The Gene as the Basis for Life," *Proceedings of the International Congress of Plant Sciences, 1* (1926), pp. 892–921; H. J. Muller, "Variations Due to Change in the Individual Gene," *American Naturalist, 56* (1922), pp. 32–50; and E. A. Carlson, "An Unacknowledged Founding of Molecular Biology: H. J. Muller's Contributions to Gene Theory, 1910–1936," *Journal of the History of Biology, 4* (1971), pp. 149–170.

26. Some of the important papers are P. A. T. Levene, "The Structure of Yeast Nucleic Acid," *Journal of Biological Chemistry, 40* (1919), pp. 415–424; and P. A. T. Levene and E. S. London, "The Structure of Thymonucleic Acid," *Journal of Biological Chemistry, 83* (1929), pp. 793–802. See also Aaron J. Ihde, "Phoebus Aaron Theodor Levene," *Dictionary of Scientific Biography, X* (New York: Scribners, 1970–1980), pp. 275–276.

27. Robert C. Olby, *Path to the Double Helix*, Ch. 6 (see Note 9).

28. John H. Northrop, "Crystalline Pepsin," *Journal of General Physiology, 13* (1930), pp. 739–780; and John H. Northrop, *Crystalline Enzymes* (New York: Columbia University Press, 1939).

29. Wendell M. Stanley, "Isolation of Crystalline Protein Possessing the Properties of Tobacco-Mosaic Virus," *Science, 81* (1935), pp. 644–645.

30. Lily E. Kay, "Stanley's Crystallization of the Tobacco Mosaic Virus," pp. 465–470 (see Note 18).

31. Gunther Stent and Richard Calendar, *Molecular Genetics* (San Francisco: Freeman, 1978), p. 578.

32. Barclay M. Newman, "Giant Molecules: The Machinery of Inheritance," *Scientific American, 158* (1938), p. 337.

33. RAC, RG2, 100, Box 170.1235, General Correspondence, Warren Weaver, August 28, 1939. Weaver cited T. R. Parsons, *Fundamentals of Biochemistry*, 5th ed. (Baltimore: W. Wood & Co., 1935).

34. RAC, RG1.1, 800D, Box 7.75, "University of Uppsala," March 15, 1940; Lisa Robinson, "The Role of the Rockefeller Foundation in the Diffusion, Use, and Design of the Ultracentrifuge, 1928–1940," unpublished paper, Department of the History of Science, University of Pennsylvania, 1982; Boelie Elzen, "Two Ultracentrifuges: A Comparative Study of the Social Construction of Artifacts," *Social Studies of Science, 16* (1986), pp. 621–662; and Boelie Elzen, *Scientists and Rotors: The Development of Biochemical Centrifuges* (Ph.D. dissertation, University of Twente, 1988), Ch. 2.

35. Pierre Laszlo, *A History of Biochemistry*, pp. 134–145, Ch. 12 (see Note 13).

36. Ibid., pp. 12–21; Robert C. Olby, *Path to the Double Helix*, pp. 7–10 (see Note 9); Olby, "Structural and Dynamical Explanations in the World of Neglected Dimensions," T. J. Horder, J. A. Witkowski, and C. C. Wylie, eds., *A History of Embryology* (Cambridge University Press, 1983), pp. 275–308. John T. Edsall, "Proteins as Macromolecules: An Essay on the Development of the Macromolecule Concept and Some of Its Vicissitudes," *Archives of Biochemistry and Biophysics, Supplement 1* (1962), pp. 12–20.

37. Both Robinson and Elzen detailed some of the disagreements surrounding Svedberg's ultracentrifuge and its later competitors (see Note 34). For the theoretical objections to Svedberg's work see Note 41.

38. RAC, RG1.1, 800D, Box 7.77, "University of Uppsala"; Svedberg's report, 1934, pp. 9–10.

39. Lily E. Kay, "Laboratory Technology and Biological Knowledge: The Tiselius Electrophoresis Apparatus, 1930–1945," *History and Philosophy of the Life Sciences, 10* (1988), pp. 51–72.

40. Joseph Fruton, "Early Theories of Protein Structure," *Annals of the New York Academy of Sciences, 325* (1979), pp. 1–15; and R. C. Olby, *The Path to the Double Helix*, pp. 11–14 (see Note 9).

41. APS, Bergmann Papers, Box 8, Link file 6, from Link's lecture "The Macro-Cellulose Molecule," January 18, 1937, pp. 6–8. There were also criticisms by R. A. Gortner, December 17, 1935 (RAC, RG1.1, 800D, Box 7.76) and by J. D. Bernal, November 10, 1939 (RAC, RG1.1, 800D, Box 7.77).

42. APS, Bergmann Papers, Box 8, Link file 6, Bergmann to Link, November 8, 1937. For a brief sketch of Bergmann's life and work see his obituary, "Max Bergmann, 1886–1944," *Science, 102* (1945), pp. 168–179; and L. E. Kay, *Cell, Molecules and Life: An Annotated Bibliography of Manuscript Sources on Physiology, Biochemistry, and Biophysics, 1900–1960, in the Library of the American Philosophical Society* (Philadelphia: American Philosophical Society, 1989), pp. 35–37, 68–74.

43. Max Bergmann, "The Structure of Proteins in Relation to Biological Problems," *Studies from the Rockefeller Institute for Medical Research, 109* (1938), pp. 35–47; Joseph Fruton, *Molecules and Life*, pp. 159–165 (see Note 14); R. C. Olby, *The Path to the Double Helix*, pp. 115–117 (see Note 9); Pierre Laszlo, *A History of Biochemistry*, pp. 240–244 (see Note 13); and R. B. Goldschmidt, "The Theory of the Gene," *Scientific Monthly, 46* (1938), pp. 268–273.

44. APS, Bergmann Papers, Box 8, Levene file, Levene to Bergmann, December 20, 1937.

45. Ibid., March 25, 1938; and Box 11, Protein Committee file, April 15, 1938.

46. Ibid., Box 8, Link File 6, Link to Bergmann, October 12, 1938.

47. Rollin D. Hotchkiss, a response to N. W. Pirie's paper, "Purification and Crystallization of Proteins," *Annals of the New York Academy of Sciences, 325* (1979), p. 34.

48. Quoted in Joseph Fruton, "Early Theories of Protein," p. 13 (see Note 40).

49. J. Fruton, "Early Theories of Protein Structure," pp. 11–12 (see Note 40); and Dorothy Crawfoot Hodgkin, "Crystallographic Measurements and the Structure of Protein Molecules as They Are," *Annals of the New York Academy of Sciences, 325* (1979), pp. 121–145.

50. J. Fruton, "Early Theories of Protein Structure," pp. 11–14 (see Note 40); P. Laszlo, *A History of Biochemistry*, Ch. 13 (see Note 13). On Wrinch's thoughts on D'Arcy Thompson see her inspired review of his *Growth and Form*, in *Isis, 34* (1943), pp. 232–234. On the professional and personal saga of Wrinch see Pnina J. Abir-Am, "Synergy or Clash: Disciplinary and Marital Strategies in the Career of Mathematical Biologist D. M. Wrinch (1894–1976)" in Abir-Am and Dorinda Outram, eds., *Uneasy Careers and Intimate Lives, Women in Science, 1789–1979* (New Brunswick, NJ: Rutgers University Press, 1987), pp. 338–394; and Anthony Serafini, *Linus Pauling: The Man and His Science* (New York: Paragon House, 1989), Ch. 6. For further discussion on Pauling versus Wrinch see Chapter 5, this volume.

51. RAC, RG12.1, Box 68; Warren Weaver Diary, October 15, 1936.

52. On Langmuir's support of the cyclol theory see J. Fruton, "Early Theories of Protein Structure," p. 12 (see Note 40); and A. Serafini, *Linus Pauling*, Ch. 6 (see Note 50).

53. APS, Bergmann Papers, Box 10, Niemann file, Niemann to Bergmann, December 5, 1937. Bergmann accused Wrinch of priority claims, of concealing "so eagerly the fact that she was already acquainted with our work in the fall of 1936." APS, Bergmann Papers, Box 10, Niemann file, Bergmann to Niemann, December 2, 1937.

CHAPTER 4

From Flies to Molecules: Physiological Genetics During the Morgan Era

Jack Schultz: A Bridge to the Phenotype

Jack Schultz's work constituted probably the earliest linkages between genetics, physiology, and the physicochemical studies of proteins and nucleic acids. When Schultz announced his interest in the biochemistry and physiology of heredity during the late 1920s, his ideas floated in the margins of American genetics. "Doing genetics" meant primarily Mendelian analyses. "I am very sorry to have given you the impression that I have lost my interest in genetics," Schultz had to defend his position to T. H. Morgan in 1929 to salvage his research plans at Caltech. "This is far from the case. My primary interest is in genetic problems, and it is with them that I have been chiefly engaged in the last two years. My concern with general physiology, and in particular with the chemistry of pigments and enzymes, has been directly inspired by the reasoning and results of the genetic analysis."[1]

Charting the course of his nascent career, Schultz described a two-pronged research strategy. Rather than continue merely to map and analyze genes horizontally along the chromosomes, as his *Drosophila* research colleagues had done, Schultz suggested a vertical approach, integrating classical genetics with the study of genes' effect on a single developmental process: the formation of eye pigment in *Drosophila*. "The first type of experiment tells us something about the relations of the genes to each other. The addition of experiments of the second type ties this information up with a tangible mechanism."[2] When he proposed this integrated approach, Schultz departed from the paradigm of American genetics. At the same time his physiological conception of gene action fit well with the new biological enterprise.

Schultz had traversed an unconventional path to his destination. Nurtured in an intellectual socialist milieu of Russian-Jewish immigrants in New York City, Schultz had been immersed in workers' political action and social philosophy, as well as in the art world: music, theater, literature, and languages. His early years at Columbia University reflected his devotion to the humanities. Mounting expenses for books, theater, and concerts led him to a bottle-washing job in Morgan's "fly room," where Schultz's intellectual kinship with Bridges and Sturtevant soon culminated in his graduate work under Morgan and Edmund B. Wilson. His close association with Columbia biophysicist Selig Hecht sharpened Schultz's conviction and skills in applying physicochemical methods to biological research.

Shortly before proposing his new research project to Morgan, Schultz had completed his doctoral thesis in which he showed that the large class of "minute" mutations in Drosophila all produced nearly identical somatic effects. That these mutations occurred at many different loci demonstrated that numerous independent mutations could lead to similar phenotypic effects on development. Schultz's project, sponsored by the Carnegie Institution, sought to investigate these mechanisms in greater detail, to explain the physicochemical basis for gene action by focusing on the processes linking genotype to phenotype.[3]

He tackled these problems at the new biology division, where the presence of Dobzhansky, George Beadle, and such visiting European biologists as David G. Catcheside, J. B. S. Haldane, and Boris Ephrussi helped broaden the scope of discussions, both scientifically and politically. With the ravages of the Depression spreading worldwide and the growing disillusionment with the alliance between business and science, politicized scientific factions in Europe and United States had begun calling for a socially responsive science. Schultz sympathized with the political commitments of J. B. S Haldane, a central figure in the British science movement. "You do political things much better in England than we do here," Schultz wrote to Haldane in 1933. "In America however I begin to feel that the only politics scientists go in for are as conservative as they can be (e.g. Millikan). The rest is silence."[4] This ideological bond with British activism, lasting beyond Haldane's American sojourn, undoubtedly did not endear Schultz to Caltech's establishment.

Schultz continued to work within the framework of classical genetics, collaborating with Bridges, Sturtevant, and Dobzhansky; but he increasingly veered toward physiological and biochemical genetics. He collaborated with embryologist Albert Tyler on fertilization experiments; and five years before Beadle's and Ephrussi's noted transplantation studies in Drosophila on the genetic mechanisms of eye pigment formation, Schultz had analyzed the absorption spectra of eye pigments of various mutant Drosophila stocks. His work strongly suggested that the formation of pigments was linked to metabolic processes. By the mid-1930s Schultz's interest in the gene as a chemical unit converged on the study of nucleic acids, a rather marginal topic for American biochemists and geneticists.[5]

The European scene, however, looked quite different. Schultz admired the physiological and biochemical approach to genetics in Holland and Germany and had planned on working in Europe. In 1937 he received a Rockefeller Fellowship to travel to the Karolinska Institute in Stockholm to investigate the role of nucleic acids in the chromosomes. Their biophysicist Torbjörn O. Caspersson had been by then studying the cytogenetics of nucleic acids for at least five years. The first to combine knowledge of cell biology and biochemistry with ultraviolet microscopy and precise spectroscopic measurements of nucleic acids in living cells, Caspersson concluded that nucleic acids were somehow crucial for protein synthesis. Caspersson's group was the first to provide microspectrophotometric interpretations of T. S. Painter's photographs of the salivary gland chromosomes in *Drosophila*, which Morgan presented at his 1933 Nobel lecture in Stockholm. Schultz's training in cytogenetics and biophysics was ideal for Caspersson's project. In fact, when planning on extending their two-year collaboration, Caspersson arranged in 1939 for a Rockefeller Fellowship at the Carnegie Institute of Washington. Though these plans dissolved with the eruption of World War II, the Stockholm experience led to a permanent intellectual partnership and a personal bond between the two.[6]

Several papers accompanied their collaboration. Combining cytology and classical genetics with absorption spectrophotometry of nucleic acids, Schultz and Caspersson demonstrated that there is a close relation between the metabolism of nuclear and cytoplasmic nucleic acids on the one hand, and the replications of genes on the other. "These considerations," Schultz and Caspersson wrote in 1938,

> have an especial interest in the case of the other self-reproducing molecules—the viruses and the bacteriophage—all of which have been shown to contain nucleic acid. It seems hence that the unique structure conditioning activity and self-reproduction, possibly by successive polymerization and depolymerization, may depend on the nucleic acid portion of the molecule. It may be that the property of a protein which allows it to reproduce itself is its ability to synthesize nucleic acid.[7]

Schultz communicated the significance of his work to Sturtevant, his Caltech mentor, arguing that "only by measurements of the nucleic acid present could one get unequivocal evidence whether the nucleic-acid-containing band [on the *Drosophila*'s salivary gland chromosomes] were the loci of the genes."[8] Since the 1930s Schultz had conceptualized the gene as an integrated unit comprised of proteins and nucleic acids, but the precise role of each component and the biochemical interactions between the two was a matter of speculation.

If establishing credit for greatest proximity to "correct" solutions is of historical urgency, then it is true that in assigning the primary replicative capacities and biological specificities to proteins Schultz's and Caspersson's formulation situated them within the protein paradigm. If other criteria are considered, however, their work gains broader cognitive and disciplinary significance. Their findings not only focused attention on the biological role of nucleic acids but emphasized the general importance of interactive mechanisms between the nucleus and the cytoplasm. Their biophysical approach, coupled with observations on the cellular level, implicitly challenged the categorical reduction of heredity to problems of nuclear proteins; and it did so at a time when scientific authority, institutional resources, and disciplinary power were heavily skewed in favor of the protein chemistry establishment clustered along the Rockefeller–Uppsala axis.

That this redressing of intellectual balance entailed career risks and a professional price is evident from the anxieties Caspersson shared with Schultz.

> There is no use trying to hide that the arrangement we made with publications from America is the worst disappointment I have ever had, without much over-words [referring to an aborted monograph]. I have almost cried in letters about the cytoplasmic nucleotides being written as I have always felt that they would be the basis for my continuation and as I badly needed paper merits. That has had absolutely no results, and also nothing appeared from your genetical side. . . . The last days the question is still more acute as Torsten Teorell [enzymologist and Svedberg's associate] has got a professorship in Uppsala. . . . I find it difficult to write a large nucleic acid monograph now, in view of the protein data, which in my mind put the role of the nucleic acids on a lower shelf.[9]

The fear of confronting Svedberg's protein establishment was well grounded, though Caspersson's fears were eventually dispelled. He was appointed professor of medical cell research and genetics in 1944 and within a couple of years became the director of the newly created Nobel Institute for Medical Cell Research and Genetics, as well as the Wallenberg Laboratory of Experimental Cell Research.[10]

Schultz's career trajectory was far more turbulent. The five years following his return from Sweden were burdened by searches for a permanent position and for financial support for the sophisticated apparatus required for his physicochemical cytogenetics. The period 1939–1943 ended with a harried year spent partly as visiting professor with Lewis Stadler at Missouri, working on variegation in corn, partly at Caltech, and partly at Woods Hole. In 1942 Schultz received an offer from the newly established Institute for Cancer Research in Fox Chase, Philadelphia.[11] Apparently neither Morgan nor Millikan voiced strong interest in Schultz's case, though Sturtevant did expend some effort to retain him at Caltech. Even though Schultz had already decided to accept the Philadelphia offer, Sturtevant tried to persuade Millikan to make a counter-offer. The weak terms proposed (an assistant professorship and a salary of $3500) only underscored Caltech's lack of commitment to Schultz's project and career. "I need scarcely tell you that I don't take it seriously as a counter-offer," he wrote to Morgan when he accepted the new position with its expanded research opportunities.[12] It was one of the recurrent ironies of Caltech's molecular biology program that its missed opportunities often coincided with institutional decisions that were out of step with its cognitive aims.

During the following years Schultz brought his expertise in genetics, cytology, embryology, physiology, and biophysics to bear on the study of basic mechanisms underlying replication, mutation, development, and growth—the very interests and skills that defined the intellectual core of Caltech's biology division. Throughout the 1930s, in professional addresses and grant proposals to the Rockefeller Foundation, Morgan trumpeted the broadening scope of genetics. The rhetoric surrounding the cooperation ideal stressed the need for building intellectual and disciplinary bridges between studies of heredity, development, and growth through physicochemical methods. Schultz's project, its content and style, fit squarely within that agenda and contributed to its goals, especially during the division's intellectual nadir during the war years.

By the early 1940s, the department had deteriorated considerably. Morgan, now retired, complained that there was virtually no genetics at Caltech. Dobzhansky had left for Columbia after years of bitter feuds with Sturtevant; and Bridges had passed away. The division pinned its hopes on Beadle, whose research and personal style (his work in biochemical genetics and his midwestern roots) made him ideal for Caltech's molecular biology program and its conservative social milieu. Beadle, however, resisted Caltech's overtures, in part because of the division's scientific decline.

Beadle, Ephrussi, and the Physiology of Gene Action

George W. Beadle, born, raised, and educated in the farmlands of Nebraska, arrived at Caltech as a National Research Council Fellow in 1931 after completing his doctoral research in corn genetics under Rollins A. Emerson. The excitement generated by Morgan's group had reached Cornell's Agricultural College during the late 1920s, and *Drosophila* mapping techniques were now being applied to corn genetics, principally by the maverick graduate students Beadle and Barbara McClintock. They had already distinguished themselves in 1927 through the discovery that some cytological phenomena, such as pollen sterility, were gene controlled. Beadle's subsequent analyses of genetic linkages in corn continued to exploit results derived from calculating mapping distances.[13]

In addition to thorough training in corn genetics, Beadle audited courses in physical chemistry and biochemistry. These were particularly exciting times for biochemistry, he recalled. The mid-1920s were the golden age of enzymology, and Cornell was at the center of ferment, where biochemist John B. Sumner conducted his Nobel Prize-winning work, which led to crystallization of the first enzyme (urease) in 1926 and to its identification as a protein. Having followed Emerson's call for cooperation among plant physiologists, biochemists, and geneticists, Beadle was predisposed to an interdisciplinary approach well before his move to Caltech. His familiarity with some of the debates surrounding the relations of genes to enzymes and to general physiology broadened his thinking on heredity.[14]

With significant publications to his name, Beadle's career as a corn geneticist was well on its way, but his plans were unorthodox. Visiting the *Drosophila* center in Austin, Texas on his way to Pasadena, Beadle shocked colleagues by announcing that he would devote his Caltech Fellowship to becoming an expert on *Drosophila* genetics. Corn work, he explained, was just too slow, and the major problems— how genes produce their effects on the phenotype—could be attacked more efficiently by using an organism that produced a new generation every 10 days. He was prepared to forsake the species on which he had cut his teeth genetically and start all over again, learning the intricacies of a new biological system.[15]

With the exception of graduate student Carl Lindegren, whose doctoral project focused on the genetics of the fungus *Neurospora*, Caltech's group in 1931 was still a place for *Drosophila* and corn genetics. With the encouragement and collaboration of Dobzhansky, Sturtevant, and Sterling Emerson (son of Beadle's mentor), Beadle plunged into *Drosophila* research hoping to learn more about the nature of gene action. Within a few months of his arrival at Caltech, Beadle published two articles on cytogenetics (one on corn and one on *Drosophila*); both addressed the relation between genes and their cytoplasmic expression.

There was a kind of circularity about the gene problem, as Schultz aptly put it. If one knew what a gene was, one could probably find out how it worked; if one understood the mechanisms of gene action, one could begin to predict what the gene was. Put differently, given enough knowledge about development and the cellular events leading to the adult character, one could utilize such knowledge to understand the role genes played in development and thus learn something about the nature of the gene itself. If, on the other hand, properties of the gene could be understood from independent evidence, it would be possible to unravel developmental processes of characters. The problem was to solve both puzzles at once, knowing the answer to neither.[16] Beadle continued to explore the mechanisms of cytogenetics; his publications over the next two years were a tour de force, not only of volume but of focus as well. His detailed genetic analyses of crossing-over mechanisms of meiosis in corn and *Drosophila* persistently aimed at elucidating the relation between genes and cellular process. These studies, however, were approaching a point of diminishing returns.

A visit to Caltech in 1934 of Rockefeller Fellow Boris Ephrussi, the Russian-born French biologist, infused Beadle with intellectual adventure. Outgoing and witty, Ephrussi was particularly enthusiastic about learning genetic methods. Having specialized in tissue culture techniques at the Institute de Biologie Physico-Chimique in Paris, he now hoped to combine these methods with transplantation experiments aimed at understanding the developmental aspects of gene action.[17] Beadle and Ephrussi spent long hours and drank much beer while discussing the interrelation between genetics and embryology and lamenting the gulf separating the two fields. To a large extent the problem was a structural one: lack of a suitable biological system, a species simultaneously compatible with genetic crosses and developmental observations. The geneticists' favorite organism *Drosophila* was not well suited for embryological studies; and frogs and sea urchins, the embryologists' classical specimens, were inappropriate for studying genetics.[18]

Beadle exemplified the "problem-oriented" trend of the new biology. His approach to physiological genetics and his search for a biological system were strikingly evident in a 1934 paper that conceptualized the central problem in terms of the "ideal organism." Reviewing more than a dozen studies in corn, *Drosophila*, flowering plants, mosses, and fungi, Beadle compared the relative merits of these biological systems, already then singling out *Neurospora* for its methodological merits (based on some of Lindegren's findings). Yet *Drosophila* had a clear advantage over most other organisms. Its genetic makeup was well understood, and some work had already been done on the physiological aspects of several characters, especially eye color.[19]

With Morgan's encouragement, Beadle spent the following year in Paris with Ephrussi learning tissue culture and transplantation techniques in *Drosophila*. Morgan arranged to continue Beadle's salary of $1500, which had been reduced by 33 percent because of the Depression. Only years later did Beadle learn that, because of Caltech's financial problems, his stipend was almost certainly provided personally by Morgan.[20]

Although their initial attempts at tissue culture promptly failed, Morgan considered their joint project to be of prime importance and had a great deal of confidence in Beadle's ability and scientific judgment. Beadle and Ephrussi switched to larval transplantations in *Drosophila*, a technical feat that turned out to be successful despite unfavorable odds. The knowledge gathered from these experiments established a conceptual framework for future work in biochemical genetics, forming new links between American geneticists and biochemists.

Research in physiological or biochemical genetics, of course, did not originate at Caltech with Beadle and Ephrussi. Since the early years of genetics and through the first two decades of the twentieth century, European geneticists had conducted experiments aimed specifically at understanding the relation between genes and the formation of color pigment. These researches ranged from the inheritance of color in flowering plants to studies of coat color in animals and from eye color experiments in insects to analyses of pigment differences, such as albinism and color blindness in humans. Beadle acknowledged the influence of these early investigations on shaping his own thoughts on the subject.[21]

He was familiar with the work of Muriel Wheldale (*The Anthocyanin Pigments in Plants*, 1916), which dealt with the genetic as well as the biochemical aspects of color development of flowering plants, work that was later extended by Rose Scott-Moncrieff. J. B. S. Haldane and C. D. Darlington, visiting scholars at Caltech during the early 1930s, promoted this line of physiological genetics, emphasizing its general importance to the problem of gene action. Some of these studies showed that albinism in flowering plants was correlated with an absence of oxidative enzymes, an observation that agreed with results obtained from experiments on recessive whiteness in rabbits. Haldane's own prolific research in physiological genetics included studies of inheritance of color-blindness and hemophilia. By 1925, the classic 1909 work that informed Haldane's own studies, *Inborn Errors of Metabolism* by the English physician Archibald Garrod, was reprinted in England. The work, which correlated enzymatic deficiency with a recessive Mendelian trait leading to the metabolic disease alkaptonuria, highlighted a connection between genes, chemical reactions, enzymes, and the development of characters in organisms. In general, the prevailing trend during the first three decades of the century was to view development as a complex chemical process—a series of progressive stages in which genes themselves acted as enzymes governing chemical reactions.[22]

Beadle also drew substantially on the works of geneticists in Holland and Germany, who had always situated problems of heredity within a broader physiological context. The school of Carl Correns in Germany had a strong research tradition of studying chemical processes of color formation in plants. The research project of the German group led by Alfred Kuhn and Ernst Caspari was of particular urgency to Beadle. By the 1930s they had adapted the older organ transplantation techniques in the meal moth to the study of diffusible substances that might control pigment production. Their rapid progress posed serious competition to Beadle and Ephrussi, as did the work of Richard Goldschmidt at the Kaiser Wilhelm Institute in Berlin. The vocal and controversial proponent of biochemical genetics had always aimed his genetic research toward the relations between heredity and developmental processes, viewing the gene as essentially enzymatic in nature.[23]

However, in an important respect, Beadle's and Ephrussi's approach to the problem of gene action differed fundamentally from that of their predecessors engaged in physiological or biochemical genetics. By selecting *Drosophila* as their organism, they broached the problem within a well-defined biological system, the genetics of which was exceptionally well understood. Despite the obvious technical difficulties of working with diminutive insects, with only small amounts of pigment visible as eye color, no other system, with the possible exception of corn, possessed such well-standardized and meticulously analyzed mutant stocks. With more than two dozen genetically characterized eye color mutants at their disposal, Beadle and Ephrussi possessed powerful physiological probes. They generally followed Morgan's functionalist approach. In directing their inquiry toward "how the gene worked" rather than asking "what the gene was," they relied on Mendelian crosses that correlated a specific biochemical event with a precise location on the chromosome. Beadle and Ephrussi thus steered clear of theoretical premises and the ongoing debates that had begun during the first decade of the century over the question whether genes *were* enzymes or only *made* enzymes. They, in fact, heeded Morgan's warnings that until rigorous experimentation was done one was not justified in assuming that genes were enzymes.[24]

Time and politics were on Beadle's side. By the late 1930s, as the situation in Germany worsened and the conditions for scientific research deteriorated, investigations in many biological fields were curtailed and lines of communications among researchers were broken. The research of Kuhn and Caspari was interrupted and eventually eclipsed by the success of Beadle and Ephrussi and later by the Nobel Prize-winning research of Beadle and Tatum. However, without these earlier studies, especially without some of the techniques, Beadle and Ephrussi could hardly have succeeded in performing their own transplantation experiments during a brief period of two years.

The theoretical background and the underlining principles behind Beadle's and Ephrussi's transplantation experiments in *Drosophila* derived primarily from Sturtevant's earlier work on naturally occurring mosaic flies, which displayed a mixture of genetic traits (two or three eye colors, for example). Sturtevant had shown in 1920 that the recessive vermilion eye color (the absence of brown component; normal or wild-type red is comprised of red and brown pigments) was nonautonomous (if one eye and a small portion of its surrounding tissue were vermilion and the remainder wild type, the genetically vermilion eye would produce both pigment components). This observation suggested that the brown pigment might be determined not by the genetic constitution of the eye pigment itself but by some other portion of the body, moving to the eye during development.[25]

Sturtevant's 1920 findings had only peripheral relevance during the period dominated by the Mendelian paradigm. In 1932, however, as attention was shifting to physiological genetics, Sturtevant stressed the potential of mosaic flies for the study of biochemical and developmental effects of genes. He concluded:

It is clear that in most cases there is a chain of reactions between the direct activity of a gene and the end-product that the geneticist deals with as a character. One may surmise that any valid generalizations about these reactions are more likely to concern the initial links than the terminal ones. However, it is the terminal ones that are usually more open to experimental attack, since the only index to the effectiveness of given experimental technique is the condition of the end product. Looked at from this point of view, the type of experiment that I have described may be considered as a beginning in the analysis of certain chains of reactions into their individual links.[26]

Sturtevant's conclusion, which implied a hormonal or humeral step-wise link between gene and eye color formation, was soon to be demonstrated by Beadle and Ephrussi in a set of experiments in which they transplanted genetically vermilion embryonic eyebuds into wild-type host larvae. These seminal studies began to illuminate the murky area of gene function and enzyme action; they also focused the attention of biologists and the Rockefeller Foundation on Beadle as a rising star in the life sciences.

Beadle's and Ephrussi's transplantation experiments were a technical feat. In contrast to the relatively straightforward theories behind the experiments, the surgery and injection procedures in embryonic *Drosophila* seemed forbidding. The transplantation method used in earlier studies of larger insects was understandably never attempted in *Drosophila* because of the small size of the flies. The dissection of donor larvae and injection of host larvae, performed under binocular microscopes, had to be perfectly synchronized: one person excising imaginal disks (embryonic eyes) from mutant donor larvae, and the other preparing host wild-type larvae, etherizing and placing them on slides in a convenient position for injection.

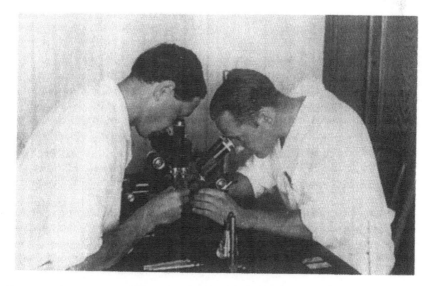

Figure 8 Boris Ephrussi (*left*) and George Beadle (*right*), ca. 1930s. Courtesy of the California Institute of Technology Archives.

The imaginal disks would then be injected into the abdominal cavity of the larvae, using specially constructed micropipettes barely large enough to accommodate the implants. Speed and efficiency were crucial factors. On "good" days about 160 transplantations were made at a rate of 30 per hour, with a success rate of 80 to 90 percent.[27]

As a result of these implantations of vermilion into wild-type larvae, Beadle and Ephrussi found that vermilion disks differentiated into adult structures of wild-type color. Because the implanted eyes developed in the hosts' abdomens with no connection to surrounding tissues, Beadle and Ephrussi concluded that the lymph of the wild-type hosts had to contain diffusible substances responsible for correcting the deficiency in brown pigment that led to normal red color. They also concluded that the postulated hormone-like substance was produced under the control of the wild-type allelomorph of the vermilion gene, that is, under the control of the normal counterpart of the eye color gene in the homologous chromosome. That substance, they reasoned, represented the link connecting the gene with its corresponding character, as Sturtevant's studies had suggested.[28]

These positive results led to a search in more than two dozen *Drosophila* eye color mutants tested by similar transplantation experiments. The search turned up one additional nonautonomous mutant called cinnabar (vermilion-like but browner eye color). The cinnabar eye disks, when transplanted into wild-type larvae, also differentiated into normal red eye color. The question was whether the same diffusible substance was involved, probably representing a single chemical reaction; or were there two diffusible substances and more than one reaction? Reciprocal transplantations made it obvious that there were two substances; and additional manipulations accompanied by educated guesses led to the conclusion that somehow eye color genes regulated a chain of chemical reactions leading to pigment formation through the action of substances circulating in the organism, loosely defined as "hormones."[29]

Some of their findings even made their way into corn genetics. Dutch plant physiologist at Caltech Jan van Overbeck relied on a similar rationale for his experiments with mutant dwarf races of corn, investigating the genetic control of the plant growth hormone auxin. Although he did not develop that line of inquiry, he did show in 1935 that there was a chain of intermediate reactions between the mutant dwarf gene and decreased growth. (The chain involved increased oxidation level, partial auxin destruction, and decreased plasticity of the young cell wall.) Though sketchy, his findings suggested that, in principle, physiological mechanisms governing genetic control of growth and development were similar in animals and plants.[30]

Yet, in general, there was little coordination between Beadle's project in biochemical genetics and Caltech's biochemistry group, reflecting some of the problems in the biology division that held back the plans for cooperation. Borsook's group was interested in the thermodynamics and bioenergetics of food combustion, but they also worked on amino acid and nitrogen metabolism, enzymes, and vitamins. Some of these studies could have been focused to complement Beadle's work in biochemical genetics. Instead, Borsook's projects remained scattered, serving mainly to promote the importance of nutrition and to train graduate students in biochemical techniques and physicochemical analyses.[31] Some of his students would later join Beadle at Stanford, but during the late 1930s Beadle looked elsewhere for developing the biochemical side of his research.

At the end of their Paris collaboration in 1936, Beadle and Ephrussi returned to America. Ephrussi, now on his third Rockefeller fellowship, brought with him to Caltech in 1936 another Rockefeller Fellow, a young biologist from the Sorbonne named Jacques Monod, whose interest in the genetic regulation of cellular processes grew out of the discussions at the laboratory of Beadle and Ephrussi. These early ideas, reinforced by Monod's collaborative research at Caltech in an environment that stressed the genetic aspects of biochemical processes, strongly influenced Monod's thinking, later reflecting in his research on the genetic regulation in bacteria. Beadle's collaboration with Ephrussi tapered off. Following a year at Harvard, Beadle, presumably due to budget constraints at Caltech but also because of the division's weaknesses, accepted a position at Stanford, a move that was a major blow to Caltech's biology division.[32]

During the last four years of collaboration, 1936–1940, Beadle and Ephrussi focused mainly on the postulated diffusible substances in the larval lymph. These experiments led to the extraction and isolation of these substances.[33] For isolating and identifying the active substances and for linking them to the gene on one hand and to pigment on the other, Beadle depended on the cooperation of biochemists. In 1930s America, where there existed no common institutional ground for biochemistry and genetics research and virtually no dialogue between the two communities, Beadle's project demanded intellectual dexterity and administrative flexibility. His deftness at leading cooperative projects had become an intellectual and institutional asset. He was courted at Caltech and by the Rockefeller Foundation.

By 1940 Beadle's collaborations with British plant biochemist Kenneth Thiemann (who had left Caltech for Harvard), Dutch plant physiologist Arie J. Haagen-Smit (who joined Caltech in 1938), and the microbial biochemist Edward L. Tatum from Wisconsin led to the identification of one of the diffusible substances in the *Drosophila* larvae with kynurenine, a derivative of the amino acid tryptophan. The second diffusible substance, derived from the first, was shown to be a direct precursor of the wild-type pigment. The next, and most important, task would be to connect these intermediate links with the two opposite ends of the chain: the pigment product and its regulating gene.[34]

As Sturtevant pointed out in a 1941 review article, there appeared to be no major conceptual or even technical obstacles to reaching the pigment end of the chain. It was bridging the gap between the hormone-like substance and the gene that was a challenge. That an oxidizing enzyme was involved in pigment formation, as the new transplantation experiments suggested, was itself no news, especially to European biologists. Beadle and Ephrussi went further, however, by showing that somehow a specific gene intervened in tryptophan metabolism through the action of specific oxidizing enzymes. Superficially, the conceptual problem of primary gene activity could have been reduced to a technical problem of enzyme chemistry. Such an approach to reproduction and growth was indeed widespread, especially at the Rockefeller Institute, where the heavy emphasis on enzymology paralleled a disregard of the genetic aspects of physiology. Even with an appre-

ciation of the role of the gene in development, the difficulties appeared insur-
mountable. As Sturtevant lamented in 1941, "the chain of developmental reactions
may be traced back to the gene, but there is no way of determining when one has
reached the gene."[35]

Despite Sturtevant's skepticism, Beadle and Ephrussi had made considerable
progress in filling in some of the blanks in the representational scheme that linked
the intermediate steps between the gene and its biochemical products. Their work
helped break the circularity between what a gene was and what it did, a circularity
that confused most previous studies in physiological genetics. As a result, Beadle's
and Ephrussi's transplantation experiments in *Drosophila* came to be regarded by
the international community of biologists as one of the three major advances in
genetics during the 1930s (the other two being Painter's work on salivary chromosomes and the studies of radiation effects on germ cells). Beadle emerged during
the late 1930s as an exemplary model of cooperative individualism—a man whose
individual initiative inspired group interest. His accomplishments soon brought him
to the attention of the Rockefeller Foundation. When scouting for key team leaders
for the molecular biology program, the Foundation officers closely scrutinized
Beadle. After a 1936 visit to Woods Hole they noted: "Beadle makes an exceptionally fine impression and is undoubtedly one of the most promising men of his
age in Biology—a man to be watched."[36]

The Riddle of Life:
Max Delbrück and Phage Genetics

In 1937, just as Beadle left for Stanford, theoretical physicist Max Delbrück
arrived at Caltech's biology division from the Kaiser Wilhelm Institute in Berlin.
Delbrück was another candidate closely watched by the Rockefeller talent scouts.
He had been investigating gene structure and mutations in *Drosophila* for five
years, using x-ray techniques and theories of quantum physics. He too demonstrated
an ethos of cooperation; his research style blended the prized qualities of individual
initiative and team work. His family history—the great grandson of Justus von
Liebig, son of the noted military historian Hans Delbrück, and nephew of Adolph
von Harnack, founder of the Kaiser-Wilhelm Gesselschaft—placed Delbrück at
the hub of Germany's academic and ruling elite. When he was invited to apply for
a Rockefeller fellowship in 1936, Delbrück stood near the intellectual and social
summit of science. He could have chosen other institutions, but he selected Caltech's biology division for its cooperative research program and its sympathetic
attitude to physicists. The reputation of Morgan's division as a place where classical
genetics joined forces with physics and chemistry, and the Rockefeller Foundation's
promotions of such projects, had reached Europe.[37]

On the surface, Delbrück's research program seemed to match Weaver's cognitive goals closely: physics applied to biology. Delbrück's epistemological objectives, however, were in fact somewhat different. Whereas Mason and Weaver
sought to escape the threat of nondeterministic physics, Delbrück's *weltanschauung*
centered around acausality. Mason's and Weaver's molecular approach to biology

was predicated on applying sophisticated technologies from the physical sciences to vital processes. Delbrück's approach, on the other hand, was one of minimal physicochemical intervention. Whereas the Rockefeller architects of the new biology hoped to recapture in the new biology the security of the old mechanistic physics, Delbrück intended to force yet another uncertainty—to discover through the riddle of life new laws of nature.

A child of the "physics revolution" of the 1920s, Delbrück entered biology in 1932 with profound appreciation for the complexity and organization of living matter and for the parallels between acausal quantum physics and biological explanations. This approach to biology was inspired by his close association with Niels Bohr during the years 1931–1932, an intellectual and psychological bond that continued throughout Delbrück's career. Bohr's suggestion that the complementarity principle in quantum physics, which imposed a finite limit on the measurement process of atomic phenomena, had a counterpart on the atomic level in biology became the main impetus for Delbrück's commitment to biology.[38]

According to the strictures imposed by the complementarity principle in biology, one could never hope to attain a complete atomic account of an organism, no matter how simple, as such investigations would necessarily interfere with the structural properties of life. Within the world view of the new quantum physics, where an observer's experimental tools became an integral part of the observed phenomenon, physicochemical manipulations were seen as distorting forces. Just as with inanimate matter, when probing the molecular constituents of animate matter Delbrück intended to use mainly the tools of theoretical physics, or mathematics.[39]

In 1932, upon terminating his studies with Bohr, Delbrück chose a position at the Kaiser Wilhelm Institute for Chemistry in Berlin. While working on the problem of nuclear fission with Lise Meitner and Otto Hahn, Delbrück intended to find a suitable research project at the neighboring Kaiser Wilhelm Institute for Biology. Within a few months Delbrück became part of a lively genetics group headed by N. W. Timoféeff-Ressovsky. In part due to H. J. Muller's influence, the team had been investigating gene structure by inducing mutations in *Drosophila* with x-rays. In fact, in 1932 H. J. Muller was a Guggenheim fellow in Timoféeff's laboratory.[40]

The gene problem was considered then to be the cutting edge of life science. Delbrück was intrigued at first by the challenge of explaining the interaction of ionizing radiation with reproducing matter, and he applied theories of quantum mechanics to genetics enthusiastically. His mathematical skills, command of physics, and theoretical interpretations of the problems were applauded by the biology group. By 1936, when invited to apply for a Rockefeller fellowship, Delbrück had shown great promise through his work in radiation genetics and stood out as a potential leader in the new molecular biology program. He had organized seminars at the Kaiser Wilhelm Institute on the relations between quantum physics and genetics and had led group meetings of biologists, chemists, and physicists. His joint publication with physical chemist K. G. Zimmer and Timoféeff-Ressovsky in 1935 "On the Nature of Gene Mutation and Gene Structure"—his debut in biology—was well received by physicists and geneticists and became a basis for a 1936 interdisciplinary conference in Copenhagen organized by Bohr and supported by the Rockefeller Foundation. The paper, according to Delbrück, also reached Morgan's group at Caltech. Although they found the mathematical language incomprehensible, they were nevertheless intrigued by the new approach. His first task at Caltech would be to give a seminar explaining his paper.[41]

Drosophila research and the radiation genetics project posed obstacles to Delbrück's mission. Although his entry into biology at the Kaiser Wilhelm Institute was facilitated by finding a research program that capitalized on his skills as physicist, his deeper epistemic motivation had not been fulfilled. He did not merely wish to convert a question in biology to a physics problem. Delbrück detested biophysics and biochemistry. He disapproved of the reductionist research programs of Hermann von Helmholtz and Jacques Loeb, which extended the old mechanistic physics to complex life phenomena. His quantum analysis sought to avoid mechanistic models, and for that he required the simplest system, the purest case of replication as a counterpart of the hydrogen atom in biology. From Delbrück's philosophical and methodological perspective, *Drosophila* was too complex an organism and radiation too disruptive a probe for studying replication in its natural state.[42]

Delbrück also found the technical vocabulary of *Drosophila* genetics too cumbersome for the elegance of his mathematical models. In 1937, after spending a couple of months visiting America's major genetics centers, he could not identify a single project compatible with his objectives. At Caltech, Sturtevant and Bridges introduced him to the *Drosophila* literature and to recent findings in genetics. Accustomed to the abstractions of mathematical language, though, Delbrück pored disconsolately over scores of forbidding papers containing long and detailed genotypic maps. Failing to grasp the material, he was overwhelmed by the complexities of even the relatively well-defined *Drosophila* system.[43]

He seemed to have had some novel ideas in the back of his mind. By the time of his arrival at Caltech, Delbrück, like most life scientists, had been captivated by the 1935 crystallization of the tobacco mosaic virus by Wendell M. Stanley of the Rockefeller Institute. By converting an organism, a reproducing and mutating entity, into inert crystals, Stanley seemed to have scored a victory for the proponents of the physicochemical view of life. Equally important, these self-replicating viral proteins were believed to be essentially autocatalytic enzymes. Stanley seemed to have confirmed the autocatalytic theory of life and to support the view that viruses, or "naked genes" as Muller called them, were nothing but giant protein molecules. As Weaver heralded in his 1939 review:

> One of the most interesting aspects of protein research, and one which has only recently emerged, is the indication that these huge molecules exhibit phenomena that we ordinarily consider possible only to living organisms. Thus viruses "reproduce" when in a suitable environment; and yet the brilliant researches of W. M. Stanley and others have shown that certain viruses which show this property so characteristic of life are nothing more than huge protein molecules.[44]

For many life scientists this viral portrait reduced genetics to the study of protein chemistry.

Initially, Delbrück too responded to Stanley's research with great enthusiasm. In his unpublished essay "Preliminary Exposition on the Topic 'Riddle of Life,'" written shortly before leaving Germany, Delbrück mused about a new genetics based on viruses. He envisioned viruses as giant living molecules that lent themselves to studies of growth and reproduction in the most simplified form and without the complexities of Mendelian recombinations. A short visit to Princeton, however, dampened his premature enthusiasm. The conceptual simplicity of viruses contrasted with the sophisticated and invasive technology required for their analysis, and the life cycle of these giant molecules was complicated by the life cycle of their host. The tobacco mosaic virus was not the minimalist system he sought for his mathematical mill.[45]

Nevertheless, Delbrück had been irreversibly captivated by the framework that equated life with reproduction and reproduction with molecularity. He came to Caltech predisposed to virus research. When he found out that he had missed a seminar on bacteriophage while on a camping trip with plant physiologist Fritz Went, he immediately went to talk with the speaker, Emory Ellis, at Borsook's department. Ellis, whose work on phage had sprung from his interest in tumor growth, had just developed an infectivity assay for phage. Delbrück remembered being impressed with the austerity and effectiveness of Ellis's laboratory arrangements. With only rudimentary knowledge of microbiology and with primitive equipment consisting of an autoclave, a few dozen pipettes, flasks, petri dishes, and some agar, Ellis had managed to set up an impressive experimental system and was generating intriguing results: discrete visible plaques formed within 20 minutes of lysis (the bursting open of a bacterium).[46] "I was absolutely overwhelmed that there were such very simple procedures with which you could visualize particles," recalled Delbrück. "I mean, you could put them on a plate with a lawn of bacteria, and the next morning every virus particle would have eaten a macroscopic one-millimeter hole in the lawn. You could hold up the plate and count the plaques. This seemed to me just beyond my wildest dreams of doing simple experiments on something like atoms in biology."[47]

That analogy of phage to "atoms in biology" corresponded closely to Delbrück's goal of finding a simple and efficient model of replication requiring little physicochemical intervention and amenable to numerical analysis. He conceptualized the riddle of life (replicating phage) within the black box of the cell (host bacterium). The box did not need to be pried open in order to study replication; biochemistry was irrelevant to his epistemological program. He would vary strictly controlled initial and boundary conditions, such as dilutions, temperatures, and phage strains. The relations between the input (infecting phage) and the output (newly replicated phage) could then be expressed as mathematical equations, revealing by inference the hidden mechanisms of replication. What were the variations in time for different replication cycles, or variations in burst sizes? At which part of the cycle did new phage appear? It was these elementary questions that Delbrück hoped to answer during the few months remaining of his Rockefeller fellowship. Given its technical and conceptual simplicity, phage appeared to be an amusingly simple project, "a fine playground for serious children who ask ambitious questions."[48]

Despite the influence of Bohr's philosophy of biology on Delbrück and his deep appreciation for life phenomena and their complexities, Delbrück did not anticipate the scope of the work required to answer his deceptively simple questions: growing cultures, preparing stocks, collating microscopic observations from thousands of petri dishes, and tabulating endless columns of plaque counts. It took months just to establish a rigorous plaque assay and, with the aid of refined mathematical analysis, to redetermine and reinterpret the phage growth curve. The change in the traditional characterization of that curve from a continuous S-shape to a discontinuous one-step growth curve would form the first step in a series of studies challenging the bacteriophage literature of the 1920s and 1930s, especially the work of J. H. Northrop and his explanations of bacteriophage action in terms of enzyme autocatalysis.[49]

The experimental precision and clarity of interpretation of Delbrück's work won him Morgan's admiration. Impressed with the physicomathematical approach to biology—though he could only vaguely follow the arguments—Morgan presented an enthusiastic account to the Rockefeller Foundation recommending the extension of Delbrück's fellowship, pointing out that "it is not often that a competent physicist is interested in applying his knowledge of physics to problems in biology." Morgan was not merely invoking the appropriate rhetoric for Rockefeller support; he emphasized Delbrück's unusual gift for biological thinking, describing him as "one of the few men we have known who is a mathematician and to whom we can go with our biological problems and find that he has a real understanding of what we are trying to say."[50]

Morgan, however, could not create a new position in the division and reluctantly had to recommend Delbrück to Vanderbilt University. After two years at Caltech as a Rockefeller fellow, Delbrück moved to Nashville, where the Rockefeller Foundation arranged for an appointment in Vanderbilt's physics department. These brief two years at Caltech were of immense importance for Delbrück's career and for the molecular biology program. The handful of papers on absorption, growth, and lysis of phage established phage as a powerful model for gene replication and demonstrated the potential effectiveness of mathematical reasoning as an analytical tool in the new genetics.[51] Just as Beadle forged permanent links between genetics and biochemistry, Delbrück established ties between genetics and physics; and as with Beadle, Delbrück's own blend of cooperative individualism would prove crucial to the building of a research school, an ethos A. A. Noyes had promoted from the start. Six years later, Delbrück's dual citizenship in physics and biology and his leadership skills would be prime factors in his return to Caltech as professor of biology.

Nascent Trends: Toward Giant Protein Molecules

Delbrück's phage project signaled changing directions in genetics. By 1940, at the end of Morgan's tenure, the new physicochemical genetics had evolved a long way from Morgan's original conception, increasingly focusing on viruses, antibodies, and proteins. His 1932 projections of future directions in gene research, the 1933 Nobel Prize address on the relation of genetics to physiology and medicine, and his 1933 appeal to the Rockefeller Foundation made no mention of immunology or virus research. These areas, which had little in common with genetics, were traditionally associated with the study of disease, a connection that Morgan had sought to avoid. Unexpectedly, however, virus research and immunology became Morgan's research priorities by 1940. Impressed, the Rockefeller Foundation officers noted that "if he [Morgan] had the power to do it (which he does not) he would turn the entire biology group on to immunity and virus work as a new field where the most important advances could be made."[52]

This rather sudden shift of priorities reflected some of the important discoveries and technological innovations of the 1930s. These developments, which would shape the mainstream of molecular biology research during the following two decades, were based, and in turn focused attention, on the primacy of proteins in reproduction, growth, and physiological regulation. The so-called giant protein molecules, especially viruses, became targets for researches that aimed to unravel the biological specificities involved in self-replication or autocatalysis. The synthesis of antibodies promised to explain heterocatalysis, the process by which a protein specified the formation of another (different) protein. The capabilities and knowledge generated during the 1930s by the new technologies of protein chemistry, especially the analytical ultracentrifuge and the Tiselius electrophoresis apparatus, had direct impact on the study of viruses and antibodies and, in turn, on the research plans in molecular genetics.

In 1930, when Morgan first mapped out his new program, researchers—mainly bacteriologists—studied viruses largely in relation to pathology. Hundreds of diseases of man, animals, and plants were known by then to be caused by viruses. Many of these viruses, which could pass through bacteria-retaining filters, had been identified and characterized based on size; the shape of some viruses was inferred from experiments with flow birefringence apparatus. Viruses, however, being of the same order of magnitude as genes, were inaccessible through ordinary methods of cell culture and, like genes, escaped the best of microscopes. No virus had ever been isolated in pure form. After crystallization of the tobacco mosaic virus in 1935 (and, soon after, other plant viruses), virus research was driven and shaped by the new powerful technologies.[53]

These new technologies also played a crucial role in immunology. Immunology work during the early 1930s, still tied to medical research, was increasingly becoming quantitative and physicochemical. By then, Paul Ehrlich's side-chain theory of antibody formation had been effectively challenged by the experiments of the Austrian physician Karl Landsteiner, strongly suggesting that antibodies were proteinaceous immunoglobulins, newly formed in response to antigens. Ehrlich's "lock-and-key" model, which explained various processes of agglutination (clumping) in terms of interlocking covalent bonds, was still the dominant metaphor in immunology (a metaphor coined by Emil Fischer). Later studies in the physical chemistry of solutions and the emphasis on the role of weak physical forces eventually displaced Ehrlich's conception. The properties and mode of action of antibodies entered the domain of biophysics and physical biochemistry.[54]

The tenuous ties between immunology and genetics were strengthened during the 1930s. As early as 1910 Landsteiner, discoverer of the four blood groups, had shown that a mix of some blood types—a combination of different human sera or sera of different species—produced agglutination. The degree of agglutination, in turn, became an indicator of differentiation between biological types. Landsteiner also established some of the earliest links between immunology and genetics by showing that blood types were inherited according to simple Mendelian rules. By 1930 his work at the Rockefeller Institute, buttressed by serological research in other centers, had demonstrated that other factors (antigens) in erythrocytes and in sera were genetically determined.[55]

Serological studies thus became increasingly important tools in genetics during the 1930s. The degrees of agglutination were being interpreted by mapping biological relations between animals, by studying the identity or difference in oxy-hemoglobins of various species, by defining gene-linked antigens of avian and mammalian red blood cells, and by studying animal systematics. By the mid-1930s there was little doubt that antibodies were sensitive indicators of physiological variance and of differences in genotypes.[56] Not surprisingly, these probes of serological genetics held out a new promise for eugenics. The refined techniques could now be exploited for pinning down heritable "markers" in the blood—the identification of antigens linked to genes responsible for mental defects.

Since around 1910, the strategies for segregating and sterilizing the feeble-minded were flawed by the problem of partial dominance, by genes responsible for mental defects being hidden in normal carriers. Reproductive restrictions on the mentally deficient phenotypes would thus work slowly. With the availability of serological tests, however, some geneticists during the 1930s hoped that the recessive genes causing mental defects in their homozygous state could be detected directly. Perhaps even in the absence of direct serological effects closely linked genes on the same chromosomes would reveal genetic divergence. Such studies—in Denmark, Germany, and the United States—promised to circumvent the limitations of mental tests and bring higher levels of rigor, detail, and resolution to genetic testing.[57] Immunology also seemed to provide new tools for embryology. Preliminary studies by Albert Tyler at Caltech in 1939 demonstrated similarities between the agglutination that occurred during fertilization and the immunological process of phagocytosis, suggesting relations of biological specificity.[58] These cognitive potentials held out promises of a reliable social technology.

These developments in immunology and virology, and their extension in serological genetics, co-evolved with innovations in laboratory technology. Until the late 1930s, when the ultracentrifuge and electrophoresis apparatus became available to leading American researchers, antibodies and viruses were not amenable to physicochemical analyses. The analytical ultracentrifuge, a sophisticated instrument that could sort and weigh macromolecules as large as viruses and antibodies, arrived in the United States in 1937. This molecular balance was crucial to the characterization of many viruses—their size and shape—and an essential tool for determining the molecular weight and shape of antibodies. The actual isolation of these globulins took place in 1937, when Arne Tiselius built his powerful electrophoresis apparatus in Uppsala. In fact, his first paper, announcing the apparatus, also reported the isolation of alpha-, beta-, and gamma-globulins from blood serum. The Tiselius, as it was called, became a principal tool for studying the physical properties of proteins in solution. Electrophoresis technology complemented ultracentrifugal analyses, and in some ways it was even more effective for the gentle separation of labile proteins.[59]

During the years 1937–1940 papers on antibodies, enzymes, and viruses cascaded from well-endowed laboratories that possessed these expensive technologies. The Rockefeller Institute was at the vanguard of these developments; and Linus Pauling, who had been in close contact with members of the Rockefeller Institute, was well aware of these innovations. His interest in protein chemistry, which began with the study of hemoglobin during the mid-1930s, soon extended to research in immunochemistry. By the late 1930s the chemistry division under Pauling would construct an ultracentrifuge and by 1941 a Tiselius electrophoresis apparatus.

Pauling's collaborations with members of the Rockefeller Institute and his involvement in immunology had a decisive influence on Morgan's group and on the direction of the molecular biology program at Caltech. The strong intellectual emphasis on immunochemistry in the chemistry division shaped the research directions at the biology division toward serological genetics and chemical embryology. From an institutional standpoint, the new trends at the biology division were a reflection of the massive support by the Rockefeller Foundation for the chemistry division during the 1930s and of Pauling's increasing preoccupation with the protein problem.

Notes

1. APS, Schultz Papers, Morgan File 1, Schultz to Morgan, January 3, 1929, p. 1.
2. Ibid.
3. Thomas F. Anderson, "Jack Schultz, May 7, 1904–April 29, 1971," *Biographical Memoirs of the National Academy of Sciences*, 47 (1975), pp. 393–422. On Schultz's dissertation, see J. Schultz, "The minute reaction in the development of *Drosophila melanogaster*," *Genetics*, 14 (1929), pp. 366–419.
4. APS, Schultz Papers, J. B. S. Haldane File, Schultz to Haldane, July 9, 1933. See also Peter J. Kuznick, *Beyond the Laboratory: Scientists as Political Activists in 1930s America* (Chicago: University of Chicago Press, 1987), Ch. 2; Gary Werskey, *The Visible College: The Collective Biography of British Scientific Socialists of the 1930s* (New York: Holt, Reinhardt, Winston, 1978); and William McGucken, *Scientists, Society, and State: The Social Relations of Science Movements in Great Britain, 1931–1947* (Columbus: Ohio State University Press, 1984).
5. For his seminal article on the topic, see J. Schultz, "The Gene as a Chemical Unit," in J. Alexander, ed., *Colloid Chemistry, Theoretical and Applied*, Vol. 5 (1944), pp. 819–850.
6. Thomas F. Anderson, "Jack Schultz," pp. 399–400 (see Note 3); Robert C. Olby, *The Path to Double Helix* (London: Macmillan, 1974), Chs. 7 and 8; Jan Sapp, *Beyond the Gene: Cytoplasmic Inheritance and the Struggle for Authority in Genetics* (New York: Oxford University Press, 1987); Jonathan Harwood, "National Styles in Science: Genetics in Germany and the United States Between the World Wars," *Isis*, 78 (1987), pp. 390–414; and Richard Burian, Jean Gayon, and Doris Zallen, "The Singular Fate of Genetics in the History of French Biology, 1900–1940," *Journal of the History of Biology*, 21 (1988), pp. 357–402. On Caspersson's Rockefeller Fellowship see APS, W. H. Lewis Papers; Caspersson to Lewis, June 5, 1939; Lewis to Caspersson, June 17, 1939; Caspersson to Lewis, July 29, 1939.

7. T. Caspersson and J. Schultz, "Nucleic Acid Metabolism of the Chromosomes in Relation to Gene Reproduction," *Nature, 142* (1938), p. 295.

8. APS, Schultz Papers, Sturtevant File, Schultz to Sturtevant, undated but ca. 1939.

9. APS, Schultz Papers, Caspersson File No. 1, Caspersson to Schultz, August 12, 1940. A. H. T. Teorell's work on the oxidation of enzymes won him the Nobel Prize in Medicine in 1955. He also contributed to the development of protein electrophoresis.

10. On Svedberg disciplinary and institutional influence, see Interlude I, this volume. On Caspersson see also Lily E. Kay, *Molecules, Cells, and Life: An Annotated Bibliography of Manuscript Sources on Physiology, Biochemistry, and Biophysics, 1900–1960 in the American Philosophical Society Library* (Philadelphia: American Philosophical Society, 1989), pp. 59–60, 66.

11. T. F. Anderson, "Jack Schultz," pp. 401–402 (see Note 3).

12. APS, Schultz Papers, Morgan File No. 2, Schultz to Morgan, August 22, 1942; and Sturtevant File, Sturtevant to Schultz, July 22, 1942 and October 20, 1942.

13. CIT, Beadle Papers, Box 3.13, "The Uncommon Farmer" (1981) and a "Farmer at Heart" (undated). Also G. W. Beadle, "Recollections," *Annual Review of Biochemistry, 43* (1974), pp. 1–13. For descriptions of corn genetics at Cornell see Evelyn Fox Keller, *A Feeling for the Organism* (San Francisco: W. H. Freeman, 1982). See also G. W. Beadle and B. McClintock, "A genetic disturbance of meiosis in *Zea mays*," *Science 68* (1928), p. 433; and G. W. Beadle, "A gene for supernumerary mitoses during spore development in *Zea mays*," *Science 70* (1929), (pp. 406–407). See also Lily E. Kay, "Beyond the Organism: G. W. Beadle's Approach to the Gene Problem," paper presented at the 1989 Summer Conference in History, Philosophy, and Social Studies of Biology, London, Ontario.

14. G. W. Beadle, "Recollections," pp. 1–13 (see Note 13).

15. B. H. Glass, personal communication, letter, January 30, 1988.

16. G. W. Beadle, "Recollections," p. 6 (see Note 13); and Jack Schultz, "Aspects of the Relation between Genes and Development in *Drosophila*," *American Naturalist, 69* (1935), pp. 30–31. For Beadle's publications, see G. W. Beadle, "Possible influence of the spindle fibre on crossing-over in *Drosophila*," *Proceedings of the National Academy of Sciences USA, 18* (1932), pp. 160–165; and G. W. Beadle, "Genes in maize for pollen sterility," *Genetics, 17* (1932) pp. 413–431.

17. RAC, RG10, R7, Natural Sciences, France, "Fellowships—Boris Ephrussi," 1931–1968.

18. G. W. Beadle, "Recollections," p. 6 (see Note 13).

19. G. W. Beadle and S. Emerson, "Further studies of crossing-over in attached X-chromosomes of *Drosophila melanogaster*," *Proceedings of the National Academy of Science USA, 21* (1935), pp. 192–206.

20. G. W. Beadle, "Recollections," p. 6 (see Note 13).

21. Ibid., p. 7.

22. R. C. Olby, *The Path*, Ch. 8 (see Note 6); Joseph S. Furton, *Molecules and Life* (New York: John Wiley & Sons, 1972), pp. 225–254. For Haldane's contributions to physiological genetics, see "John Burdon Sanderson Haldane," *Biographical Memories of the Royal Society, 12* (1966), pp. 219–249, a review that emphasizes his familiarity with Garrod's work.

23. R. C. Olby, *The Path*, pp. 137–141 (see Note 6). On the works of Kuhn, Caspari, and their German colleagues, see Jonathan Harwood, "History of Genetics in Germany," *The Mendel Newsletter, 24* (1984), p. 3; and J. Harwood, "National Styles in Science," pp. 390–414 (see Note 6).

24. On the gene-enzyme problem and Morgan's position see Interlude I, this volume.

25. Alfred H. Sturtevant, "The Vermilion Gene and Gyandromorphism," *Proceedings of the Society of Experimental Biology and Medicine, 17* (1920), pp. 70–71.

26. A. H. Sturtevant, "The Use of Mosaics in the Study of the Developmental Effects of Genes," *Proceedings of the 6th International Congress of Genetics, 1* (1932), pp. 304–307.

27. B. Ephrussi and G. W. Beadle, "A Technique of Transplantation for *Drosophila*," *American Naturalist, 50* (1936), pp. 218–224.

28. B. Ephrussi, "Chemistry of 'Eye Color Hormones' of *Drosophila*," *Quarterly Review of Biology, 17* (1942), p. 328.

29. Ibid., p. 330.

30. RAC, RG1.1, 205D, Box 5.74; Morgan's Report to Weaver, March 12, 1935.

31. Ibid.

32. CIT, Oral History, James Bonner, p. 15; Beadle, "Recollections," p. 7 (see Note 13); and J. Monod, "From Enzymatic Adaptation to Allosteric Transitions," in *Nobel Lectures in Molecular Biology* (New York: Elsevier, North-Holland, 1977), p. 261. On Ephrussi and Monod see also, Horace F. Judson, *The Eighth Day of Creation* (New York: Simon & Schuster, 1979), pp. 355–358.

33. B. Ephrussi, "Chemistry of Eye Color," pp. 331–332 (see Note 28).

34. Ibid., pp. 332–336. Undoubtedly, Adolph Butenandt's priority when analyzing kynurenine must have been a major disappointment to Beadle and might have contributed to his decision to abandon the project.

35. A. H. Sturtevant, "Physiological Aspects of Genetics," *Annual Review of Physiology, 3* (1941), pp. 41–56.

36. RAC, RG1.1, 205D, Box 7, File 88, F. B. Hanson Diary, September 4–5, 1936.

37. CIT, Delbrück Papers, Box 40.32, Fellowships, Delbrück to Miller, December 12, 1936. For a more detailed account of Delbrück's activities see CIT, Oral History, Delbrück, 1978, Part 1; and Lily E. Kay, "Conceptual Models and Analytical Tools: The Biology of Physicist Max Delbrück," *Journal of the History of Biology, 18* (1985), pp. 207–246.

38. For Bohr's interest in biology see Niels Bohr, "Light and Life," *Nature, 131* (1933), pp. 457–459; and Finn Aaserud, "The Redirection of the Niels Bohr Institute in the 1930s: Response to Conditions for Basic Science Enterprise," Ph.D. dissertation, Johns Hopkins University, 1984, Ch. 4. On the relations between Delbrück and Bohr in connection to the new biology see Lily E. Kay, "The Secret of Life: Niels Bohr's Influence on the Biology Program of Max Delbrück," *Rivista di Storia delle Scienza* (1985), pp. 487–510.

39. Max Delbrück "A Physicist Looks at Biology," in J. Cairns, G. Stent, and J. Watson, eds., *Phage and the Origins of Molecular Biology* (New York: Cold Spring Harbor Laboratory of Quantitative Biology, 1966), pp. 9–22; and Lily E. Kay, "Conceptual Models," pp. 209–213 (see Note 37).

40. Lily E. Kay, "Conceptual Models," pp. 213–227 (see Note 37).

41. Ibid.; and N. W. Timofeeff-Ressovsky, K. G. Zimmer, and M. Delbrück, "Über die Natur der Genmutation und der Genstruktur," *Nachrichten Gesamte Wissenschaftliche Göttingen, Mathematischen-phys. KL.6* (1935), pp. 190–245.

42. Lily E. Kay, "Conceptual Models," pp. 209–213 (see Note 37).

43. Ibid., p. 230.

44. RAC, RG2, 100, Box 170.1235, General Correspondence, Warren Weaver, statement for Review, August 28, 1939, p. 7. For a more complete account of Stanley's discovery and its impact see Lily E. Kay "W. M. Stanley's Crystallization of the Tobacco Mosaic Virus," *Isis, 77,* no. 288 (1986).

45. CIT, Delbrück Papers, Box 36.1, "Preliminary Write-up on the Topic Riddle of Life" (Berlin, August 1937); also published in *Science, 168* (1970), pp. 1314–1315, in Delbrück's Nobel Prize address.

46. Lily E. Kay, "Conceptual Models," pp. 230–231 (see Note 37).

47. CIT, Oral History, Delbrück, 1978, p. 57.

48. M. Delbrück, "Experiments with Bacterial Viruses," *Harvey Lectures, 41* (1945–1946), p. 161.

49. Emory L. Ellis and M. Delbrück, "The Growth of Bacteriophage," *Journal of General Physiology, 22* (1939), pp. 365–384.

50. RAC, RG10, R7, Fellowships, Morgan to Miller, April 27, 1938; and RAC, RG1.1, 200D, Box 164.2015, Morgan to Weaver, September 27, 1939.

51. Lily E. Kay, "Conceptual Models," p. 234 (see Note 37).

52. RAC, RG1.1, 205D, Box 7.91, Morgan to Hanson, April 1–18, 1940.

53. Sally Smith Hughes, *The Virus: A History of the Concept* (New York: Science History Publications, 1977); Lisa Wilkinson, "The Development of the Virus Concept as Reflected in Corpora of Studies on Individual Pathogens," *Medical History* (1979), pp. 1–28, and Lily E. Kay, "W. M. Stanley's Crystallization of the Tobacco Mosaic Virus," part I. (See Note 44.)

54. A. M. Silverstein, "History of Immunology," in W. E. Paul, ed., *Fundamentals of Immunology* (New York: Raven Press, 1984), pp. 23–40; and P. H. M. Mazumdar, "The Antigen-Antibody Reaction and the Physics and Chemistry of Life," *Bulletin of the History of Medicine*, *48* (1974), pp. 1–21.

55. Karl Landsteiner, *The Specificity of Serological Reactions*, 2nd ed. (Cambridge, MA: Harvard University Press, 1945).

56. H. H. Standskot, "Physiological Aspects of Human Genetics, Five Human Blood Characteristics," *Physiological Reviews*, *24* (1944), pp. 445–466; M. R. Irwin and R. W. Cumley, "Immunogenetics Studies of Species Relationships," *American Naturalist*, *57* (1934), pp. 211–233; and in the same volume A. Boyden, "Serology and Animal Systematics," pp. 234–249.

57. Diane B. Paul, "The Rockefeller Foundation and the Origins of Behavior Genetics," in Keith Benson, Jane Maienschein, and Ronald Rainger, eds., *The Expansion of American Biology* (New Brunswick, NJ: Rutgers University Press, 1991), pp. 262–283.

58. A. Tyler, "An Auto-Antibody Concept of Cell Structure, Growth and Differentiation," *Six Growth Symposium* (1947), pp. 7–19; and M. Heidelberger, "Immunological Approaches to Biological Problems," *American Naturalist*, *57* (1943), pp. 193–198. These researches began in 1939 and are discussed in Chapter 5, this volume.

59. See Interlude I, this volume.

CHAPTER 5

Convergence of Goals: From Physical Chemistry to Bio-Organic Chemistry, 1930–1940

Gates Chemical Laboratory, 1930

When Mason approved the first Rockefeller grant to T. H. Morgan in 1930, it was agreed that the cooperative attack on vital processes was predicated on the development of chemistry; the thriving chemistry division would have to expand even further to include bioorganic research. During the 15 years of Arthur Amos Noyes's leadership of chemical research at Caltech, the Gates Chemical Laboratory had attained international acclaim. It was "the most forward looking Department of Chemistry with respect to physical chemistry in the world," according to Linus Pauling.[1] As early as 1922, when he first arrived in Caltech from Oregon Agricultural College, Pauling was struck by the progressive chemistry program at the Gates Laboratory. There was little formal instruction; instead, European-style seminars emphasized new topics on the borderland between physics and chemistry: chemical thermodynamics, relativistic thermodynamics, quantum theory, and mathematical physics.

In addition to being recognized for its emphasis on physical chemistry, Noyes's specialty, by the mid-1920s Caltech had acquired a reputation for being the main American center for work with x-ray crystallography. That program too was masterminded by Noyes. Having arranged for a generous grant from the Carnegie Institution, in 1916 Noyes employed C. L. Burdick (trained by W. H. Bragg) to build Caltech's first spectrometer, "probably the best spectrometer of its day."[2] When Linus Pauling embarked on x-ray diffraction studies in 1922, new photographic methods and sophisticated techniques of data analysis at Caltech had yielded the first structural interpretation of an organic molecule, followed within a few years by crystal structure determinations of several inorganic compounds.[3]

Noyes's vision was not limited to physical and structural chemistry; he encouraged other lines of research in borderland chemistry. Thus when local physician Bernhardt Smith proposed in 1922 to commit his personal funds to develop an improved process for insulin synthesis at the Gates Laboratory, Noyes seized the opportunity. He wrote to Hale about the new prospects.

> A rather unusual opportunity has offered itself to make a start in biomedical research; and it seems to me it was worthwhile to improve this opportunity because the problem was readily one of transcendent human significance to which we might reasonably hope to make a useful contribution, and because it would probably make much easier future development in these directions. Thus this might be true if we can show Prichette and perhaps the Rockefeller people that we are already doing effective work in a small way in this field.[4]

The insulin venture turned out to be an institutional success, generating in turn several schemes for a large-scale development of biomedical research at Caltech. Sizable grants from the Carnegie Corporation enabled Noyes to follow up on his scheme and bring John J. Abel from Johns Hopkins in 1925 to coordinate the work on insulin, hoping that Abel's position would eventually develop into a permanent program.[5]

Although the original plan did not materialize and Abel did not remain at Caltech, Noyes's strategy had other dividends. When Mason, then President of the Rockefeller Foundation and director of the Natural Sciences Division, approved $4 million in 1930 to Caltech's natural sciences curriculum, with $1.14 million earmarked for biology, he could point to the early interest in insulin research, the enthusiastic local support for biology, and Caltech's cooperative spirit. A program in bioorganic chemistry in Noyes's department, complementing the work in biology in Morgan's division, would be cultivated on fertile grounds.[6]

Rockefeller Foundation officers noted that the chemistry division had a remarkable record. By 1931 ten full-time faculty members and their assistants and students had produced 269 research papers. The Carnegie Institution had supported 20 research projects in physical chemistry and between 1924 and 1930 had provided funds for at least seven projects during each academic year. Sixteen National Research fellows in chemistry chose to go to Pasadena during the 1920s, a total exceeded only by established centers such as Harvard and the University of California. It was clear that distinguished faculty members such as A. A. Noyes, Richard C. Tolman, Roscoe Dickinson (the division's first Ph.D.), and Linus Pauling were attracting bright young chemists to Caltech. Since 1915 the two chemistry rooms in the General Science building had grown into the Gates Laboratory, which had filled to capacity by 1931. Despite its budget having almost tripled, Noyes complained that there was "literally no space for another research man" and that "the funds available have made it increasingly difficult to provide adequately for the salaries of the research men and for the instruments required in their investigation."[7] Like Harry Chandler's designs for Los Angeles expansion, Caltech's "think big" plans entailed constant upward revision.

If one had to single out the primary reason for the division's remarkable growth, it would surely be the leadership of A. A. Noyes, his academic insights, administrative policies, and his central position in America's science and business elite. Although not a brilliant researcher, Noyes was an inspiring leader and teacher, possessing deep understanding of critical problems in chemistry and a sense for the fruitful directions to be taken. Deliberately charting a course that would distinguish the Gates Laboratory from other chemistry centers, Noyes in effect established a new research school. By all accounts, there was no other place in America where one could investigate the constitution of matter by combining traditional physical chemistry, the principles of quantum physics, with the powerful methods of x-ray diffraction; Caltech alone owned that field. Strong cooperative ties with the physics division and a close relationship between Noyes and Millikan strengthened the borderland program in physical and structural chemistry.[8]

This feature, of course, was not unique to the Gates Laboratory. It exemplified Caltech's "cooperation spirit," mirroring its institutional structure. The German model of an institute centered around a powerful personality—a model that guided science departments in older American universities—was incompatible with Caltech's cooperative ideal. Like Millikan and Hale, Noyes advocated a decentralized but coordinated multiteam research enterprise based on the modern business corporation. Acting as a corporate executive, Noyes delegated the administrative responsibilities and running of the projects to the staff; a management team set policy and directed resources, and a standing committee had purchasing authority, planned undergraduate curricula, and guided graduate students. Noyes promoted the cooperative effort; he had

> very little sympathy with extreme individualism in scientific research, especially in educational institutions, for although individualists have made some of the important contributions to science, they commonly fail to create a spirit of research in their institutions, to develop a school of students who will follow in their footsteps, and to secure cooperative efforts in research by which in the long run the largest results are attained.[9]

As chief executive, Noyes of course insisted that the general line of research would be subject to his approval, but he was definitely an "institution man" guided by the principles of corporate liberalism.

Noyes's influential position within the establishment network of business and science was crucial to the growth of the division. A former president of the American Chemical Society and member of the National Academy of Sciences and the National Research Council, Noyes had powerful friends on the East and West Coasts— corporate leaders, industrial magnates, financiers, politicians, academic administrators, directors of foundations—who shared his ideology and visions of Caltech's destiny. John C. Merriam, Noyes's colleague on the Research Council and after 1920 president of the Carnegie Institution, and Max Mason are just two examples of Noyes's primary links with the foundations.[10] Through California associates Henry Robinson, Henry O'Melveny, and Herbert Hoover, Noyes's influence extended to the spheres of business and politics. By 1930 both foundations and the

Research Council were closely intertwined with Caltech, which in turn was enmeshed in California's industrial sector.

Success had its side effects, however. The rapid growth of the Gates Laboratory generated its own problems. Noyes's complaint of overcrowding and inadequate facilities reflected a tension in the division between the push to expand and the pull to contain. Mirroring the conflicts within the region, the Institute was caught between its own Dionysian forces exerting their will to dominate, while the Apollonian spirit yearned for an intimate community of scholars.

Noyes's pursuit of power was somewhat paradoxical. Despite his political stature, he was a rather shy and retiring man. Unmarried, with no family of his own, Noyes, who lived only a block away from the laboratory, was deeply involved in the activities of Caltech; its community was his surrogate family. He was a strong man, who in his quiet manner worked hard behind the scenes to see his ideas implemented by the Institute's administrators. According to Pauling, although Millikan was Caltech's great public figure, it was Noyes who was largely responsible for the policies announced by Millikan, policies such as the stress on pure rather than applied research and especially the emphasis on a limited-enrollment policy.[11]

Noyes's check on academic growth was designed to avoid mass education, generate close contacts between faculty and graduate students, and foster the cooperative spirit. Throughout the 1920s, Noyes cultivated at Caltech a tradition of inviting new graduate students in chemistry to join him on camping trips to the desert or to stay at his beach house at Corona Del Mar (where Morgan set up his marine laboratory during the early 1930s). Evenings by the campfire, Noyes would recite poetry and inspire discussion on diverse topics, including chemistry. The swelling population of the division endangered these activities, threatening the intimacy that had permitted informal administration and casual collaborations. Above all, expansion could weaken the interaction between faculty and students and interfere with the nurturing of novices to scientific maturity. These close encounters and outings were in part intended to help size up new students and single out talent. As a new graduate student, Pauling participated in several such camping trips and was quickly selected by Noyes to be groomed for leadership.[12]

Pauling's background (born 1901) fit well with Caltech's community. Raised in the remoteness of Condon, Oregon and the son of the town's pharmacist, Pauling's milieu embodied the Rooseveltian myth: rugged "God-fearing" folk taming the austere wilderness of the Pacific Northwest with self-mastery, work ethic, and prayer. In most respects Pauling's nativist roots were congruous with those of the leaders of American science, especially Warren Weaver. Like Weaver, Pauling's early years were shaped by the ambience of a Protestant village and by an involvement in his father's chemical business; but unlike Weaver, religious worship played an insignificant role in Pauling's adult life, though the work ethic remained. He excelled in chemistry throughout his school years, and his enrollment in 1917 in Oregon Agricultural College reflected his commitment to some kind of career in chemistry; a couple of years later this commitment meant graduate school. Berkeley's chemistry department, headed by G. N. Lewis, was Pauling's first choice, but their delayed response tilted the balance in favor of his second choice, Caltech.[13]

The young man from Oregon soon became Noyes's favorite pupil. During the years he spent under Noyes's tutelage, he never failed to dazzle his mentor and colleagues with his phenomenal memory and intellectual originality, especially in chemistry. Pauling was so imbued with the physical aspects of chemistry that he first considered specializing in atomic physics. As was customary at Caltech, Millikan and Noyes saw to it that promising students received fellowships and grants to train in top European laboratories. Pauling was encouraged to go to Europe as a National Research Fellow and a Guggenheim Fellow. He studied in Munich (1925–1926) with the leading theoretical physicist Arnold Summerfeld, continuing his training in atomic physics and quantum mechanics the following year with Erwin Schrödinger in Zurich and with Niels Bohr in Copenhagen.[14]

Combining the pull toward atomic physics with a fascination with physical chemistry and the structure of compounds, Pauling integrated the two fields to form one of the most important scientific theories, linking atomic structure with chemical properties. His great work, "The Nature of the Chemical Bond" (1931), succeeded in explicating the forces operating between atoms and molecules in terms of the principles of quantum mechanics. That work, together with Pauling's papers on crystal structure and quantum physics, numbering nearly 50 by 1931, began to revolutionize concepts of chemical structure and molecular architecture.[15] Hailed as a prodigy of American science, Pauling at the age of 30 was inundated with honors and professional acclaim, becoming an associate editor of the *Journal of the American Chemical Society* in 1930. He was offered lucrative professorships by MIT, the University of Michigan, and Harvard, all of which he declined in order to assume a full professorship at Caltech in 1931. That same year, Irving C. Langmuir's brother established the Langmuir Prize in Chemistry. Awarding the first prize to Pauling, Langmuir predicted that the young star at the threshold of his career might yet win the Nobel Prize. A proud Noyes announced that Pauling was "the most promising young man with whom I have ever come in contact in my many years in teaching."[16] There was little doubt that in due course Pauling would become Noyes's successor. Here too, however, a conflict was brewing. Although Pauling could lead and inspire, the young maverick was also a strong individualist. His powerful personality was better suited for leadership of a German-style institute than an academic corporation, and in practice his personal ambitions would tend to rise above the cooperative ideal.

Vital Processes: Pauling and Weaver

Having established himself by 1931 as a first-rate theoretical chemist, Pauling, now heading a large research team, had the luxury of indulging various intellectual curiosities and diversifying his research interests. With the arrival of T. H. Morgan and the biology team in 1928 Pauling began to take an interest in genetics and occasionally participated in the Tuesday afternoon biology seminars.[17] Although no collaborative projects between the Gates and Kerckhoff Laboratories developed during the 1930s, at least the implicit encouragement of cooperation removed obstacles for interdisciplinary dialogue for the adventurous few. For example, even though biochemist Henry Borsook did not get along or collaborate with Pauling, he reminisced nostalgically about the early days when Caltech was still small and intellectually cohesive. In those days, Borsook recalled, the conversations around the long table in the dining hall and the casual social gatherings kept faculty broadly informed about the researches in the Institute.[18]

It may have been Borsook's studies in biochemistry or Noyes's early interest in developing bioorganic chemistry that motivated Pauling to begin investigating organic molecules. It might have been just a reasonable progression—from the simple to the complex—to extend the studies of chemical bonding and molecular structure from simple inorganic compounds to the more challenging organic substances. Certainly Mason's plans for a psychobiology program and the presence of Morgan's team played a pivotal role in shaping Pauling's new interest. At any event, in 1932 Pauling began applying the concept of molecular resonance to account for the variations in interatomic distances in organic compounds by investigating the thermodynamics and bond configurations of urea, oxamide, oxamic acid, and carboxylic acids. Within two years, Pauling and his collaborators demonstrated the existence of a resonance structure in the amide group (the basic molecular configuration of amino acids), a crucial clue for elucidating the secondary structure of proteins. A few years later (1938) Pauling and his new collaborator, Robert Corey, would mount a precision attack on the atomic architecture of proteins.[19]

Pauling began expanding the scope of his research just as the Rockefeller Foundation converged toward a definite policy, when Warren Weaver was summoned by his mentor and Foundation president Max Mason to join the Rockefeller staff. Soon after assuming the directorship of the natural sciences division in February 1932, as part of a comprehensive orientation for launching the new biology program Weaver traveled extensively throughout Europe and the United States. Familiarizing himself with research centers where borderland biology had already made a start and where there existed a framework for cooperation, Weaver carved new territory and cultivated old grounds. Morgan's division of physicochemical biology was by then designated as ideal for Weaver's cooperative program. After learning from Morgan at Woods Hole in September 1933 about the crisis generated by the Kerckhoff endowment, and that the biology division had been hard-hit by the Depression and needed emergency funds, Weaver planned an October trip to Pasadena.[20]

Both for Weaver and his hosts the visit to Caltech meant more than an item on a business agenda. Having spent three happy years (1917–1920) as a member of Caltech's physics faculty, Weaver retained his respect and sentiments for Noyes, Millikan, and the Institute that had offered him his first academic post. Because of his fondness for Weaver, Millikan had encouraged him to return to Caltech. Never officially accepting Weaver's resignation, Millikan had hoped that Weaver would "continue to be a professor of the California Institute of Technology, on leave until your return."[21] Although Weaver did not return as a physics professor, he returned as a philanthropic agent and was now in a position to express his appreciation to his senior colleagues at Caltech.

The visit, Weaver's first since leaving Caltech, comprised a lavish display of mutual enthusiasm. Weaver was especially impressed with Pauling and his 20-

member group, which, according to Weaver, essentially formed an institute of theoretical chemistry in the European sense. Although not quite in keeping with the group spirit Noyes would have preferred to project, Pauling's dominance was nevertheless perceived as an asset. "P. has a speculative mind of the first order," wrote Weaver in his report, "great analytical ability, and the genius to keep in close and inspiring touch with experimental work . . . nearly universally rated the leading theoretical chemist in the world."[22] His observations were amply reinforced by Caltech's own boosters.

Noyes, a bit apologetic for his excessive pride and promotion of his department, hoped that Weaver would not think it was the "normal California enthusiasm when he [Noyes] said that, were all the rest of the Chemistry Dept. wiped away except P., it would still be one of the most important departments of chemistry in the world."[23] Although leadership abilities and organizational skills were not explicitly mentioned, it was clear from the size of Pauling's team and from its intellectual vigor that the young chemist was seasoned in the ways of politicoscientific management—the kind of man the Rockefeller Foundation liked to court. Weaver encouraged Pauling to direct his theoretical methods and technical skills toward biological research.

Pauling in turn was well prepared for Weaver's visit. Anticipating the new opportunities for his research, he submitted to Weaver a detailed report on the chemistry work supported by the Rockefeller Foundation. He described the new methods of x-ray and electron diffraction and their importance for researches of organic substances, "with the hope that ultimately an attack can be made in this way on the purpurins, chlorophyll, hemoglobin, and other substances of biological importance."[24] A formal application for a three-year grant, $10,000 per year, in conjunction with Morgan's proposal for physiological genetics and neurophysiology followed shortly.

Back in New York at a December conference, Weaver and his staff assessed Caltech's chemistry program. "So fundamental is the nature of the problems under investigation by Professor Pauling," they concluded, "that they necessarily underlie in a most significant manner the vital processes which constitute the present major interest of the Natural Sciences."[25] According to Mason, "C.I.T-Research in Chem. [was] at the center of the program of study of vital processes as furnishing aid in the sciences underlying human behavior."[26] For the present the Foundation would act conservatively, obligating itself for one year only, until Pauling had demonstrated his commitment to the Rockefeller program. Recommending an appropriation of $10,000 for the academic year 1934-1935 (Morgan received $50,000), Weaver stressed to Pauling that favorable consideration of the chemistry application depended on the fact that the work had developed to the point where it promised application to substances of biological importance.[27]

Pauling wasted no time. He mounted an immediate attack on the structural features of hemoglobin. With 10^{20} hemoglobin molecules circulating in the blood he considered it to be the most ubiquitous and important substance. As if guided by the Aristotelian association of blood with human qualities, he broached the study of human molecular structure with hemoglobin research. He began by surveying earlier work on the biochemistry of porphyrins to which he intended to apply his methods of theoretical and structural chemistry. Eight months later, in time to apply for a grant renewal, he reported to Weaver that preliminary progress had been thoroughly satisfactory. Through the use of electron diffraction techniques, Pauling and his assistants found that incorrect formulas had been assigned to some of these substances by all previous investigators, and that a newly discovered x-ray method promised a powerful means of attack on the structure of the iron environment of the heme group. He was planning an intensive attack on the hemoglobin problem and was contemplating applying for a larger grant—perhaps for three years—so the essential basic research on simpler substances, which had spawned the new techniques, might continue. The Foundation, however, preferred a limited commitment for an additional year and decided to forego the support of inorganic research.[28]

Self-assured and impatient, Pauling did not tolerate rejections. Extracting a commitment from Millikan and the trustees to partially support the research in inorganic chemistry, Pauling used the promise as a vote of confidence. He dispatched two letters simultaneously, one to the Rockefeller Foundation and one to Weaver, informing them of the Institute's support. Despite the Foundation's judgment to the contrary, he did apply for a three-year grant (1935–1938), forcefully contesting the limited renewal. "The hemoglobin problem is a very difficult one," Pauling wrote to Weaver, "and I consider it unlikely that we can obtain results providing a real test of our methods in one year. On the other hand, I am confident that within three years we could obtain very valuable information regarding the structure of the hemoglobin substances, the nature of the bond to globin, the process of addition of oxygen, etc."[29] He emphasized to the Foundation Millikan's readiness to match their grant (on a 50 percent basis) by supporting Pauling's basic research in inorganic chemistry during the three-year period.[30] A month later Pauling's grant was renewed for three years at the requested $10,000 per year, thus providing his program with long-range stability in terms of staff, resources, and choice of scientific problems.

By April 1935, when he visited Weaver in New York, Pauling had submitted his first paper on hemoglobin to the *Proceedings of the National Academy of Sciences*. His premature conclusion on the structure of hemoglobin was wrong, as later studies by Max F. Perutz would show, but at least he mastered the hemoglobin literature, which he believed had been inadequately utilized. He spent part of the summer at the marine station at Corona Del Mar working on the hemocyanin of the keyhole limpet, developing sensitive magnetic methods for determining free oxygen. A similar method of determining the magnetic moment for hemoglobin— based on the iron content—supplied information on the nature of the chemical bond in the heme molecule.[31]

Here Pauling and his collaborators were remarkably effective. Using their new magnetometer, they established that the dark venous blood (unoxygenated) was attracted to a magnet, whereas the bright arterial blood (oxygenated) was repelled. This clue to the electronic structure of iron and its relation to the heme molecule would prove crucial for his landmark studies of sickle cell anemia a decade and a half later. Furthermore, the hemoglobin work had two immediate consequences: Pauling became involved in the problem of the structure of proteins; and through his association with the Rockefeller Institute, especially Karl Landsteiner, he became interested in the problem of immunology—the structure of antibodies and the nature of serological reactions.[32]

Alfred E. Mirsky of the Rockefeller Institute, one of the world's leading protein chemists, came to collaborate with Pauling during the following academic year 1935–1936. Having made the important discovery several years earlier that denaturation of proteins such as trypsin and hemoglobin could be reversed, Mirsky was now interested in explaining the physicochemical and physiological mechanisms of protein denaturation. The collaboration resulted in a seminal publication, "On the Structure of Native, Denatured and Coagulated Proteins" (1936), in which they presented a general theory of protein structure. For the first time hydrogen bonds—weak but flexible connections between molecules—were assigned a major physiological role. They postulated that a polypeptide chain was coiled in a specific configuration in the native molecule, stabilized mostly by hydrogen bonds between one part of the chain and another; denaturation, then, implied a loss of such configuration to a more random form. In suggesting that hydrogen bonds determined the three-dimensional configuration of proteins—and thus their biological specificity—Pauling and Mirsky enunciated a fundamental relation between molecular structure and biological function. It was also one of the cornerstones of

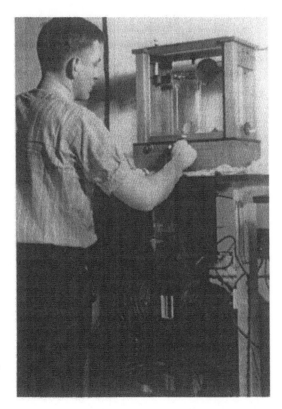

Figure 9 Dr. Harrison Davies of Pauling's group measuring the magnetic susceptibility of a hemoglobin solution. Courtesy of the Rockefeller Archive Center.

Pauling's conception of molecular architecture, a metaphor and method for explaining life in health and disease, which would lend legitimacy to the molecular biology enterprise.[33]

Having proved its effectiveness in the borderland between physical chemistry and biology, and with the concurrent development of Morgan's division, the chemistry division was now in a favorable position to execute Noyes's original plan of building a laboratory of bioorganic chemistry. During Weaver's visit to Caltech in March 1936 Noyes, feeble and riddled with illness, expressed his concern at a conference held in Millikan's office to see his plans realized—to build, equip, and staff as soon as possible the connecting link between the existing chemistry building and the biology laboratory.[34]

Pauling, conscious of his brilliance and reputation, was well aware that his recent accomplishments in hemoglobin research constituted a primary justification for the proposed center of bioorganic chemistry. He was thus in a position to make demands and apply strong pressure on Caltech and the Foundation. According to Weaver, Pauling implied that he would leave Caltech unless the Foundation furnished him with able colleagues in organic chemistry so he could develop his new research properly. To speed up the process of building the center, Noyes informed Weaver that he had already persuaded Edward Crellin, a retired steel magnate in Pasadena, to make available in 1936 a fund of $350,000 for constructing and equipping the new chemistry laboratory. With that decisive move, Weaver, heavily committed to the expansion, stated that "It was my own idea, before receiving Noyes's letter, that the development could and should be considerably more pretentious than he indicates."[35] Noyes's mission was now completed. He died in June 1936. "He was of rare mould; a civilized man in a world whose soul is still largely barbaric," eulogized Noyes's friend and successor at MIT. "The ancestry of A. A. Noyes was unexcelled. Pioneer stock of English origin, there has been no breed more versatile, more adherent to the tenets of common sense," he noted, lamenting the passing of an exemplar of the great race.[36]

With Noyes's death, the conflicts in the chemistry division, which had been brewing over recent years, bubbled to the surface. Apparently Pauling had always thought that his mentor was too institutionally minded, always willing to sacrifice the interest of the department for broader institutional goals. Although Pauling appreciated the importance of such an attitude, he also thought that each special branch of science in the Institute should have its own champion and guardian. According to Weaver's impression during his visit to Caltech the following January (1937), Pauling's aggressive politics and pursuit of power had generated resentments in some circles, giving him "with Millikan and others, the reputation of being something of a young dictator because he has tried to force M. to be very explicit about his promises for support of chemistry."[37] Millikan did not like explicit or written agreements; he always preferred implicit understandings, an informal administration by gentlemen's agreement. Moreover, the Caltech community did not prize noncooperative individuals.

The rest of the chemistry groups, Weaver was told, thought that Pauling was excessively aggressive. During the last year or two of Noyes's life, when his physical and mental energies were low, Pauling worried and pestered Noyes in what they thought to be an unwarranted and even unforgivable way, trying to force him into an "intolerable activity of leadership for chemistry." Some of the group members went so far as to believe that Pauling's attitude made Noyes's last years definitely unhappy, if it did not hasten his death. In fact, when arrangements were made for Noyes's funeral, the list of honorary pallbearers included the entire Executive Council and every member of the chemistry division except Pauling.[38]

This episode precipitated a bitter confrontation between Pauling and Millikan, accentuating an already strained relationship between the two men. When the control of the chemistry division was vested in a divisional council including Pauling, and the council in turn offered him the council's chairmanship, Pauling refused because the offer did not include the title "Director of Laboratory" and it explicitly stated that the council and not the chairman would control the division. Weaver openly and amiably discussed the whole matter with Pauling and later with Max Mason, now a member of Caltech's Executive Council. Weaver told Pauling that Millikan was shocked by Pauling's curt and impudent letter of refusal of what was generally viewed as a great honor for a young man.[39]

Pauling, however, was not impressed with mere honorific titles. He did not approve of what he considered to be an inefficient mode of running a division by committee and was not willing to accept responsibility for the growth and development of the division unless he were given the authority that should accompany it.[40] Pauling's demand was met. He was appointed "Director of the Gates Laboratory and Chairman of the Division of Chemistry and Chemical Engineering," a title commensurate with his authority. In 1937, with the Crellin building under construction, Pauling was also appointed director of the new laboratory; his newborn son was named Edward Crellin.

Crellin Laboratory: Nascent Trends

The development of bioorganic chemistry at Caltech was indeed considerably more expansive than Noyes had envisioned. After an informal agreement had been reached, Millikan submitted to the Rockefeller Foundation a grant application for $300,000. A $50,000 portion would be used at a rate of $10,000 per year for Morgan's biology program—biochemistry, biophysics, and physiology—and $250,000 for the work in bioorganic and structural chemistry over a period of five to seven years. The construction of the Crellin Laboratory was nearly completed; and if the application were approved, as expected, the program would be launched July 1, 1938.[41]

As the program's chief architect at Caltech, Pauling envisioned a grand research center, with state-of-the-art technologies elucidating the molecular processes of life. Three prestigious appointments were planned, including that of the English biochemist Alexander R. Todd of the Lister Institute, now spending six months at Caltech as visiting lecturer; but he would decline the offer at the last minute. Protein chemist Carl Niemann of the Rockefeller Institute was to take up residence July 1938, to be joined shortly after by Edwin R. Buchman, a specialist in vitamin syntheses. A protégé of Karl Link at the University of Wisconsin, and having

Figure 10 Crellin Laboratory of Chemistry. Courtesy of the Rockefeller Archive Center.

established his scientific reputation at the Laboratory of Max Bergmann, Niemann came highly recommended by Weaver as one of the most promising organic chemists in America. At least three additional appointments were planned on the level of associate and assistant professors, supplemented by postdoctoral positions and research fellowships. Several "human computers," specialists in Fourier analysis of x-ray diffraction patterns, would be employed for carrying out the laborious calculations, a task that two decades later would be performed electronically within hours.[42]

The organic chemists occupied the second and third floors of the Crellin Laboratory and the auxiliary rooms on the roof, their modern chemical benches equipped with vacuum suctions, ventilated hoods, and electrical outlets. Pauling also requested from the Rockefeller Foundation funds for specialized apparatus and supplies for organic analyses and syntheses: analytical and microbalances, apparatus for catalytic high-pressure reduction, table-top centrifuges, pH meters, electrometric titration instruments, and gas analyses apparatus. These instruments appear rudimentary when considered in light of present-day chemical armamentaria, but they represented the technical excellence of the times; pH meters, for example, were first commercially produced during the early 1930s.[43] In fact, Associate Professor of Chemistry Arnold O. Beckman, a Caltech graduate specializing in the design of physicochemical apparatus, had just then marketed his new sensitive pH meter. Founder of Beckman Instruments (1935), a company that by the 1950s would become a leading manufacturer of biochemical apparatus, Beckman played a central advisory role in the technological planning of the Crellin Laboratory. He has remained associated with Caltech as a permanent liaison between the realm of scientific imagination and world of the instrument shop.[44]

Pauling's wish-list included also a wide array of expensive optical equipment for chemical analyses: microscopes and accessories, precision polarimeters for measuring optical rotation by isomers of plane-polarized light, refractometers for determining substance density, a microphotometer, and a spectrophotometer.[45] The spectrophotometer, a complex and expensive apparatus for determining the nature of substances through analyses of their absorption spectra, was Beckman's area of expertise. A decade later, the Beckman DU spectrophotometer would become a landmark in his company's ascendancy to biotechnological superiority.

The details of the research in organic chemistry in the new laboratory were still somewhat unclear, depending on final academic appointments. There was an extreme shortage in America of able organic chemists, especially those interested in biological problems, Pauling pointed out. He used this point to promote the Crellin enterprise as a vehicle for building organic chemistry in the United States. Should Todd accept Caltech's offer, his group would continue the investigations on the structure of vitamins, hormones, and other natural products of plants and microorganisms, work that would complement Buchman's organic research on vitamins. Niemann's appointment was certain, and thus an extensive attack was planned on the central protein problem, separating polypeptides, and developing new methods for ordering amino acid residues.[46]

The bioorganic approach to protein chemistry would be complemented by an attack from the physical-structural side, led by Pauling and his group. Situated close to the organic chemists, the physical chemists occupied the first floor, basement, and subbasement of the Crellin building, which housed the sophisticated apparatus of photochemistry, magnetochemistry, spectroscopy, and x-ray and electron diffraction. New methods of attack on organic molecules were in progress. One of Pauling's associates had just developed a powerful spectroscopic technique of infrared photography for determining the presence of NH-O hydrogen bonds; when quantitatively and qualitatively refined, it could reveal the hydrogen bonds in different proteins. Pauling was planning to use this method, he reported to Weaver, to distinguish between conventional polypeptide structure and the cyclol structure proposed by Dorothy Wrinch. Her cyclol hypothesis, postulating that proteins were constructed from hexagonal rings, was coming under attack from several quarters, even including Bergmann and Niemann. With Niemann's aid, Pauling planned to strike a fatal blow to the cyclol theory, an attack that eventually would end Wrinch's career.[47]

The Crellin Laboratory, a physical link between Caltech's biologists and chemists, epitomized the Rockefeller Foundation's program, by now renamed molecular biology. Even though there had been little collaboration between the two groups (plant physiologist Fritz W. Went briefly collaborated with two chemists on the structure of plant hormones) Pauling boasted about a cohesive cooperative attack. "The presence at the Institute of such leaders in the field of genetics as Professors T. H. Morgan and A. H. Sturtevant and Dr. C. B. Bridges," he wrote to Weaver, "makes obvious the appropriateness of a chemical attack on the structure of chromosomes."[48] By this statement, of course, Pauling meant that the methods of protein chemistry would be applied to explicate the properties of the protein gene.

Millikan too stressed in his report to Weaver the cooperative nature of the venture: "[O]ur bio-organic development spills over into the biological department quite as much as into the chemical in that the Kerckhoff building, having a floor space of 48,000 square feet, which we have been designating as for physiology, is in a very real sense, however, part of the bio-organic development."[49] That report, with its clear programmatic statements accompanying a comprehensive review of Caltech's growth in the natural sciences over a decade and a half, had been actually solicited by Weaver. Having come under attack from Herbert S. Gasser (Foundation Trustee and President of the Rockefeller Institute) for an excessive financial commitment to a single "project," Weaver too needed to persuade the holders of the purse strings of the viability of the new venture.[50]

The Foundation's support and confidence were amply acknowledged by Caltech's leadership. At the dedication of the Crellin Laboratory on May 16, 1938, Millikan singled out the long-term commitment of the Rockefeller Foundation "which has watched so carefully over this young and vigorously growing institution, and has offered to help it substantially at a number of critical junctures. Indeed, its total contributions to the development of the Institute, without including its support of the 200-inch telescope project, have now amounted all told to about $4 million."[51] The synergy between intellectual capital and economic resources buttressed the technocratic vision of progress. With the Foundation's support and the generous help of prominent Pasadena families, Millikan predicted that the Institute could "scarcely fail to win the race for human betterment" through chemical and biochemical advances.

The term "human betterment" must be viewed within a politics of meaning with its own historicity. "The race for human betterment" had a specific linguistic meaning during the 1930s, grounded in eugenic discourse. As the *New York Times* announced, the Rockefeller gift to Caltech was aimed at "the biological improvement of the race." For Caltech's community, including members of the Human Betterment Foundation, the physicochemical attack on vital processes was part of the same mission that guided the eugenic sterilization campaign, then at its height in California. Although there is no written record that during the 1930s Pauling was directly motivated by the social goals of the Rockefeller Foundation's agenda "Science of Man" or by the eugenic campaign of the Human Betterment Foundation, his interests in human applications of biochemical research are documented. From the 1940s on his medical interests and growing politicization engaged him in problems of human behavior and a commitment to the genetic purification of the germ plasm.[52]

At the center of Pauling's program was the problem of protein structure in relation to diverse vital processes, ranging from gene replication to the oxygenation of hemoglobin. Even though bioorganic chemists in the new laboratory would study several physiologically active substances manufactured in the body or ingested in foodstuffs, the proteins, with their thousands or tens of thousands of atoms, were the single most urgent problem, Pauling stressed. He predicted that the crucial steps toward the solution of that great problem would occur during the next 10 years.

There were two possible modes of attack on the protein problem. The first, the British crystallographers' approach, was direct x-ray analysis of an intact crystalline protein. The second method was an indirect approach, a model-building method based on an exact knowledge of the structures of a protein's constituent amino acids and small peptides. The second method attempted to first build up the protein and then check the structure by x-ray methods. J. D. Bernal remembered "very well discussing the problem with Pauling just before the war. He [Pauling] was in favor of the second method, which I thought indirect and liable to take a very long time."[53] Discounting the war years 1940–1945, during which time chemists accomplished little work on protein structure, Pauling's estimate of 10 years was accurate. By 1951 Pauling and Robert Corey would work out the helical structure of the protein alpha-keratin.

Corey's move to Caltech in 1937 was a fortuitous turn of events for Pauling's protein project. Corey had worked for 10 years with Ralph W. G. Wyckoff at the Rockefeller Institute on the physical chemistry of proteins, but he lost his post when the latter left. Perhaps because of Wyckoff's ties to Caltech, having spent the years 1920–1922 there as Visiting Research Associate, Corey decided to work in Pasadena during the last year of his Rockefeller appointment. During a single year, using his own x-ray equipment, Corey made great strides into the protein puzzle. He showed that in the crystalline dipeptide diketopiperazine (a simplified analogue of amino acids), the amide bonds were coplanar, strongly suggesting the

Figure 11   Caltech's x-ray diffraction apparatus for studying the crystal structure of amino acids and peptides, ca. 1939. Courtesy of the Rockefeller Archive Center.

presence of a resonance structure—observations that fit precisely with Pauling's studies of the amide bond in urea during the early 1930s.[54]

Pauling communicated to Weaver, by now his confidant and adviser, his profound respect for Corey's accomplishments. Even though there were no plans for additional positions, Pauling thought that appointing Corey as a senior research fellow would be a great asset to the new chemistry program, despite Corey's limitations. Badly crippled by infantile paralysis, he was unlikely ever to attain a higher academic rank. Weaver's recommendation to retain Corey stood Pauling in good stead. Within a year (1939) Corey and his collaborators worked out the principal features of the structure of the first and simplest amino acid glycine.[55]

In 1939 Pauling also requested from the Foundation a substantial increase in that year's budget to build a chemistry instrument shop: bench lathe, milling machine, and other smaller equipment. Up until 1938, x-ray and other physicochemical apparatus had been built in the astrophysics shop, but the backlog of astrophysical apparatus impeded the rate of construction of the x-ray apparatus. The new x-ray spectrograph and the electron diffraction apparatus with their specialized photographic gear would be coupled to an electrical Fourier synthesizer, permitting two-dimensional Fourier calculations to be made in less than a week. With Weaver's approval, a fully equipped chemistry shop was set up, substantially accelerating the x-ray program. By 1940 Corey and his collaborators had tackled several amino acids and peptides, and published a structure of DL-alanine.[56]

While Corey's group focused on the problem of protein structure with x-ray techniques, Pauling, in collaboration with Niemann, consummated their plan—the

Figure 12   V. Schomaker and D. P. Stevenson studying the structure of gas molecules with the electron diffraction apparatus at Crellin Laboratory, ca. 1939.

final blow to Wrinch's cyclol theory, published in their 1939 paper "The Structure of Proteins."[57] Although bearing a broad and general title, the paper was a specific and unambiguous attack on Wrinch's theory. "It is our opinion," they opened their attack,

> that the polypeptide chain structure of proteins, with hydrogen bonds and other interatomic forces (weaker than those corresponding to covalent bond formation) acting between polypeptide chains, part of chains, and side-chains, is compatible not only with the chemical and physical properties of proteins but also with the detailed information about molecular structure in general which has been provided by the experimental and theoretical researches of the last decade. . . . It has been recognized by workers in the field of modern structural chemistry that the lack of conformity of the cyclol structures with the rules found to hold for simple molecules makes it very improbable that any protein molecules contain structural elements of the cyclol type.[58]

A heap of refutations cascaded behind this opening statement: x-ray results regarding protein structure that derived from the works of British crystallographers, recent determination in Pauling's laboratory of the N-H . . . O bond angles, and thermochemical data regarding protein structure. Pauling concluded that the great energetic disadvantage of cage structures relative to polypeptide chains makes the cyclol structure highly unlikely.

Figure 13 Carl Niemann at work, 1939. Courtesy of the Rockefeller Archive Center.

It was ironic that in this paper Niemann contributed to the eventual refutation of the work he had performed with Max Bergmann only three years earlier, research that had brought him scientific visibility and his appointment at Caltech. Although Bergmann and Niemann had opposed Wrinch's methods and personal style, their results lent support to her theory, which defined cyclol as consisting of 288 amino acid residues. Their own experiments—postulating that a protein was composed of series of multiples of 288 amino acid residues—turned out to be flawed because of samples of questionable homogeneity and faulty analytical methods.[59] The paper "The Structure of Proteins" did not attack the Bergmann-Niemann theory directly. Rather, it showed only that Wrinch's cyclol did not agree with the Bergmann-Niemann multiple rule. By integrating Corey's recent results from the studies of peptides and amino acids with Pauling's own calculations of bond angles and energies in amides, together with objections advanced by other crystallographers, the paper sowed the seeds of destruction of Niemann's prior accomplishments. The same techniques of x-ray diffraction, coupled with chromatographic methods of ordering amino acids, would establish a decade later that a pattern of 288 amino acid multiples did not exist.

Their strategy worked. The paper was an overnight success. Pauling and Niemann received numerous requests for reprints and congratulatory letters on the timeliness and rigor of the paper. Indeed, "The De-bunking of Wrinch" by Pauling and Niemann, as A. R. Todd referred to it, became biochemistry's cause célèbre, adding yet another trophy to Pauling's expanding scientific kingdom. Pauling's pleasure over the outcome of the controversy radiated from his report to Weaver, but Weaver's sober response suggested that such polemics were unnecessary, even inappropriate. He thought that it would be unfortunate to be drawn into a scientific controversy in which the emphasis fell upon the controversy rather than upon the sciences. "[N]othing ever spoke so convincingly as a quiet presentation of facts," Weaver pointed out in his understated manner. Both Weaver and Pauling agreed, however, that the debate had served a useful purpose by focusing a great deal of attention on the fundamental problem of protein structure.[60]

This assessment may be true if human factors are ignored. Although Niemann's career within Pauling's powerful sphere of influence suffered little damage as a result of the refutation of his earlier work, Wrinch paid a high price. Largely due to Pauling's personal efforts and his vociferous campaign, not only her project but even the soundness of her judgment were categorically discredited, leading to the demise of her career. As it turned out, Wrinch was not entirely wrong. In 1951 a cyclol structure was discovered in a class of primitive proteins (ergot alkaloids), demonstrating that Wrinch's cage structure, though uncommon, was at least thermodynamically viable. Pauling's chain theory of proteins won out, but Wrinch's judgment proved to be sound.[61] Science does not proceed merely by sober intellectual negotiation; it is a socially charged process with high professional stakes.

By mid-1939 the Crellin Laboratory, in conjunction with the expanded Kerckhoff Laboratory, represented, at least on paper, a comprehensive joint effort in chemistry and biology. Lumping together all those scientists supported by the Rockefeller Foundation whose research interests apparently matched the aims of its new program in physicochemical biology—faculty, staff, fellows, graduate students—Weaver prepared a report for the Foundation. Lest any doubt linger regarding the merit of the Caltech project, he pointed with pride to the group working on problems that involved the application of chemistry to biology; they represented training or research experience obtained in 38 institutions including 15 foreign universities: elite American institutions such as MIT, Columbia, Harvard, Princeton, Johns Hopkins, Yale, and Cornell, as well as such European centers as Berlin, Göttingen, Oxford, Cambridge, Paris, Zurich, Utrecht, and Uppsala. Some scientists had also worked at the Rockefeller Institute, the U.S. Department of Agriculture, and the Food and Drug Administration. "This widely trained and broadly experienced group constitutes an impressive proof, I would say, of the fact that this development at the California Institute is already well underway."[62]

Just at that time the war broke out. Neither Pauling nor most biochemists in America or Europe could devote time to basic research during the war years. However, the fundamental knowledge accrued during the 1930s formed a framework within which researchers could apply the rudimentary knowledge of protein structure to new practical problems relevant to the war effort. Pauling's interest in biologically active substances brought him face to face with one of the most challenging problems in physiology: the genesis and structure of antibodies. He began applying theories of protein structure and intermolecular forces to immunology. Much of this research, some of which was considered controversial even at the time, was later discredited. Throughout the 1940s, however, immunochemistry in alliance with serological genetics formed a central project within the molecular biology program at the Crellin and Kerckhoff Laboratories. Although many other research projects were curtailed during the war, the work on antibodies and serological genetics flourished because of its medical significance for the war effort.

Notes

1. J. W. Servos, "The Knowledge Corporation: A. A. Noyes and Chemistry at Cal-Tech, 1915–1930," *Ambix, 23* (1976), p. 179, where he referred to Pauling's address at the dedication of the Noyes Laboratory of Chemical Physics. See also J. W. Servos, *Physical Chemistry from Ostwold to Pauling: The Making of Science in America* (Princeton: Princeton University Press, 1990), especially Ch. 6.

2. From R. J. Paradowski, "The Structural Chemistry of Linus Pauling," Ph.D. dissertation, University of Wisconsin, 1972, p. 171. A quote from C. L. Burdick, "The Genesis and Beginnings of X-ray Crystallography at Caltech," in *Fifty Years of X-ray Diffraction*, Peter Paul Ewald, ed. (Utrecht: International Union of Crystallography, 1962), p. 557.

3. J. W. Servos, "The Knowledge Corporation," p. 177 (see Note 1).

4. Ibid., p. 178; Noyes to Hale, December 22, 1922, Roll 28, *Hale Papers*.

5. RAC, RG1.1, 205D, Box 5.66, November 8, 1930. See also Chapter 2, this volume.

6. Ibid.

7. J. W. Servos, "The Knowledge Corporation," p. 179 (see Note 1). CIT, Box 25, Robert A. Millikan Papers, memo of Noyes dating from sometime in 1930.

8. J. W. Servos, "The Knowledge Corporation," pp. 180–181 (see Note 1). Excellent discussion on the significance of research schools and disciplinary development, including Noyes's school is given by Gerald Geison in "Scientific Change, Emerging Specialties and Research Schools," *History of Science, 19* (1981), pp. 20–40.

9. J. W. Servos, "The Knowledge Corporation," p. 182 (see Note 1); Noyes to Loeb, February 1916, Box 2, Scherer Papers.

10. J. W. Servos, "The Knowledge Corporation," p. 181 (see Note 1).

11. R. J. Paradowski, "The Structural Chemistry," p. 46 (see Note 2).

12. Ibid., p. 47.

13. Anthony Serafini, *Linus Pauling: The Man and His Science* (New York: Paragon House, 1989), Ch. 1.

14. G. W. Gray, "Pauling and Beadle," *Scientific American, 180*, No. 5 (1949), p. 16.

15. L. Pauling, *The Nature of the Chemical Bond, and the Structure of Molecules and Crystals* (Ithaca: Cornell University Press, 1939); based on a series of articles by Pauling in the *Journal of the American Chemical Society*: "The Nature of the Chemical Bond, I," *53* (1931), p. 1367; "The Nature of the Chemical Bond, II," *53*, p. 3225; "The Nature of the Chemical Bond, III," *54* (1932), p. 988.

16. From "Linus Pauling," *Scientific American, 145* (1931), p. 293.

17. L. Pauling, "Fifty Years of Progress in Structural Chemistry and Molecular Biology," *Daedalus, 99* (1970), p. 1002.

18. CIT, Oral History, Borsook, pp. 12–13.

19. L. Pauling, "Interatomic Distances in Covalent Molecules and Resonance between Two or More Lewis Electronic Structures," *Proceedings of the National Academy of Science USA, 18* (1932), pp. 293–297; and R. C. Olby, *The Path to the Double Helix* (London: Macmillan, 1974), pp. 273–275.

20. RAC, RG1.1, 205D, Box 5.71, Weaver's report from Woods Hole, September 8–10, 1933. For Weaver's account of his connection to Mason and to the Rockefeller program see Warren Weaver's autobiography, *Scene of Change* (New York: Charles Scribner's Sons, 1970), Chs. 2, 4, and 5. See also R. E. Kohler, "The Management of Science: The Experience of Warren Weaver and the Rockefeller Foundation Programme in Molecular Biology," *Minerva, 14* (1976), pp. 249–293.

21. W. Weaver, *Scene of Change*, Ch. 3 (see Note 20); Millikan's letter is quoted on p. 48.

22. RAC, RG1.1, 205D, Box 5.71, Weaver's report on chemistry, October 23, 24, 25, 1933.

23. Ibid.; also Noyes to Weaver, November 7, 1933.

24. Ibid., "Research in Chemistry," October 24, 1933.

25. Ibid., "Officer's conference," December 18, 1933.

26. Ibid., cross-reference regarding Mason's letter on S. M. Gunn, December 18, 1933.

27. Ibid., Weaver to Pauling, December 19, 1933.

28. RAC, RG1.1, 205D, Box 6.73, Pauling to Weaver, September 25, 1934.

29. Ibid., Pauling to Weaver, November 26, 1934.

30. Ibid., Pauling to the Rockefeller Foundation, November 26, 1934.

31. L. Pauling, "The Oxygen Equilibrium of Hemoglobin and its Structural Interpretation," *Proceedings of the National Academy of Science USA, 21* (1935), pp. 186–191. Also RAC, RG1.1, 205D, Box 6.75, Report of H. M. Miller, September 25–27, 1935. L. Pauling and C. D. Coryell, "The Magnetic Properties and Structure of the Hemochromogens and Related Substances," and "The Magnetic Properties and Structure of Hemoglobin, Oxyhemoglobin, and Carbonmonoxy Hemoglobin," *Proceedings of the National Academy of Sciences USA, 22* (1936), pp. 159–163 and 210–216, respectively.

32. L. Pauling, "Fifty Years," p. 1002 (see Note 17).

33. RAC, RG1.1, 205D, Box 6.74, Report of H. M. Miller, September 25–27, 1935. A. E. Mirsky and L. Pauling, "On the Structure of Native, Denatured, and Coagulated Proteins," *Proceedings of the National Academy of Science USA, 22* (1936), pp. 439–447; and L. Pauling, "Fifty Years," p. 1002 (see Note 17).

34. RAC, RG1.1, 205D, Box 6.74, Weaver's report, March 6, 1936.

35. Ibid., "CIT Chemistry Project," April 22, 1936.

36. OSU, Pauling Papers, A. A. Noyes correspondence file, 1926–1938; and Frederick G. Keyer, "Arthur Amos Noyes" (reprinted from *The Nucleus*, October 1936).

37. RAC, RG1.1, 205D, Box 6.76, Weaver's report, January 31, 1932.

38. Ibid., January 30, 1937.

39. Ibid., January 31, 1937.

40. Ibid.

41. RAC, RG1.1, 205D, Box 6.77, Millikan to Weaver, August 7, 1936.

42. Ibid., Pauling's "Outline of Program," August 7, 1937. On Niemann's collaboration with Bergmann, see Interlude I, this volume.

43. RAC, RG1.1, 205D, Box 6.77, Pauling's "Outline of Program," August 1937, p. 4; also L. Pauling, "The Future of the Crellin Laboratory" (dedication address), *Science*, 87 (1938), pp. 563–565.

44. CIT, Oral History, Beckman, pp. 37–43. Also Arnold Thackray and Jeffrey L. Sturchio, "The Education of an Entrepreneur: The Early Career of Arnold Beckman," and Paul F. Cranefield, "The Glass Electrode, the pH Meter, and Ion-Selective Electrodes," in Carol L. Morberg, ed., *The Beckman Symposium on Biomedical Instrumentation* (New York: Rockefeller University Press, 1986), pp. 3–26.

45. RAC, RG1.1, 205D, Box 6.77, "Outline of Program," p. 4, August 7, 1937.

46. Ibid., pp. 6–7.

47. Ibid., p. 8. See Interlude I, this volume; and Anthony Serafini, *Linus Pauling: The Man and His Science* (New York: Paragon House, 1989), Ch. 6.

48. RAC, RG1.1, 205D, Box 6.77; "Outline of Program," August 7, 1937, p. 9.

49. RAC, RG1.1, 205D, Box 6.78, "Millikan's report to Weaver," November 6, 1937, p. 2.

50. Ibid., Weaver to Millikan, November 1, 1937.

51. R. A. Millikan, "The Development of Chemistry at the California Institute of Technology" (dedication address), *Science*, 87 (1938), pp. 565–566.

52. See discussion in Chapter 3, p. 84–85, and 98, and Epilogue pp. 274–276, this volume.

53. J. D. Bernal, "The Pattern of Linus Pauling's Work in Relation to Molecular Biology," in Alexander Rich and Norman Davidson, eds., *Structural Chemistry and Molecular Biology* (San Francisco: W. H. Freeman), p. 372.

54. R. B. Corey, "The Crystal Structure of Diketopiperazine," *Journal of the American Chemical Society*, 60 (1938), pp. 1598–1604; and R. C. Olby, *The Path*, pp. 275–278 (see Note 19).

55. RAC, RG1.1, 205D, Box 6.79, Pauling to Weaver, February 23, 1938.

56. RAC, RG1.1, 205D, Box 6.81, Pauling to Weaver, April 6, 1939; R. B. Corey, "Interatomic Distances in Proteins and Related Substances," *Chemical Reviews*, 26 (1940), pp. 385–406.

57. L. Pauling and C. Niemann, "The Structure of Proteins," *Journal of the American Chemical Society*, 61 (1939), pp. 1860–1867.

58. Ibid., p. 1.

59. J. Fruton, "Early Theories of Protein Structure," *Annals of the New York Academy of Sciences*, 325 (1979), pp. 10–14.

60. RAC, RG1.1, 205D, Box 6.82, Pauling to Weaver, October 20, 1939.

61. A. Serafini, *Linus Pauling*, chapter 6 (see Note 13).

62. RAC, RG1.1, 205D, Box 6.81; Weaver to Pauling, November 1939; Weaver's Report, April 10, 1939.

CHAPTER 6

Spoils of War: Immunochemistry and Serological Genetics, 1940–1945

Terra Incognita: Shift to Immunology

Immunology was traditionally tied to medical research and clinical practice. Medically trained immunologists shared little common scientific vocabulary with geneticists or with researchers in the botanical and zoological traditions. Intellectual links between the sciences of immunity and heredity were weak during the early 1930s. Immunology therefore had not been an active interest of the Rockefeller Foundation's natural sciences division, nor did it seem to have a useful function within the medically divorced biology program at Caltech. Thus when Pauling and Sturtevant submitted an application to the Rockefeller Foundation in 1940 for support of immunology—immunochemistry and serological genetics—their plans represented a surprising departure from the molecular biology program as initially formulated at the Rockefeller Foundation and at Caltech. From 1940 until the mid-1950s, immunology (the biology and chemistry of antibodies) became a common research area for the chemistry and biology divisions. Though consistently omitted from the historical constructions of both immunology and molecular biology, this program in immunochemistry and serological genetics comprised a principal chapter in the study of the biological and chemical specificities of the "giant protein molecules," a key project within the cognitive structure of the molecular biology program.

From an institutional standpoint as well, immunology was of critical importance, benefiting Caltech immensely during the war years. Because of potential practical applications, research projects in immunology not only received lavish support from the Rockefeller Foundation but were sponsored by the federal government's Committee on Medical Research (CMR) of the Office of Scientific Research and Development (OSRD) as essential to the war effort. Ultimately, Sturtevant's "new" serological genetics and Pauling's "revolutionary" work in immunochemistry did not fulfill their promise to shed light on questions of heredity and growth. During and after the war, however, the research on the structure and biological action of antibodies provided continuity, prestige, and resources to Caltech's molecular biology program.

* * *

Pauling's ideas about the structural basis of biological specificity, which underlay the work in immunology, developed during the years 1936 to 1939 while working on hemoglobin.[1] His ideas on molecular interactions of proteins were an extension of the collaboration with A. E. Mirsky of the Rockefeller Institute, who had spent the academic year 1935–1936 at Caltech. Their joint project, a study of the structural differences between native, denatured, and coagulated proteins, had led in 1936 to a seminal publication. The authors attributed the structural and functional differences in proteins to differences in their hydrogen bonds and postulated that these weak bonds defined the three-dimensional molecular configuration of proteins, which in turn determined their biological specificities.[2] Pauling's conceptualization of protein specificity in terms of spatial folding—regardless of the ordering of its constituent amino acids—became the basis for his program in immunochemistry; the Rockefeller Institute was his social link to medical research.

According to Pauling, in May 1936, after he had given a seminar on hemoglobin at the Rockefeller Institute, Karl Landsteiner asked him to come to his laboratory to discuss immunology. Landsteiner, seeking explanations for the mechanism of antibody formation, challenged Pauling to account for that phenomenon in terms of molecular structure. A novice to immunology, Pauling failed to construct an explanation, but his curiosity was piqued. After surveying the literature and reading Landsteiner's new book, *The Specificity of Serological Reactions*, he attended to the problem seriously. During the fall of 1937, when serving as a guest lecturer at Cornell, Pauling frequently met with Landsteiner to discuss problems in immunology.[3]

The problem of antibody formation had occupied physiologists, bacteriologists, and physicians since World War I and was a topic of much debate during the 1930s. Until 1917 immunology had been firmly grounded in Paul Ehrlich's (1897) theories of antigen–antibody reaction. Ehrlich had defined an antibody as a discrete preexisting molecular entity, originating as a side chain or receptor on the cell surface. Accordingly, the antibody, by possessing a discrete chemical configuration, determined its specific interaction with a complementary configuration on the antigen molecule. Drawing on the analogous process of enzyme-substrate interaction, Ehrlich had postulated the presence of functional domains in both antigen and antibody. These chemical domains in turn allowed specific chemical interactions to take place through the formation of what he believed to be irreversible covalent bonds, the two fitting together like a "lock and key," a metaphor coined by Emil Fischer. Ehrlich's concept of antibody–antigen interaction was well grounded in the organic chemistry of the day.[4]

Although specific aspects of Ehrlich's explanation of the antibody–antigen reaction had been challenged by various researchers, the idea of preexisting antibody molecules covalently bonded to antigens was generally favored until the second decade of the century, when physical chemistry had gained wide acceptance. By the early 1930s, the application of physicomathematical methods of physical chemistry to immunology led to several influential studies, notably those of J. R. Marrack in England and M. Heidelberger at Columbia University (formerly at the Rockefeller Institute). These researches strongly suggested that weak physical interactions such as Van der Waals forces and ionic attractions, not covalent bonds, effected the combination of antigens with antibodies in various proportions, permitting the formation of a "lattice," or framework.[5]

The greatest challenge to Ehrlich's theories, however, arose from Landsteiner's landmark studies on the nature of serological reactions. Beginning in 1917, Landsteiner and his collaborators at the Rockefeller Institute prepared several artificially conjugated antigens by coupling simple inorganic compounds (called haptens) to protein carriers and injecting these artificial antigens into animals. Under normal physiological conditions the organism could have never encountered these synthetic molecules. Yet, as Landsteiner showed, the animals' antisera contained antibodies against these synthetic antigens. Clearly, so it seemed, the specific antibodies to these nonphysiological substances could not have preexisted in the cell surface, as Ehrlich had postulated. Instead, there had to be a chemical mechanism for the de novo synthesis of antibodies, a direct response to an injected antigen. It was that physicochemical mechanism that Landsteiner sought to explain when he approached Pauling.[6] Landsteiner's work on conjugated haptens and Heidelberger's lattice theory supplied Pauling with a conceptual framework for his new research in immunochemistry. The Rockefeller Institute linked him to the medical establishment.

Given Landsteiner's authority and the weight of the experimental evidence, the construction of a molecular mechanism for the de novo genesis of antibodies appeared to be a promising project. However, Landsteiner studied antibodies in isolation from their cellular milieu, focusing primarily on chemical mechanisms, an approach reflecting the Institute's emphasis on the physicochemical aspects of the life phenomena. From the start, Landsteiner's conceptualization of the problem ignored the biological complexities inherent in the production of antibodies, questions that had occupied physicians and bacteriologists for decades. It glossed over such unexplained phenomena as the persistence of antibody production in the apparent absence of antigen and failed to account for an enhanced response on a second exposure to the same antigen. No attention was given to the biological riddle of how the organism distinguished self from nonself in terms of antigenic response.

In addition to the investigations in immunochemistry, various researches had emerged during the 1930s on gene-linked antigens in blood. Serological studies of animal systematics demonstrated a direct relation between the formation of antibodies and heritable genetic "markers"; serological genetics also held promise for eugenic manipulation of heterozygous traits. The advent during the late 1930s of the analytical ultracentrifuge and the Tiselius electrophoresis apparatus generated new projects for research on antibodies and other giant protein molecules. A cooperative venture in immunology between chemistry and biology could lay the groundwork for a major program in biomedical research at Caltech funded by the Rockefeller Foundation. Pauling's playful curiosity and casual collaboration mushroomed into an ambitious plan.

By the end of 1938, when the Crellin Laboratory (of bioorganic chemistry) opened—attached to the adjacent new Kerckhoff biology building—immunology already occupied a prominent place. Carl Niemann, who had just joined Pauling's group, devoted the introductory essay in Caltech's bulletin to the recent advances in immunology, pointing to the key role that serological reactions played in the immunization process. In view of the recent trends and Niemann's promising project, there was even discussion of bringing Landsteiner (and possibly P. A. T. Levene) to Pasadena after his retirement at the Rockefeller Institute.[7] Pauling's enthusiasm for immunology soon spilled over to the biology group even though they had had no prior interest in the subject. Pauling's international reputation and his growing influence at the Rockefeller Institute and the Foundation lent authority to the new venture, promising large grants to the biology division.

Indeed, Pauling's aggressive management and scientific territoriality went virtually unchallenged by the floundering division of biology. By 1940 the most productive and dynamic researchers in genetics were gone. Calvin Bridges had died in 1938 of a heart attack, Jack Schultz had left for Philadelphia, and Theodosius Dobzhansky, after 11 years punctuated by feuds with Sturtevant, had accepted an offer from Columbia University. Rockefeller Fellow Max Delbrück defaulted to a physics position at Vanderbilt University. According to Morgan, Caltech's biology division had no resources for creating a new position for this researcher who had started the new field of virus genetics and who had been the primary liaison between the biological and physical sciences at Caltech. Pauling, who generally did not show high regard for the biology division, singled out Delbrück as the one person who could grasp the problems in biology with the tools of mathematical physics. In fact, a paper published jointly by Pauling and Delbrück in 1940 was one of a handful of collaborative publications between the biology and chemistry divisions produced over two decades.[8]

The loss of George Beadle to Stanford dealt a major blow to the biology division. After spending five years (1931–1936) at Caltech, Beadle had broad research experience in the classical genetics of corn and *Drosophila* and since 1935 had been exploring the links between mutations and biochemical pathways. Beadle was an exemplary product of Caltech's program. His cooperative style, wide network of institutional connections, and effective management of research projects exemplified Caltech's ideal of scientific leadership. Having been groomed as successor to Morgan, an all-out effort was made in 1940 to lure Beadle to Caltech. "As you probably have heard," Morgan explained to Beadle, "Dobzhansky has accepted a call to Columbia and this gives us a chance to carry out a plan that we have had in mind for some time—viz. to try to get you to come back. Sturtevant is writing to you to use whatever persuasion is in him to welcome you to our group." The offer consisted of a full professorship at a salary of $5000 (a competitive salary for that rank) and technical assistance. "You can be assured," Morgan stressed, "that we are ready to do anything within our power to make the offer attractive. . . . I should like to add my personal plea to you and Mrs. Beadle to come home."[9]

The offer did not persuade Beadle. Having gathered at Stanford a dynamic research team including a few Caltech graduates, Beadle had just embarked on a project of biochemical genetics of *Neurospora*, studies that would later win him the Nobel Prize. In all likelihood he declined the offer for several reasons. According to statements made in 1945, the Caltech biology division was fragmented, its cooperative spirit stifled by internal conflicts; and a pernicious trend toward applied research emerged as the division increasingly assumed a service role to California's agribusiness. Beadle was well informed about the Caltech situation through his close contact with Sterling Emerson. Possibly the bitter feud between Sturtevant and Dobzhansky discouraged Beadle from entering into a potential rivalry with Sturtevant over the division's leadership.

Based on seniority, Sturtevant was rightly considered to be Morgan's successor. He had accompanied Morgan to Caltech 12 years earlier and assisted him in building the division. A dignified and reserved southerner, Sturtevant, though deeply involved in departmental affairs, preferred the seclusion of the laboratory. He was at a high plateau of productivity even during the 1930s. Although his scientific conservatism did not lend itself to new ventures in physiological genetics or to experimenting with new biological systems such as microorganisms, he did follow closely the new developments in molecular genetics and often offered valuable insights to others—to Beadle, Schultz, Ephrussi, and Emerson.[10]

Sturtevant, however, did not "think big." He had no special talent for administration, no leadership skills. His austere manner and dry personality did not readily attract collaborators. He possessed neither Morgan's academic clout nor his political connections, nor did he share his mentor's aspirations for building Caltech's biology division to world prominence. More importantly, he did not know how, or perhaps did not care, to cultivate relationships with officers and presidents of foundations and did not seem to appreciate the complex web of corporate science. Certainly, Sturtevant was no equal partner to a strong-willed manager such as Pauling. Yet it was only proper that after Morgan's retirement Sturtevant should be offered the chairmanship of the division's council, an offer Sturtevant did not wholeheartedly desire but also did not refuse.[11]

With Pauling's aggressive promotion of immunology as a cooperative venture, Sturtevant willy-nilly began exploring new research avenues by applying serological techniques to *Drosophila* genetics. In 1939, under the guidance of R. C. Lancefield of the Rockefeller Institute, Sturtevant began studying the antisera of rabbits injected with various *Drosophila* mutants. Emerson began collaborating with Roscoe R. Hyde of the Johns Hopkins School of Hygiene and Public Health, attempting to raise rabbit antisera to plants carrying specific combinations of self-sterility genes. Based on differences in the specificities of serological reaction, these sera were tested against antisera extracts from pollen carrying the same or different allelomorphs.[12]

The new immunochemical methods also seduced the embryologists. Albert Tyler, Morgan's protégé, had been working at Corona Del Mar on classical problems in embryology since the 1920s: fertilization and development of marine invertebrates. He now joined the new physicochemical bandwagon and began viewing

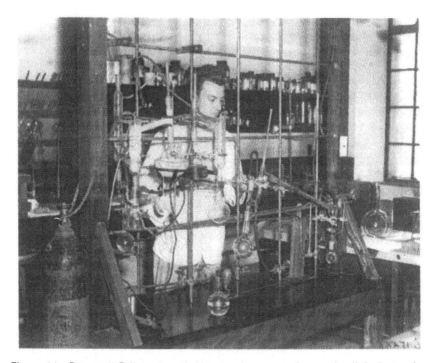

Figure 14 Research Fellow Joseph Nye conducts experiments in clinical genetics. Courtesy of the Rockefeller Archive Center.

the old questions of fertilization and embryonic development through the new lens of serological reactions. Within a mere few months Tyler converted from traditional experimental embryology to the new "chemical embryology," adopting an entirely new scientific vocabulary. In his 1939 publication "Extraction of an Egg Membrane-Lysine from Sperm of the Giant Keyhole Limpet," a study conducted in the traditional style, Tyler never referred to serological reactions or to agglutinations occurring during fertilization, nor did he offer explanations in terms of antibody reactions.[13] Although two decades earlier Frank R. Lillie had called attention to parallels between specificities of serological reactions and fertilization, the analogy went largely unappreciated, primarily due to Jacques Loeb's mechanistic impact on embryology. Tyler had not referred to Lillie's work earlier. In 1940, however, Tyler's follow-up paper "Agglutination of Sea Urchin Eggs by Means of a Substance Extracted from the Egg," prepared in consultation with Pauling, was well seasoned with new chemical terminology and concepts from immunochemistry.[14]

By adopting a biochemical stand, the paper also unwittingly incorporated some of the dominant themes of the Rockefeller Institute contingent, Pauling's primary source of influence. Thus when explaining the agglutination process in fertilization, Tyler, evidently swayed by Pauling's interpretations, adopted the protein paradigm of Mirsky, Northrop, Landsteiner, and Heidelberger. Tyler concluded that the specificities involved in the serological reactions of fertilization were analogous to the autocatalytic reactions of crystalline enzymes. Echoing the views of Northrop, the leading proponent of the autocatalytic theory of replication, Tyler went on to compare the specificities of agglutination reactions to the mechanisms of bacteriophage replication. Applying Heidelberger's and Marrack's framework theory, Tyler concluded that all cells were composed of alternate layers of substances that were capable of reacting with one another in a serological, or chemically specific, manner.

The products of Tyler's swift conversion were ephemeral. Under Pauling's influence, Tyler's new chemical approach was largely window dressing, borrowing disjointed concepts that did not serve as an experimental framework. This conversion unfortunately seemed to confirm Simon Flexner's argument that Weaver's program might tempt scientists into areas of research mostly out of desire for grants and that new tools do not constitute a new biology.[15] The newly funded physicochemical embryology at Caltech combined the desire for grants with the lure of scientific fashion; in its new chemical garb embryology fit well within the molecular biology program.

Despite the flaws in Tyler's work, his ideas stimulated Sturtevant to explore the chemistry of mutations in a novel way. While Tyler was constructing immunochemical mechanisms of fertilization, Sturtevant and his genetics group designed experiments in serological genetics. Tyler's interpretations, in fact, suggested to Sturtevant parallel mechanisms for gene action. If the autocatalytic reactions of gene replication and the heterocatalytic reactions of gene mutations involved protein specificities analogous to fertilization and serological reactions, perhaps controlled serological manipulations could effect protein mutations. In his 1940 paper "Can Specific Mutations Be Induced by Serological Methods?" Sturtevant combined Tyler's findings with earlier researches in immunological genetics and protein chemistry. According to Sturtevant, J. B. S. Haldane and M. Irwin had demonstrated during the 1930s that there was a one-to-one correspondence between the presence of specific single genes and their specific antigens: For instance, erythrocyte antigens of birds and mammals were direct gene products. Based on Landsteiner's work, it seemed plausible that antigens could induce the direct genesis of antibodies and then, because of their specific chemical configuration, react serologically with those antibodies. "These considerations led to the supposition," wrote Sturtevant, "that if a particular gene is responsible for the formation of a given antigen, then there is a possibility that the antibodies induced by this antigen may react with the gene. If these possibilities exist, there is a series of consequences that are of interest to the geneticist."[16] Later, with Sturtevant's influence and Beadle's recommendation, Emerson would embark on exploring the experimental consequences of these ideas by attempting to study antibody-induced mutations in *Neurospora*.

The preliminary tries of Sturtevant, Emerson, and Tyler sufficed to impress the officers of the Rockefeller Foundation. When the first grant application for immunology was submitted in 1940, the officers judged the findings to be promising. Repeating the claims of several prominent scientists, the officers argued that the kind of specificities studied by immunologists and geneticists were so similar that shared materials and methods were sure to yield valuable results. Caught in the web of their own directive strategies and advisory network, the officers triumphantly concluded that "it seems likely that in this field lies the best hope of attacking the general problem of gene action."[17] Morgan, cheering from the sidelines, endorsed the new venture, a trend reinforced with a grant of $12,000 for equipment, animals, and chemicals.

These excursions by the biology staff into the *terra incognita* of immunology, however, could hardly have won the confidence and support of the Rockefeller Foundation were it not for Pauling's intellectual clout and institutional power. In addition to imposing his ideas on the floundering division of biology, Pauling initiated his own venture in immunology, a program born largely out of his 1940 article, "A Theory of the Structure and Process of Formation of Antibodies," a paper perceived by leading immunologists and chemists at the time as a classic.[18] A pure thought piece, the paper integrated Landsteiner's and Heidelberger's works with the theoretical insights of the Pauling–Delbrück article.[19] Together these ideas formed the conceptual foundation for the immunology program at Caltech, a framework that would dominate research in molecular immunology for more than 15 years. The cognitive promise was amplified by potential commercial applications (controversial to be sure), forecasting a revolution in the practice of biology and medicine.

Pauling seemed to be fully aware of the future and immediate consequences of the new project. A program of fundamental research in immunology would enable Caltech to forge strong links with California's medical establishment, broadening the service role of molecular biology and expanding its resource base in accordance with Noyes's vision. A medical research program would also receive a priority status during the war emergency.

Problem of Antibody Formation

The United States entered World War II in December 1941, but the preparedness machinery had been set in motion nearly two years earlier. When the question of science mobilization for war resurfaced during the spring of 1940, Frank Jewett, President of the National Academy of Sciences (1939–1947), native of Pasadena and graduate of Caltech, argued that the Academy lacked the authority and power to mobilize science quickly and effectively. A vigorous campaign launched by leaders of America's scientific establishment, Frank Jewett, Vannevar Bush, Karl Compton, and James Conant led, during the summer of 1940, to an executive order by President F. D. Roosevelt to establish a National Defense Research Committee (NDRC). Its purpose was to contract with educational institutions, scientific organizations, individuals, and industry for the purpose of coordinating scientific research on the problems underlying the development, production, and deployment of war devices. A second executive order culminated the following summer in the creation of the Office of Scientific Research and Development (OSRD) under Bush's directorship, endowing it with resources and power beyond any previous coalition of science, industry, and the military.[20]

Four elite universities—MIT, Caltech, Columbia, and Harvard (in that order)—spearheaded the war mobilization. Just as in peacetime, these institutions received the lion's share of grants and contracts. Their leaders increasingly divided their time between the nation's capital and their academic home constituencies, setting a pattern for the management of postwar "big science." Caltech's leading physical chemist R. C. Tolman, an active organizer of the NDRC, "just packed up and moved to Washington to be at the center of things."[21]

In October 1940, following a meeting in Washington, DC, Pauling embarked on several war projects that would eventually strengthen Caltech's position as a primary research and development center of chemical warfare. It was through his initiation into war mobilization that Pauling became aware of the strategic place of medical research and of the potential importance of immunology in the war effort. The Committee on Medical Research (CMR) was just assembled under the direction of A. N. Richards, a leading pharmacologist at the University of Pennsylvania School of Medicine and an influential consultant for Merck and Company. The war projects, coordinated between the government, universities, and pharmaceutical companies, encompassed research on malaria; infectious, venereal, and tropical diseases; convalescence; neuropsychiatry; various aspects of surgery; and aviation medicine. Medical chemistry work focused on the biochemistry of adrenal and cortical hormones and the production of penicillin; physiology projects included nutrition, acclimatization, water sterilization, shock physiology, the development of blood substitutes and agents for boosting resistance to disease (drugs and vaccines), and projects related to biological warfare. Pauling's casual excursions in immunology turned into a vigorous war project. Timing and content constituted in effect a protective strategy for molecular biology during a period when basic research on proteins was in danger of displacement.[22]

* * *

The concept of antibody synthesis was broached as an element of a broader theoretical problem. In July 1940 Pauling, together with Max Delbrück, published in *Science* the article "The Nature of the Intermolecular Forces Operative in Biological Processes," introducing ideas that linked the process of antibody formation to enzyme synthesis, virus replication, and gene action.[23] Later Pauling would attach a great deal of weight to the paper for its supposed prophetic insights. However, Delbrück, for good reasons, downplayed his own role, limiting the claims of his contribution to reading Pauling's draft and to offering criticisms. Delbrück presumably merely agreed to lend his name or, rather, his title of theoretical physicist to the grand vision—a unified molecular theory of biology. The entire argument of the paper was implicitly grounded in the primacy of proteins as biological determinants of heredity, growth, and cellular regulation.

The mechanisms involved in protein synthesis and the three-dimensional folding of highly complex molecules in the living cell, the authors argued, were only in part determined by covalent bonds. A major role was played by the intermolecular interactions of van der Waals attractions and repulsions, electrostatic interactions, and hydrogen bond formations. These weak physical interactions were such as to give stability to a system of two molecules with complementary structure in juxtaposition, rather than of two molecules with necessarily identical structures. Accordingly, the Pauling–Delbrück paper proclaimed complementarity to be the primary factor determining specific molecular attractions and guiding the enzymatic synthesis of molecules. This structural complementarity, of course, had nothing to do with Niels Bohr's complementarity principle, the leitmotif of Delbrück's biology program, nor did it anticipate the 1953 discovery of complementary nitrogenous bases in the DNA molecule. Rather, the Pauling–Delbrück argument extended Emil Fischer's popular "lock and key" model beyond enzyme–substrate interactions.

The physical basis for such interactions was relatively simple. Attractive forces between molecules varied inversely with a power of the distance, and maximum stability of a complex was achieved by bringing the molecules as close together as possible. In order to reach such optimal stability, the molecules had to possess complementary surfaces such as die and coin. Identity and complementarity were not mutually exclusive, however. Structurally identical surfaces could also be complementary, as in the case of autocatalysis, genetic reproduction, or growth of bacteriophage, where a protein molecule supposedly acted as a template for its own replication. The phenomenon of antibody formation, on the other hand, exemplified a case of nonidentical complementarity, where an antigen acted as a template for synthesis of a complementary but structurally different antibody.[24]

The import of these ideas was stressed in Weaver's 1939 review of the Foundation's protein-centered program, presumably to reinforce the value and progress of his project. "In their simpler forms, proteins are as obviously 'dead' as any powder which ever filled a bottle. In their most complex forms they are the chief constituent of the pulsating protoplasm which is the very stuff of life."

Recent studies validated that image. Borrowing from Pauling's and Delbrück's terminology, Weaver reported that the immense molecules

> have only recently been shown, on theoretical grounds, to exert long-range forces which seem of the sort necessary to explain the ability of one parent pattern of atoms to seek out of a mixture the necessary units and charm them into arranging themselves into duplicate offspring pattern—for it is in some such terms as these that the molecular scientist describes the biological process of reduplication or reproduction. On the other hand, the complexity of the protein molecule appears to furnish, when viewed in terms of atomic forces, a sufficiently intricate, detailed pattern to make understandable the precise specificity of protein reaction.[25]

The template hypothesis of antibody formation was fully articulated in October 1940 in "A Theory of the Structure and Process of Formation of Antibodies," published in the *Journal of the American Chemical Society*. Though remarkably creative, the argument was strictly a theoretical construction, a digest of several published researches in immunochemistry, with no new experimental procedures or data. It was revealing of Pauling's scientific philosophy. Probably best understood as Piercian pragmatism, Pauling's approach to natural phenomena centered on the premise that theories were convergent approximations, explanatory models designed to encompass diverse scientific observations. In a rare moment of self-reflection, Pauling described his method through a contrast with that of Landsteiner: "I found that Landsteiner and I had a much different approach to science: Landsteiner would ask, 'What do these experimental observations force us to believe about the nature of the world?' and I would ask, 'What is the most simple, general, and intellectually satisfying picture of the world that encompasses these observations and is not incompatible with them?' "[26] This approach would guide Pauling's model-building studies of protein structure; and this statement, in fact, comprised nearly his exact words when he introduced the new theory of immunochemistry.

To be sure, Pauling did not claim originality for the idea of structural complementarity in antibody synthesis. He cited several biochemists, including Felix Haurowitz, Stuart Mudd, and Jerome Alexander, all of whom had suggested variants of these ideas. He also adopted Heidelberger's framework theory to arrive at the insightful conclusion that antibody molecules were, at least, bivalent. Beyond integrating and interpreting previous studies, Pauling's own preliminary contribution to immunochemistry consisted of proposing an elegant mechanism, accounting graphically in six steps for the process of antibody formation.[27]

Contrary to the views of several researchers, Pauling simply assumed that all antibody molecules contained the same polypeptide chains as normal globulins. Based on the protein-folding theory published by Pauling and Mirsky in 1936, Pauling concluded that antibodies differed from normal globulins only in the configuration of the chain, in the way the two end parts of the globulin polypeptide chain were coiled. These small ends, as a result of their amino acid composition and order, could assume many configurations with nearly the same stability. Under the influence of an antigen molecule they assumed configurations complementary to surface regions of the antigen, thus forming two active ends. The central portion of the chain would fold up, freeing the oppositely directed ends to attach to two antigen molecules.[28]

Pauling admitted that there was no direct evidence supporting his basic assumptions. The assumptions, he explained, were justified because they constituted the simplest and most reasonable mechanism that could account for diverse experimental data. His proposed mechanism claimed to explain the heterogeneity of sera, the bivalence of antibodies and multivalence of antigens, the ontogeny of the framework structure, the role of antigens as templates for antibodies, and the various criteria of antigenic activity. These diverse experimental observations, however, reflected a strictly chemical view of immunology. In his attempt to reconcile theory with accumulated knowledge, Pauling, like Landsteiner, paid no attention to significant biological patterns: the enhanced response on second exposure, or the persistence of antibody production in the absence of antigen. In general, he ignored the range of interactions between antibodies, cells, and the organism.[29]

Beyond its theoretical promise—apparent coherence and explanatory power—Pauling's antibody model carried revolutionary implications for physiology and medicine, suggesting that any antibody derived from serum or globulins could be manufactured in vitro. As Pauling explained, the globulin could be treated with denaturing agents sufficiently strong to cause the chain ends to uncoil; the agents then slowly removed as an antigen or hapten (synthetic antigen) was introduced into the solution in high concentration. The chain ends of the denatured globulin would then coil up to assume the three-dimensional configuration around the antigen by forming hydrogen bonds. These configurations, representing maximum stability under given reaction conditions, would be complementary to the antigen or hapten.[30]

The new technology must have seemed limited only by the imagination. Not only would it enable humans and animals to ward off deadly diseases simply with an infusion of artificial antibodies, it could endow scientists with the ability to alter the immune system. Artificial antibodies could even supply lasting interventive power when coupled with the projected gene manipulations effected through mutations induced by antibodies, as proposed in Sturtevant's 1940 paper. Potential products of immunological manipulations could be harnessed for applications in germ and biological warfare. Moreover, if the procedure for the artificial production of antibodies could be patented, a windfall of profits to Caltech and pharmaceutical companies would render Pauling's immunochemical research one of the most scientifically successful and commercially lucrative projects in history.[31]

The paper on antibody formation captured the scientific imagination. With several hundred requests for reprints, Pauling estimated, the enthusiasm exceeded by far the interest in any of his other publications. Soon after, in January 1941, Pauling submitted a detailed proposal to the Rockefeller Foundation for support of a major program in immunochemistry; the Foundation's support for Pauling's research by then amounting to nearly $200,000. Confident and ambitious, Pauling applied for a five-year grant for $100,000, an enormous investment for a new venture based largely on one theoretical paper. In addition to several research fellows, assistants, and at least one visiting professor, he also requested expensive equipment and supplies, without which, he argued, the promise of the new research would not be fulfilled.[32]

Researchers were handicapped by their inability to determine with sufficient accuracy the amounts of antigen in small precipitates. With adequate financial resources, the problem could be easily solved by utilizing the new technology of radioisotopes, made to order in the cyclotron laboratory at Berkeley. Similarly, the separation of various antibodies from heterogeneous sera had been a major obstacle for decades. With the construction of a Tiselius electrophoresis apparatus (at a cost of about $5000) the difficulty could be readily overcome. Different kinds of antibodies complexed with charged haptens could be separated under the influence of an electrical force. These tools, Pauling pointed out, would also benefit the biology division.[33]

Pauling's lavish scheme prompted a consultation during January 1941 between the Foundation's officers and immunochemists Landsteiner and Heidelberger regarding the support of Pauling's proposed work. Both scientists expressed a nearly identical opinion. Pauling, they concurred, was one of the greatest chemists in the world and should definitely be encouraged to develop his program, provided he collaborated with a competent immunologist. They recommended Dan Campbell from the University of Chicago, with whom Pauling had consulted while preparing the paper on the formation of antibodies and whom he hoped to add to the staff of the Crellin Laboratory. Both Heidelberger and Landsteiner agreed that if Pauling could obtain anything significant on the in vitro reproduction of artificial antibodies, it would indeed be of revolutionary importance. As a result of the consultation,

Weaver planned a visit to Caltech the following month for meetings and on-site assessments.[34]

Landsteiner or Heidelberger may have informed Pauling about the review, or Weaver himself could have intimated his reservations regarding the grand scope of the new venture. Either way, a week later Pauling apologized for his excessive demands, admitting to Weaver that he got somewhat carried away by his enthusiasm for the new field and agreed that it would be wise to set up the project on a smaller scale and attack the problem more slowly. Indeed, during Weaver's visit to Caltech, the two agreed that a grant of $11,000 per year, over two or three years, would be adequate. It would cover the salaries of new investigators, including that of Campbell, and pay for radioisotope counters, a Tiselius apparatus, and for the expert technical assistance around the new technologies.[35]

In May 1941 the Foundation appropriated $33,000 for 1941–1944 to develop immunochemistry at Caltech under Pauling in conjunction with serological genetics

Figure 15   Dan H. Campbell and Linus Pauling, 1943. Courtesy of the Rockefeller Archive Center.

in the biology division, thus initiating a long-term commitment to what would turn out to be a controversial scientific project. In justifying their support for the new venture, the officers recounted the progress in immunology in relation to biochemistry and genetics, achieved, they pointed out, despite the absence of any clue into the chemical nature of antibodies. Pauling's new theory promised to fill that gap. "It is the opinion of several leading workers in this field that positive results in this direction will make possible revolutionary developments in immunology."[36]

By coincidence of timing, British biochemist N. W. Pirie had a chance to comment to Weaver on Pauling's plans in immunochemistry. Pauling's suggestions were stimulating, indeed, Pirie thought. Though, with obvious distaste for Pauling's flamboyance, he cautioned "that the whole situation will be excellent, just providing Pauling does not 'Wrinch' it. That is, he hopes that Pauling will not pile hypothesis on hypothesis, and will not insist on speaking of this hypothesis on every conceivable occasion, but will now quietly await experimental evidence."[37] These charges were nearly the same as those Pauling had leveled against Dorothy Wrinch in 1939, criticizing her manner of promoting the cyclol theory. Persuasion and promotion were central to Pauling's modus operandi, however. By summer 1941, the new program at the Institute was in full swing, and Pauling quietly applied for a patent on the artificial manufacture of antibodies.[38]

Science at War

The "preparedness" agenda intensified while Pauling's immunochemistry program gathered momentum. During the summer of 1941 Caltech escalated its military activities, devoting a large part of its personnel and facilities to the war effort. To the wide variety of OSRD contracts for research and development of warfare devices the institute added a number of special instructional programs for members of the armed forces. More than 2000 students—army meteorology cadets, navy V-12 engineers, and aeronautics officers—enrolled in these programs between 1941 and 1946, a period of great economic boom for southern California. Aircraft and ship-building industries mushroomed, and large military bases were established. The unprecedented economic growth was a boon to Caltech; in partnership with the new industries and the military, the Institute shared in the spoils of war.[39]

The intensification of war-related projects confronted the Rockefeller Foundation with new decisions: How should Rockefeller Foundation policies be modified in response to war exigencies? The Foundation could certainly not conduct business as usual. If the officers continued to bring to the trustees "nothing but long-range, 'pure' research items," Weaver predicted that "the trustees will think that we are unrealistic, impractical, and that we are living in an academic ivory tower, unaware of what is happening to the world." On the other hand, hardliners such as Rockefeller trustee John Foster Dulles warned that the Foundation would not be fooled into doing ad hoc business. Without relinquishing the normal enthusiasm for long-range activities, Weaver's conviction "that the RF cannot possibly afford to disregard some of important emergency opportunities which will appear," reflected the Foundations's eventual course of action. The support for Ernest O. Lawrence for the construction of the Berkeley cyclotron and the production of fissionable materials, represented the most significant of these opportunities.[40]

Special problems arose around annual appropriations on long-term grants awarded before the war; obviously adjustments were needed in response to the shifting priories in research. As part of a general inquiry, the Foundation thus requested from Pauling in January 1942 information concerning the impact, or projected impact, of the war on the program of bioorganic chemistry at Caltech, the availability of personnel or the acquisition of materials and equipment. The Foundation stated its preference for maintaining those basic and long-range studies that could be sustained on a high level without conflicting with the demands of defense. However, the officers did not wish to see the quality of research compromised by war demands or time constraints. Where and when disruption did occur, they thought that it might prove necessary to reduce the level of support or even terminate it.[41] Pauling reported that the bioorganic research program had only marginally been affected by the war. The principal effect so far was an unusually large turnover in personnel, especially in structural and physical chemistry. Although Corey and his protein group had been assigned to war work, Pauling stated that he did not anticipate major interruptions. The work in immunochemistry was not affected at all, and research was progressing well.[42]

A month later, February 1942, Pauling announced to the Rockefeller Foundation the sensational news: They had succeeded in making antibodies to *Pneumococcus* polysaccharides in vitro for the first time in history. The excitement over the practical applications now seemed justified, and Pauling approached A. N. Richards, hoping that CMR would accelerate and expand the immunochemistry program. Pending review by immunology experts, Richards recommended substantial funds for Pauling's work. Because of the exceptional nature of the research, Pauling also convinced the Rockefeller Foundation to increase the level of support for immunology in order to extend his experiments to toxins, viruses, bacteria, and other antigens. A grant of $20,000 for the year 1942–1943 was approved immediately; and if the promises of the new technique were borne out that year, the officers would recommend additional support.[43] The Foundation liked to point to cases where fundamental research revolutionized medical practice.

Commercial interests soon followed. Lederle Laboratories expressed an interest in collaborating with Pauling to develop the new research technology, a potentially complicated situation for Rockefeller grantees. Pauling requested advice from the Foundation, and its officers discouraged him from accepting aid from commercial firms, explaining that such ties might later place him under unnecessary obligations. The natural sciences division had no set policy regarding commercial exploitation of research, but the medical sciences division was strongly opposed to patents. Because Pauling was moving into biomedical research and the work now spilled into medical fields, it was best, the Foundation suggested, to avoid commercial entanglements. Pauling had to confess to having already applied for a patent a year earlier but expressed his willingness to follow the Foundation's advice in the future regarding involvements with commercial houses. He also cautioned that despite the promise of medical applications, the manufacture of artificial antibodies was still at a preliminary research stage.[44]

Having secured commercial interests and Foundation support, Pauling advanced to the next projected step: large-scale development of medical research at Caltech. As a committee member of the Hixon Fund, he was already advancing the cause of neurophysiology, and in December 1941 he proposed to the Institute's board of trustees a general plan for an "extensive cooperative attack by modern physical and chemical methods on problems of biology and medicine."[45] A description of a program of fundamental medical research followed on February 9, 1942, focusing on problems of hypertension, enzymology, immunology, and various branches of physiology. "I believe there is need for an Institute for Fundamental Research in Medicine in the West, and that the best location for it is at California Institute of Technology," Pauling argued, echoing his mentor Noyes's words. The proposed institute would be modeled mainly after the Rockefeller Institute for Medical Research, although it would be considerably smaller, entailing an endowment of $5 million to $10 million and an annual budget of $200,000 to $400,000. The endowments, Pauling hoped, would come from the Southern California community, as well as from the Rockefeller Foundation.[46]

In proposing the new plan, Pauling, like Noyes, intended to establish strong formal ties with clinical medicine and regional teaching hospitals. He envisioned a collaboration between the physics, chemistry, and biology divisions at Caltech and the University of Southern California Medical School, Huntington Memorial Hospital, Good Hope Hospital, and Los Angeles General Hospital. Senior and junior faculty of the proposed institute would come from a handful of elite medical centers but especially from the Rockefeller Institute, Pauling's intellectual and institutional model.[47]

Despite Pauling's assurances that he did not anticipate interruptions in basic research, the war apparently did prevent Caltech's administration from following up on Pauling's plan. By June 1942 most of the work in the chemistry division had some bearing on the war. Pauling was a principal investigator on 14 OSRD contracts, including the project of developing blood and serum substitutes. He served on the Committee on Medical Research, and headed several projects for NDRC, developing rocket explosives and propellants for the Navy.[48] Little time was left for basic research; even immunology was in danger of extinction. Pauling informed the Foundation that he was troubled by the recent War Production Board limitation order, threatening to cut off supplies to all nonessential projects, thus hampering the research program in immunochemistry. It was agreed that Pauling should sign a nominal contract for $1 with Richard's committee (CMR), formally classifying immunochemistry as a war research project. The Rockefeller Foundation would continue to support the program, but for priority purposes it received the sponsorship of the CMR.[49]

In March 1943 the chemistry and biology divisions presented the Rockefeller Foundation with a glowing account of the progress in immunology and with plans for future studies. Pauling reported on the extensive investigations carried out on the serological properties of simple substances and on the dynamics of precipitation reactions; new investigations on cell agglutination were on the way. The work on artificial antibodies, described in a recent publication in the *Journal of Experimental Medicine*, was proceeding vigorously, he reported. By 1943 the instrument makers of the chemistry shop had constructed their own Tiselius apparatus, a model superior to those at Uppsala and the Rockefeller Institute.[50] A sizable group attended to the new technologies and the research problems it generated. Graduate students in chemistry and biology were increasingly attracted to topics of immunology, an experience, Pauling predicted, that would certainly influence their future work.[51]

Tyler reported that the serological study of fertilization was, with minor war-related diversions, in good shape. Aspects of the work on serological reactions were also relevant to the CMR project of developing blood substitutes—problems related to serum sickness and determining the protective values of antisera. Citing Pauling's theory of antibody formation as a conceptual framework, Tyler outlined future investigations on auto-antibodies in cells and bacteria as an integral part of the program of serological genetics and immunochemistry.[52]

Emerson, now working on *Neurospora*, recounted the progress on experiments designed to test Sturtevant's hypothesis that mutations could be induced by antibodies directed against gene products. By 1943 Beadle's collaboration with E. L. Tatum at Stanford had demonstrated the genetic control of biochemical pathways in *Neurospora* and the one-to-one relation between a gene and its enzymatically regulated product. The relation between gene and enzyme was now clearer, and the *Neurospora* proved to be a more efficient system than *Drosophila*. Proposing *Neurospora* experiments in serological genetics, Emerson expected to show that

Figure 16 David Pressman (*left*) and Dan Campbell with the Tiselius apparatus. Courtesy of the Rockefeller Archive Center.

an enzyme and the particular gene protein necessary for its production were antigenically related. By showing that antibodies against one produced changes in the other, Emerson boasted a novel approach to the problem of the gene.[53]

The price tag on the research proposal for 1943–1944 added up to $29,000, far beyond the scope envisioned in the initial Rockefeller grant to Sturtevant and Pauling in 1940–1941. The officers stated with concern in May 1943 that both Caltech and the Foundation "would be faced in a year from now with this situation, that a large program at a $29,000 annual level would either be up for renewal or for liquidation. We would much prefer in entering (for us) a new field, to begin in a much more modest way."[54] The reticence was reinforced by the mounting skepticism among experts about the production of artificial antibodies.

Confessing lack of competence in immunology, the officers of the natural sciences division conducted a relatively formal "peer review" of Pauling's program and its links to serological genetics. This time the reviewers did not offer unqualified recommendations, and their assessments of Pauling's plan to manufacture artificial antibodies were a cause for concern. If it proved true that antibodies could be produced in vitro, repeated Landsteiner, indeed any amount of money would not be wasted, but he was critical of Pauling's work to date. He had tried to repeat Pauling's experiments on a small scale, with negative results. Landsteiner's general conclusion was that if he were a betting man he would think the chances less that 50–50 that Pauling had manufactured antibodies. He also believed that Pauling was working on an unnecessarily broad front. If Pauling solved the central problem of producing antibodies in vitro, it alone would suffice to open up great vistas of research for many workers and attract financial support from diverse sources. His private views of Pauling's work were even more damning. Evidently, Landsteiner was furious and complained that Peyton Rous had accepted Pauling's article for publication in the *Journal of Experimental Medicine* without seeking wider review. Landsteiner, however, was most eager to have Pauling contribute the theoretical chapter to the forthcoming edition of *The Specificity of Serological Reactions* and did not want to upset the collaboration. His colleague Elvin Kabat, on the other hand, reviewed Pauling's antibody work critically.[55]

Because Pauling had reported producing antibodies to *Pneumococcus* polysaccharides, Oswald T. Avery of the Rockefeller Institute, an expert on type specificity and biochemical properties of that bacterial system, was asked to comment on Pauling's work. Avery said he was not impressed with the attempts to protect mice with Pauling's materials. Like Landsteiner, Avery believed that the work should be supported to the extent that was necessary to demonstrate truth or falsity, cautioning, however, that the negative of Pauling's theory could never be proved (that is, one could not prove that it is impossible to produce antibodies in vitro). The Foundation officers also noted that Avery "happens to know Sturtevant, Tyler, and the biological group for whom part of the grant would be used and thinks very highly of this able group of geneticists working in the immunological field."[56]

Heidelberger, until then a strong supporter of Pauling, was now skeptical. In a phone conversation with Foundation officer H. M. Miller, Heidelberger still thought that Pauling's brilliant theory merited adequate support. Echoing Pirie's views, though, Heidelberger thought that the work required a critical and skeptical frame of mind, and Pauling was not always critical of his own work. The validity of the work rested on proper controls, but because Pauling had not published details of controlled experiments no one could repeat the work. In fact, Heidelberger said, Landsteiner, Kendall (at Goldwater Memorial Hospital), and a young man at the Lederle Laboratories had attempted to repeat the experiments, with negative results. While approving of Tyler's serological studies, Heidelberger expressed hope that Pauling was on the right track, though he had no evidence to support such hopes. He considered Campbell (who had come from Chicago to work with Pauling) "the weak spot in the situation."[57] Only Campbell's Chicago mentor W. H. Taliaferro judged Pauling's experiments as "beautiful work" and urged the Foundation to continue its support of the project.[58]

Given the reviews, the Foundation informed Pauling during the summer of 1943 that support for the following year would be reduced. In addition to the $11,000 already available under the long-term grant, they would appropriate only a limited amount (based on Pauling's estimate) necessary to secure a swift and clear decision on the central point of artificial antibodies. "We would do this with the hope," the officers stated, "that another year of work might result in the publication of experimental evidence in such clear and detailed form that experts could evaluate the results precisely, and that the technical procedures in their smallest details should be so described that the experiments could be repeated exactly elsewhere, should anyone desire to do so."[59]

The decision, the officers explained, was based on the unanimous judgment of advisers, who concurred on the centrality of the question of artificial antibodies. It was critical that this point be resolved unambiguously, and it should require only a modest financial setup. Should this point be positively settled, the officers argued, support would be forthcoming from many sources, and Pauling's program could expand considerably without sole dependence on Foundation support. This process, the officers pointed out, would be directly in line with the best Foundation principles, namely, taking a gamble during the initial stages of an unusually promising situation and developing it to the point where it commanded other forms of support.[60] "Taking a gamble," of course, is a relative term. Given the modus operandi of the Foundation, the temperament of the trustees, and their conservative management, research investments were seldom risky. The funding of projects and individuals was based strictly on recommendations from well-established scientific advisers. The Foundation seldom gambled; even the initial stages of new projects were prudently assessed.[61]

Pauling was displeased. His rebuttal struck at the most vulnerable areas of the Foundation's policy: support of pure versus applied science, exploitation of discoveries and commercial profits, and criteria for assessing scientific merit. The "advancement of knowledge," or the relation between fundamental research and its applications, hung in delicate balance. The precedent set by the Laura Spelmann Memorial, its philosophy of research as a means to an end, still informed the Foundation's policy. The trustees monitored their investments to ensure that projects would not be determined purely by academic consideration, and officers were under pressure to demonstrate the potential utility of knowledge. Although Weaver constantly promoted the Foundation's commitment to fundamental research, the presentation of research projects at board meetings fared better if they were associated with tangible returns.[62]

The issue of commercial profit was ambiguous, however. Weaver's molecular biology program was seen as a coordinated effort between the divisions of medical sciences and social sciences. Whereas medical research lent itself to commercial products, especially medical instruments and drugs, the products of biological research were indirect and of limited commercial potential. The natural sciences division of the Rockefeller Foundation could therefore ignore questions of profits, except in situations when the interests of the medical and natural sciences intersected, as in Pauling's immunology project.

The assessment of scientific merit was especially complicated when several disciplines converged in one project, creating tangles in the network of experts who could review such interdisciplinary work. Weaver, a physicist with little preparation in biology, had to rely heavily on a handful of scientific advisers, a dependence that made him vulnerable when the advisory opinion was challenged. Recalling the Foundation's negative reaction to potential commercial ties with Lederle Laboratories, Pauling sarcastically inquired whether the Foundation had now changed its tune when it pointed to "other sources of support" for artificial antibodies. He reminded the officers that it was the Foundation's professed commitment to fundamental research and his own interest in the theoretical aspects of immunology that had caused him to turn down the financial aid from the CMR and from commercial houses.[63]

The possibility of producing artificial antibodies, as expressed in the three final sentences of his 1940 paper, was a minor point in the immunochemistry program, Pauling argued. The other experiments, completed and in progress, were of much greater scientific interest because of their theoretical import. The program, Pauling stressed, was more than merely an effort to manufacture antibodies, more than even fundamental research in immunochemistry; it was an effort to introduce a new point of view into immunology. Judging by the enthusiasm of the followers, nearly 20 men, Pauling judged the program as successful.[64]

Pauling was indignant. He strongly disagreed with the advisers who claimed that a study of the manufacture of artificial antibodies should require only a modest financial setup. That might be true for big pharmaceutical houses, which were well equipped for such work, he countered, but at Caltech it was a costly, complicated process. Given wartime constraints on expansion and the difficulty of obtaining assistance, and in view of the letter from the Rockefeller Foundation, he would be willing to plan a reduced program corresponding to an added sum of $8000. The officers approved the request, stipulating that Pauling determine as quickly and clearly as possible whether artificial antibodies could be produced in vitro; their displeasure was evident from the tone of their lengthy response to Pauling's rebuttal.[65]

A few days later Pauling retreated. He explained that although from the point of view of pure science the other experiments in immunochemistry were more important, he could appreciate the significance of artificial antibodies in prevention and control of disease. Seen from such a vantage point, Pauling reported his renewed enthusiasm for the project, ranking it as the most important part of the

Figure 17 Linus Pauling in the laboratory. Courtesy of the Rockefeller Archive Center.

program. He conveyed disappointment with his ambiguous results, expressed confidence that clear answers would be forthcoming soon, and thanked the Foundation for its support. Whether a strategy of appeasement or an expression of sincerity, the response was a wise move to protect his long-term relations with the Foundation.[66]

Several researchers tested Pauling's theory during the following year, with negative results. However, the argument that negative results did not necessarily invalidate the theory, coupled with Pauling's clout and the importance of the project, did not substantially reduce the support for Pauling's program. By May 1944 Rockefeller support for immunology had amounted to $78,000. Despite the failure to produce artificial antibodies and the doubt that the failure might have cast upon his project, Pauling impressed Millikan with his progress and prospects. He was confident, he told the aging patriarch, that the immunochemistry program would continue to provide first-rate contributions to pure science, as well as practical results in medicine. Perceiving himself as the chief architect of Caltech's future in molecular biology and the custodian of Noyes's dreams, Pauling expressed his hope that "in the not too distant future the California Institute of Technology will extend the field of its research activities to cover not only the fundamental sciences of physics, chemistry, and biology, but also human physiology and subclinical medicine, with special attention to the application to these of new concepts and techniques in the fundamental sciences."[67]

Pauling's visions of developing molecular medicine alongside molecular biology would remain a life-long mission.

Terra Firma: 1944–1945

Pauling's stature and the Foundation's implicit confidence in his leadership served the biology division well. Pauling told Millikan that if it were not for the war a major joint program with biology, centered around immunology, would be developed under Pauling's leadership. Indeed, by linking their projects to Pauling's program, the biology division under Sturtevant received Foundation support for Tyler and Emerson throughout the war period. Though only a fraction of Pauling's grant, the support for these projects kept biology afloat. Emerson also benefited from another Rockefeller adviser, his friend at Stanford, G. W. Beadle. When asked by Sturtevant to recommend Emerson's proposal (to induce mutations in *Neurospora* with antibodies) to the Foundation, Beadle judged the theory to be a reasonable one. He too pointed out that there were plenty of possibilities of obtaining negative results, even if the theory were right. It was somewhat of a gamble in Beadle's opinion, but it was worth funding because if it did work it would "without a question be a find that will open up an entirely new approach to both genetics and immunology and accordingly should certainly be looked into."[68]

The molecular biology program in the biology division was now subsisting on the intellectual energy and managerial feats of Pauling and the advisory clout of Beadle. Beadle and his *Neurospora* research group at Stanford, in a sense, formed biology's leadership in exile. Surrounded by Caltech graduates, and closely informed on the situation in the biology division, Beadle was conspicuous by his absence. His scientific ideas were incorporated into Emerson's work and indirectly influenced the work of Sturtevant's group. Beadle, in contrast to Sturtevant, was politically savvy; the relations he cultivated with Weaver and the Rockefeller Foundation would soon propel him and Caltech to the biological vanguard.

Pauling, and indirectly Beadle, imparted the biology division with a semblance of programmatic coherence. In the absence of strong leadership in genetics, biochemistry, biophysics, and animal physiology, and with constant tension between foreign and American scientists in the division, hopes of formulating broad departmental goals slowly vanished. The power now rested mainly with the plant physiologists, especially the Utrecht contingent, led aggressively by Arie Haagen-Smit and Frits Went. Their projects flourished through the powerful links with California's agribusiness; commercial plant genetics and applied botany seem to have the most tangible assets of the division during the war years.[69]

Those who maintained a serious commitment to Noyes's institutional philosophy—the priority of pure over applied research—looked askance at the publications cascading from the Kerckhoff Laboratory devoted to commercial interests: flavor control in canned pineapple, sugar content in produce, the morphology of tomatoes.[70] Shady stories reached Max Delbrück at Vanderbilt in 1945 through his friend E. Buchman from the chemistry division, who reported that the biology situation was "going from bad to worse." The "old boys" who lived off commercial research promised to mend their ways, but the scene was not likely to improve. "The really disquieting thing," Buchman feared, "is the character of new appointments in Biology. Due to the system of 'horse trading' now in force, the head of each group has been pushing through his own appointments without check and of course, he has taken care to put his friends in regardless of caliber."[71]

The war effort intensified the sense of inferiority in the biology division. Though war-related projects infused some purpose and support to the fragmented biology division, it did not help to elevate the low morale. True, Caltech's *Bulletin* recounted biologists' involvement in various war activities. In addition to Tyler's work on blood and serum, Henry Borsook led a crusade for nutrition, publishing extensively on the nutritional status of aircraft workers in southern California and the effects of vitamin supplements on absenteeism and personnel rating. There were also publications on the effects of sugar and benzedrine on fatigue in paratroopers; investigations of motion and sea sickness and of night vision; and studies of the impact of stress (prolonged wakefulness, for instance) on physical efficiency. If one adds to these published studies classified secret projects, such as work on biological warfare, the biologists' participation appears substantial, demonstrating that they applied their skills in physiology, biochemistry, nutrition, and immunology to the war effort.[72]

These activities received little attention, however, compared with the physical sciences. Biologists at Caltech, as elsewhere, went generally unappreciated for their patriotic science, nor would they later share the honors of their colleagues in physics and chemistry. Watching from the sidelines, biologists had followed the phenomenal growth of academic and industrial chemistry during the 1920s and 1930s, World War I, the "chemists' war" serving as a springboard for a permanent campaign for "better living through chemistry." They were now witnessing the "physicists' war," recognized through the growing prestige of physicists and the government's commitment to physics and engineering. Despite vigorous participation in the war effort, biologists could claim no war of their own; their contributions went uncelebrated.[73]

The deteriorated biology division, with a staff of 32, was a structural flaw in Pauling's blueprint for postwar molecular biology. His grandiose plan was predicated on chemistry's partnership with a strong biology division. His own chemistry empire of 86 scientists was thriving. More than 50 researchers worked on two dozen projects for the NDRC and OSRD, and at least two-score participated in immunochemistry research under the CMR contract. Funds were abundant, productivity peaked, and morale was high. Though the war would soon end and with it many war projects, the benefits and lessons of wartime science, Pauling reasoned, did not have to terminate. Like most scientists, Pauling had gotten used to wartime efficiency, priority considerations of projects, and massive support for research. Reluctant to relinquish the spoils of war, Pauling began to look for ways to maintain his programs in the style to which he had become accustomed.[74]

As early as August 1944, Pauling proposed to Weaver a plan for research into the structural chemistry of proteins. The main organizational features of the program, Pauling explained, would be a system of orders and reports modeled after military projects. With large numbers of trained young OSRD scientists seeking employment, Pauling expected to mount a vigorous attack on the problem of protein

Figure 18 Henry Borsook with automatic chromatographic fraction collector for studying proteins. Courtesy of the Rockefeller Archive Center.

structure, a program "designed to operate in the same efficient way" as wartime research. He envisioned 20 full-time assistants, a couple of typists to process orders and reports, technicians, and "human computers." The estimated cost of the three-year project came to $150,000.[75]

The organization of postwar science was just emerging as an issue of national debate, and Weaver had strong opinions on the matter. Although he agreed with Pauling on the desirability of a major program in protein structure, he confessed to "a good deal of skepticism as to whether it is either possible or desirable to carry over into peacetime research, many of the elements of organization and control which properly and inevitably characterize wartime work." Weaver suspected ventures that smacked of national planning as flirtations with socialism. Though the Rockefeller advisory network propagated many structural features of interwar science into the postwar era, Weaver would remain a vocal opponent of federally sponsored research well into the 1960s.[76]

Although Pauling never again wrote to Weaver about the subject of restructuring science, his organizational scheme—the efficient coordination of large projects—animated his plans for a protein-centered molecular biology. He proposed to Weaver a six- to eight-year intensive program, deploying the diverse techniques of protein analysis, arguing that this scheme would allow more effective use of funds. Any one method alone would not provide the solution to the great protein problem, he stressed. "I am enthusiastic to learn the answers to the most interesting questions posed for us by Nature," he wrote to Weaver, "and I am afraid that unless a very intensive attack is made on these problems, the answers may not be found during our lifetime."[77]

In preparation for the comprehensive plan and in view of the large investment, Weaver initiated an informal survey within Rockefeller's advisory network in protein chemistry. By soliciting opinions on Pauling's suitability for spearheading such a program and on Caltech's standing relative to other research centers, Weaver sought to reduce risks and justify his program to the Rockefeller trustees. The reviewers, including prominent biochemists J. W. Williams from the University of Wisconsin, F. O. Schmitt from MIT, and V. du Vigneaud from Cornell, agreed that the protein problem was indeed of primary importance and Pauling was the best man for the job.[78]

On September 10, 1945, a week after the Japanese surrender in Tokyo Bay, Pauling flew to New York to set the wheels in motion for his postwar program. He met with the Guggenheim Board and called the Rockefeller Foundation office to discuss the plans for physical, organic, and structural chemistry at Caltech in relation to biological and medical problems. Though Weaver was already convinced of the intellectual viability of the venture, his final commitment would be informed not only by the particular needs of the chemistry and biology divisions but by the broader academic goals and administrative policies of Caltech.[79]

Major power blocks were now shifting at Caltech, the rumblings of which were heard at the latest board meetings. During preliminary chats at lunch, Pauling intimated to Rockefeller Foundation officer H. M. Miller that administrative affairs at Caltech had been stormy recently. The aging Millikan, who for some time had been unable to provide effective leadership, had been stalling with the resignation of his title and duties. When he finally did resign as chairman of the Executive Council, speculations surfaced regarding the concomitant resignation of Max Mason. These changes led to a reconstituted Executive Council with its chairman serving ex officio as chairman of the Board of Trustees. The Board would now consist of three trustees and several faculty members: Houston, Tolman, Clark, Millikan, Pauling, and a "younger man from the humanities." New rules, by-laws, and principles were instituted, and within the next six months the Institute expected to choose a new president.[80]

Members of the biology division fared poorly within the reconstituted administrative structure and apparently were not even represented on the board. The situation in biology had degenerated since Morgan's active leadership had ceased, Pauling informed Miller. "Sturtevant had no particular liking or capacity for administration. There is an important division within the crowd between the American biologists and the foreign contingent headed by Hagen-schmidt [sic]."[81] Pauling earnestly hoped, he told Miller, that the situation would be cleaned up. He looked forward to a strong and unambiguous leadership of biology and to a new regimen that would be completely and cordially interested in close cooperation with chemistry. The situation in biology was of course worse than Pauling had described. It made little sense to expose the extent of the damage when seeking support for joint projects with biology. To improve the situation, Pauling thought that "they ought to add a first-rate man each in enzyme chemistry, physiology, pharmacology, and perhaps viruses. This enlarged group would then, together with Pauling's group on structural chemistry and immunological chemistry, have the existing strength in genetics, etc., and form what would in effect be an institute for molecular biology."[82]

As Weaver stated, the plans for molecular biology had to wait until the resolution of the administrative situation at Caltech and the formulation of fiscal policies. With Pauling's cooperation, however, he intended to monitor closely the Foundation's investments at Caltech and protect his own special interests in the Institute. As he reported on the conclusion of the New York meeting: "It is understood that Pauling, who from now on will be fully informed on all questions of central policies of the Institute, will keep us informed. If certain obvious difficulties can be removed, WW considers this one of the most important and very possibly, the most important existing opportunity in this country for highly competent and imaginative application of modern physical and chemical techniques to basic biological problems."[83]

Evidently Pauling's failure to manufacture artificial antibodies had caused him little damage. In 1945 Pauling's work was strongly tied to medical research and his group well linked to the medical establishment through the Rockefeller Institute and the CMR. Although the fundamental work on protein structure had come to a standstill during the war, the flawed research on the structure and action of antibodies propelled Caltech's program to the vanguard of life science. Caltech's molecular biology program emerged at the threshold of the new era on even firmer grounds than before, with immunology at its center as part of the protein paradigm.

* * *

Pauling's whole scheme for molecular biology was predicated on bringing George Beadle, now considered one of the world's leading biologists, back to Caltech. In fact, as he spoke with Weaver, the biology division was already negotiating an appointment with Beadle, but it would be Pauling's forceful maneuvers that would bring Beadle back as chairman of the division. By 1945 Beadle was widely known in the community of life scientists for his outstanding contributions to biological knowledge. His research linked formalistic concepts of classical genetics with material, or biochemical, explanations. His program in biochemical genetics, which he developed at Stanford with biochemist Edward L. Tatum during the war years, replaced *Drosophila* with the bread mold *Neurospora crassa*, a simple microorganism amenable to genetic investigations on the biochemical level. By utilizing the *Neurospora* system Beadle was able to solve a central problem in heredity research, a problem that had been a focus of ongoing debate since the first decade of the twentieth century. The debate centered on the relation between genes and enzymes: whether genes *were* enzymes or they only *made* enzymes. Beadle demonstrated that one gene controlled only a single biochemical reaction, which in turn was regulated by one specific enzyme.[84]

In addition to its cognitive import, Beadle's work was recognized as a principal disciplinary innovation. *Neurospora* research brought together two areas in life science that in the United States had previously been remote: genetics and biochemistry. Because American genetics had been shaped primarily by its service role to agricultural sciences—plant and animal breeding—whereas biochemistry developed mainly within a medical context, these two fields represented very different scientific traditions, with dissimilar vocabularies and laboratory training. Beadle's program forged some of the earliest links between these two disciplines in America. His election to the National Academy of Science in 1944 reflected these cognitive and disciplinary accomplishments.[85]

What made Beadle's ascent to leadership even more remarkable was the time frame. His research in *Neurospora* genetics was launched at the end of 1940, just at the height of the "preparedness period," reaching its zenith (in terms of funding and personnel) in 1943, at a time when most fundamental researches were being cut back. Although national resources were diverted to war-related projects, Beadle's program in biochemical genetics flourished—because of its practical and commercial applications to the war effort. Although Beadle's primary commitment was to fundamental knowledge, it was mainly the practical application that gave the program its priority considerations and resources. Beadle thus had proved himself not only a first-rate mind but an effective promoter of interdisciplinary cooperation and a savvy manager of university, industry, and government interests—an equal partner in Pauling's projected enterprise.

Notes

1. L. Pauling, "Fifty Years of Progress in Structural Chemistry and Molecular Biology," *Daedalus*, 99 (1970), p. 1005.

2. A. E. Mirsky and L. Pauling, "On the Structure of Native, Denatured, and Coagulated Proteins," *Proceedings of the National Academy of Sciences USA*, 22 (1936), pp. 439–447.

3. L. Pauling, "Fifty Years of Progress in Structural Chemistry," p. 1005 (see Note 1).

4. A. M. Silverstein, "History of Immunology," in W. E. Paul, ed., *Fundamentals of Immunology* (New York: Raven Press, 1984), pp. 23–40; and P. M. H. Mazumdar, "The Antigen-Antibody Reaction and the Physics and Chemistry of Life," *Bulletin of the History of Medicine*, 48 (1974), pp. 1–21.

5. M. Heidelberger, *Chemical Review*, 24 (1939), p. 323; and J. R. Marrack, *The Chemistry of Antigens and Antibodies* (London: H. M. Stationary Office, 1934), Report No. 230 of the Medical Council.

6. A. M. Silverstein, "History of Immunology," pp. 33–35 (see Note 4).

7. C. Niemann, "A Recent Advance in Immunology," *Bulletin of the California Institute of Technology*, 47 (1938), pp. 16–19. RG 12.1, Box 68, December 31, 1938, Weaver's diary, p. 194.

8. RAC, RG1.1, 205D, Box 6.79, Pauling to Weaver, February 23, 1938, pp. 2–3.

9. CIT, Biology Division Records, Box 8.10, Morgan to Beadle, February 23, 1940.

10. A rough sketch of Sturtevant's background and personality may be found in G. E. Allen, *Thomas Hunt Morgan* (Princeton: Princeton University Press, 1978), pp. 168–169. General impressions of M. Delbrück in CIT, *Oral History* interviews, as well as T. Dobzhansky's in the *Columbia Oral History Project* regarding their experiences with Sturtevant confirm and elaborate Allen's biographical sketch.

11. This assessment is based on the examination of several letters from Sturtevant to Beadle and to Morgan in 1940 (CIT, Biology Division Records, Box 8.10), on the correspondence between Sturtevant and Beadle in 1945 (CIT, Beadle Papers, Box 2.36), and on Sturtevant's communication with the Rockefeller Foundation between 1940 and 1945, (e.g., RAC, RG1.1, 205D, Box 7.94, May 27, 1943). That opinion was also expressed by Pauling and the Foundation (RAC, RG1.1, 205D, Box 7.96, September 10, 1945).

12. RAC, RG1.1, 205D, Box 7.91, Grant in Serological Genetics, June 14, 1940.

13. A. Tyler, "Extraction of an Egg Membrane-Lysin from Sperm of the Giant Keyhole Limpet (*Megathura Crenulata*)," *Proceedings of the National Academy of Sciences USA*, 25 (1939), pp. 317–323.

14. A. Tyler, "Agglutination of Sea Urchin Eggs by Means of a Substance Extracted from the Eggs," *Proceedings of the National Academy of Sciences USA*, 26 (1940), pp. 249–256. In that paper Tyler acknowledged Pauling's contribution to the manuscript. See also A. Tyler, "An Auto-Antibody Concept of Cell Structure, Growth and Differentiation," *Six Growth Symposium* (1947), pp. 7–19; and M. Heidelberger, "Immunochemical Approaches to Biological Problems," *American Naturalist, 57* (1943), pp. 193–198.

15. RAC, RG3, 915, Box 4.41, Simon Flexner to R. B. Fosdick, November 14, 1934, pp. 2–6.

16. A. H. Sturtevant, "Can Specific Mutations be Induced by Serological Methods?" *Proceedings of the National Academy of Sciences USA, 30* (1944), pp. 176–178. The paper was written in 1940 during the collaboration with Hyde but was sent for publication only after Hyde's death in 1943, unaltered, according to Sturtevant.

17. RAC, RG1.1, 205D, Box 7.91, Grant for Serological Genetics, June 14, 1940.

18. L. Pauling, "A Theory of the Structure and Process of Formation of Antibodies," *Journal of the American Chemical Society, 62* (1940), pp. 2643–2657.

19. L. Pauling and M. Delbrück, "The Nature of the Intermolecular Forces Operative in Biological Processes," *Science, 92* (1940), pp. 77–79.

20. R. G. Cochrane, "The Academy in World War II," *The National Academy of Sciences* (Washington, DC: National Academy of Sciences, 1978), pp. 382–432; G. W. Gray, *Science at War* (New York: Harper & Brothers, 1943); and James Phinney Baxter III, *Scientists Against Time* (Boston: Little, Brown, 1946).

21. Quoted in R. G. Cochrane, *The National Academy of Sciences*, p. 396 (see Note 20). Several authors have shown how wartime planning and funding, as well as the cementing of bonds between universities, the military, and industry firmly established the patterns and institutional infrastructure for postwar science. For example see Daniel Greenberg, *The Politics of Pure Science* (New York: New American Library, 1967); and David F. Noble, *Forces of Production* (New York: Alfred E. Knopf, 1984), especially Ch. 1.

22. R. J. Paradowski, "The Structural Chemistry of Linus Pauling," Ph.D. dissertation, University of Wisconsin, 1972, p. 95; Irwin Stewart, *Organizing Scientific Research for War* (Boston: Little, Brown, 1948), Ch. 7; and Barton J. Bernstein, "Origins of the U.S. Biological Warfare Program," in Susan Wright, ed., *Preventing a Biological Arms Race* (Cambridge, MA: MIT Press, 1990), Ch. 1.

23. L. Pauling and M. Delbrück, "The Nature of the Intermolecular Forces Operative in Biological Processes," *Science, 92* (1940), pp. 77–79; CIT, Oral History, M. Delbrück, p. 72.

24. L. Pauling and M. Delbrück, "The Nature of Intermolecular Forces," pp. 78–79 (see Note 23).

25. RAC, RG2, 100, Box 179.1235, General correspondence—Warren Weaver statement of review, August 28, 1939.

26. L. Pauling, "Fifty Years of Progress in Structural Chemistry," p. 1005 (see Note 1); and L. Pauling, "A Theory of the Structure and Process of Formation of Antibodies," pp. 2643–2657 (see Note 26). For relevant discussion on pragmatism, see Ian Hacking, *Representing and Intervening*, (Cambridge: Cambridge University Press, 1983), Part I.

27. L. Pauling, "A Theory of the Formation of Antibodies," pp. 2643–2657 (see Note 26).

28. Ibid., p. 2645.

29. Ibid., p. 2644.

30. Ibid., p. 2656.

31. Only general accounts have been written about biological warfare projects. See R. G. Cochrane, *The National Academy of Sciences*, pp. 408–414 (see Note 20); and a telling review by J. R. Newman, "A Bacteriologist's Account of Biological Warfare, the Underpublicized Complement of the Atomic Bomb," *Scientific American*, *180*, No. 6 (1949), pp. 56–58 of the book by T. R. Roseburry, *Peace or Pestilence* about germ warfare during World War II. See also Susan Wright, *Preventing a Biological Arms Race* (Cambridge, MA: MIT Press, 1990).

32. RAC, RG1.1, 205D, Box 7.92, Pauling to Weaver, January 2, 1941. Pauling described to Weaver the reception of the paper on March 18, 1941.

33. Ibid., January 2, 1941.

34. Ibid., H. M. Miller's report, January 17–23, 1941.

35. Ibid., Pauling to Weaver, January 30, 1941 and Weaver's report, February 10, 1941.

36. RAC, RG1.1, 205D, Box 7.91, grant for immunochemistry, May 16, 1941.

37. RAC, RG1.1, 205D, Box 7.92, Weaver's interview with Pirie, April 3, 1941.

38. RAC, RG1.1, 205D, Box 7.93, Pauling to Weaver, June 11, 1942. Pauling disclosed that he had applied for a patent only a year later in context of the debate on commercial exploitations.

39. "Historical Sketch," *Bulletin of the California Institute of Technology*, *53–56* (1945), p. 58. For the economic development of Southern California during the 1940s see W. Bean and J. J. Rawls, *California: An Interpretive History* (New York: McGraw-Hill, 1982), pp. 364–368, 396–401.

40. RAC, RG3, 915; Box 1.4, October 26, 1942 and April 2, 1943.

41. RAC, RG1.1, 205D, Box 7.85, Hanson to Pauling, January 21, 1942.

42. Ibid., Pauling to Hanson, February 6, 1942.

43. RAC, RG1.1, 205D, Box 7.93, Memorandum to Weaver, March 16, 1942.

44. Ibid., Pauling to Hanson, June 4, 1942; Hanson to Pauling, June 8, 1942; Pauling to Hanson, June 11, 1942.

45. CIT, Millikan Papers, Box 18.6, Pauling to Balch, February 9, 1942, pp. 1–3.

46. Ibid.

47. Ibid.

48. R. J. Paradowski, "The Structural Chemistry of Linus Pauling," pp. 96–97 (see Note 22).

49. RAC, RG1.1, 205D, Box 7.93, Pauling to Hanson, June 23, 1942; Hanson to Pauling, June 26, 1942.

50. RAC, RG1.1, 205D, Box 7.94, Research in Immunochemistry, March 1943. Caltech's Tiselius apparatus was described in *Science*, *97* (1943), p. 564; and more fully by S. M. Swingle, *Review of Scientific Instruments*, *18* (1947), pp. 128–132. For a history of the Tiselius apparatus and its relation to the Rockefeller Foundation see L. E. Kay, "The Technology of Knowledge and Power: The Tiselius Apparatus and the Life Sciences, 1930–1945," *History and Philosophy of the Life Sciences, 10* (1988), pp. 51–72.

51. RAC, RG1.1, 205D, Box 7.94, Report on Serological Genetics, March 1943.

52. Ibid.

53. Ibid.

54. Ibid., Hanson to Sturtevant, May 7, 1943.

55. Ibid., Hanson's interviews, April 30, 1943. For Landsteiner's private views, I am indebted to Joshua Lederberg for his letter to me, April 19, 1988.

56. Ibid.

57. Ibid; Hanson's interviews, May 3, 1943.

58. Ibid.

59. Ibid., Hanson to Pauling, May 7, 1943.

60. Ibid.

61. A similar conclusion was reached by P. Abir-Am, "The Discourse of Physical Power and Biological Knowledge in the 1930s," *Social Studies of Science*, *13* (1982), pp. 341–382.

62. For example, RAC, RG3, 900, Box 24.180, J. H. Willits to F. B. Fosdick, August 25, 1939 emphasized the broader aims of research.

63. RAC, RG1.1, 205D, Box 7.94, Pauling to Hanson, May 17, 1943.

64. Ibid.

65. Ibid., Hanson to Pauling, May 24, 1943.

66. Ibid., Pauling to Hanson, May 28, 1943.

67. RAC, RG1.1, 205D, Box 7.95, Pauling to Millikan, May 1944.

68. RAC, RG1.1, 205D, Box 7.94, Beadle to Hanson, May 25, 1943.

69. RAC, RG1.1, 205D, Box 7.96, Pauling's interview with Miller, Weaver's report, September 10, 1945.

70. Many publications of such commercial nature are listed in the *Bulletin of the California Institute of Technology, 56* (1946), pp. 4–31.

71. CIT, Delbrück Papers, Box 4.38, Buchman to Delbrück, December 8, 1945.

72. *Bulletin of the California Institute of Technology, 56* (1946), pp. 4–31.

73. G. W. Gray, *Science at War*, passim; and D. T. Rhees, "The Chemists' Crusade: The Popularization of Chemistry in America, 1907–1922," Ph.D. dissertation, University of Pennsylvania, 1986.

74. David C. Noble, *Forces of Production*, p. 11 (see Note 21) has discussed this influence of war time organization on postwar research.

75. RAC, RG1.1, 205D, Box 7.95, Pauling to Weaver, August 18, 1944.

76. Ibid., Weaver to Pauling, September 21, 1944. For the transition to postwar science and for Weaver's role, see Interlude II, this volume.

77. Ibid., Pauling to Weaver, October 5, 1944.

78. Ibid., Interviews by Hanson, October 14 to November 20, 1944.

79. RAC, RG1.1, 205D, Box 7.96, Weaver's report on H. M. Miller's meeting with Pauling, September 10, 1945.

80. Ibid.

81. Ibid.

82. Ibid.

83. Ibid.

84. For accounts of Beadle and biochemical genetics, see J. S. Fruton, *Molecules and Life* (New York: John Wiley & Sons, 1972), Ch. 3; R. C. Olby, *The Path to the Double Helix* (London: Macmillan, 1978), Ch. 2; H. F. Judson, *The Eighth Day of Creation* (New York: Simon & Schuster, 1979), Ch. 7; F. H. Portugal and J. S. Cohen, *A Century of DNA* (Cambridge: MIT Press, 1977), Ch. 8; N. H. Horowitz, "Genetics and the Synthesis of Proteins," *Annals of the New York Academy of Sciences, 325* (1979), pp. 253–266. See also Lily E. Kay, "Selling Pure Science in Wartime: The Biochemical Genetics of G. W. Beadle," *Journal of the History of Biology* (1989), pp. 73–101.

85. On the intellectual gulf between genetics and biochemistry see, for example, H. Fraenkel-Conrat, "Protein Chemists Encounter Viruses," *Annals of the New York Academy of Sciences, 325* (1979), pp. 309–318. On the agricultural context of biology in general and genetics in particular see C. E. Rosenberg, "Science, Technology, and Economic Growth: The Case of the Agricultural Experiment Station Scientist, 1875–1914" and "The Social Environment of Scientific Innovation: Factors in the Development of Genetics in the United States," in his *No Other Gods* (Baltimore: Johns Hopkins University Press, 1978), pp. 153–172, 196–209. Also D. B. Paul and B. A. Kimmelman, "Mendel in America: Theory and Practice, 1900–1919," in R. Reinger, K. B. Benson, and J. Maienschein, eds., *The American Development of Biology* (Philadelphia: University of Pennsylvania Press, 1988), pp. 281–310. On the medical context of biochemistry see R. E. Kohler, *From Medical Chemistry to Biochemistry* (Cambridge: Cambridge University Press, 1982). We shall see later that the University of Wisconsin was an important exception: There biochemistry developed within a context that linked agriculture and medicine through nutrition and pharmacology. Also see Jan Sapp, *Where the Truth Lies: Franz Moewus and the Origins of Molecular Biology* (Cambridge: Cambridge University Press, 1990) for a critical reassessment of the received view.

CHAPTER 7

Microorganisms and Macromanagement: Beadle's Return to Caltech

New Biological System

In 1937 Beadle's path-breaking investigations, which began at Caltech in 1934 with Ephrussi, inched on uneventfully. Beadle had just moved from Harvard to Stanford and was looking for a "biochemist to work on hormone-like substances that are concerned with eye pigments in *Drosophila*."[1] Against the advice of his academic elders at the University of Wisconsin, Edward Tatum accepted Beadle's offer, foregoing the safer and more lucrative research in dairy microbiology in favor of the challenge of biochemical genetics. Given the wide chasm separating the two disciplines, attracting biochemists to genetics was no easy task; the move would entail professional risks. Beadle recalled that one day as Edward Tatum's father, the prominent Wisconsin pharmacologist Arthur Tatum, was visiting the laboratory he called him aside to tell him he was concerned about the professional future of his son. "Here you have him [Tatum Jr.] in a position in which he is neither a pure biochemist nor a bona fide geneticist," the senior Tatum observed, expressing his fear that his son "will find no appropriate opportunity in either one."[2] The Wisconsin connection would prove to be not only intellectually productive but a useful link to the food and drug industries.

Tatum's training during the 1930s in "pure biochemistry" reflected the emphasis on agricultural research at the University of Wisconsin, particularly biochemical nutrition in the context of the burgeoning dairy industry. The vital functions of several "growth factors"—hormones, vitamins, amino acids, nucleic acids, and other substances—were being elucidated; and animal feeding studies not only clarified cause and prevention in animal and human disease, they created new

194

topics in the biochemistry of metabolism.[3] The study of "growth factors" extended during the early 1930s to the terra incognita of nutrition of microorganisms, Tatum's graduate work. His project, which contributed to the identification of vitamin B_1 (thiamine) as a required growth factor in bacteria, was followed by postgraduate work at the University of Utrecht on nutrition and growth of fungi. Although there was already evidence for the universal need of several vitamins in microorganisms, the tendency at that time was to consider "growth factors" as highly individual requirements, peculiar to strains or species of microorganisms. There was little basis for linking these variations to gene mutations or to variations in higher organisms.[4]

It was only when Tatum joined Beadle at Stanford in 1937, and through the studies of eye color development in *Drosophila*, that he came to appreciate the cognitive and technical potency of genetics. Beadle, on the other hand, had come to appreciate the power of metabolic analyses. There were still many unanswered questions about the genetic control of the steps involved in pigment production, but the methods of *Drosophila* research, when used in conjunction with biochemical techniques, were unbearably cumbersome and the results erratic. With Tatum's biochemical experience with "growth factors," Beadle could exploit the newest analytical methods of nutrition. Whether Beadle had invited Tatum to Stanford with an eye to Tatum's skills in biochemical nutrition or it was a fortuitous choice is uncertain. In any event, their collaboration led to feeding experiments in *Drosophila*, testing the relation between the production of vermilion pigment and the presence of various amino acids in the diet. These experiments did establish some interesting positive links between vermilion production and fruit fly metabolism, but the results were too inconsistent to delineate meaningful patterns. The promising new methods of nutrition and metabolic analysis Tatum introduced into *Drosophila* research could not be properly exploited in that organism. These complications pointed to the urgency of finding a biological system with well-defined nutrition requirements and the biochemistry of its metabolic pathways already known.[5]

Finding the "ideal organism"—one amenable to investigations in biochemical genetics—was a challenge. In the case of eye color studies in *Drosophila* larvae, the problem had been to locate an organism suitable for genetics and embryology. The new task was to identify a biological system compatible with the methodologies of nutritional biochemistry and genetics. As Beadle liked to point out, "It is both an accident of organic evolution and an indication of man's lack of foresight that the organisms studied in most detail by biochemists have not been those on which geneticists have concentrated." Biochemists, mostly in the service of medicine, had focused on humans and bacteria; but according to Beadle, these two organisms were unsuited to the geneticist—the one because of a long life cycle and social obstacles to controlled mating, the other because of the absence of a sexual cycle (sexual reproduction in bacteria was discovered by J. Lederberg and E. Tatum in 1947). Geneticists, on the other hand, had chosen the vinegar fly and Indian corn as the classical organisms of their science. Both suffered from disadvantages to the biochemist in not lending themselves readily to culture under precisely defined environmental conditions. Neither could be grown conveniently on a medium completely known from a chemical standpoint. There had to be an easier approach to identifying genes with chemical reactions than the one they had been following.[6]

Beadle hit upon the idea of using the red bread mold *Neurospora* instead of *Drosophila* sometime in 1940 while auditing Tatum's course in comparative biochemistry at Stanford.[7] It occurred to him to reverse the experimental procedure. Instead of starting from the gene end—from a known mutation—and working toward the biochemical product (as he had done in the transplantation experiments), why not start from the biochemical end—from a known biochemical reaction—and work backward to the gene? With the biochemical end already worked out, Beadle reasoned, he could capitalize on his skills as a geneticist and stick to his specialty, as he put it. This reverse approach required working with a biochemically well-defined biological system and well-characterized genetic mechanisms that were reasonably easy to analyze and control. One could then induce random mutations to block biochemical reactions in the organism's metabolic pathways. A biochemical reaction could then be readily identified and linked to the specific mutations associated with it. *Neurospora* seemed to fit these specifications exceptionally well.[8]

Beadle's idea of using *Neurospora* for his studies in biochemical genetics was not entirely as fortuitous as his historical reconstructions imply. He had been exposed to *Neurospora* research since his Cornell days and had followed closely Carl Lindegren's *Neurospora* project during the early 1930s. Lindegren had worked out the genetics of *Neurospora* as a dissertation project under Morgan at Caltech, a project recommended to Morgan during the 1920s by his friend, New York botanist B. O. Dodge. An enthusiastic promoter of *Neurospora* as an organism for genetic work, Dodge had persuaded Morgan to take a collection of *Neurospora* cultures with him from Columbia to Caltech. With abundant support and the advice of E. G. Anderson, Bridges, Emerson, and Sturtevant, Lindegren and his wife had worked out by the mid-1930s the basic genetics of the *Neurospora*.[9]

The research of the Lindegrens demonstrated the advantages of working with the fungus. Its haploid cells (possessing only a single set of genes), in which complications associated with dominance did not arise, and its relatively short life cycle of 10 days between sexual generations made the mold attractive for genetic analyses. Building on Dodge's work, the Lindegrens established that the sexual union of two haploid cells from opposite mating types produced a zygote that, following the two meiotic divisions, produced four haploid cells, each of which then divided by mitosis. As a result, eight genetically identical spore cells were neatly lined up in a spore sac according to their closeness of lineage, a feature that facilitated an orderly analysis of the gene sequence. With a microscope, a technician—or in Beadle's words, a "spore isolater"—could isolate a spore sac, remove the eight spores in sequence, and place each into a tube with culture medium. The spores would then undergo rapid asexual reproduction, yielding a large population derived from a single chromosome set. The uniformity and rapid yield had clear advantages over the complicated pattern of reproduction in *Drosophila*. Beadle was well aware of these features; in fact, he had singled out the *Neurospora* for its genetic advantage back in 1934.[10]

Beadle obtained stocks of *Neurospora* from Carl Lindegren (now at Washington University), and Tatum performed the biochemical characterizations of *Neurospora* metabolism.[11] Applying his expertise in the biochemical nutrition of fungi, Tatum worked out within a few months the normal nutritional requirements of the organism. Its diet turned out to be exceedingly frugal; all three species of the fungus could grow on a minimal medium containing sugar, salts, and the newly synthesized vitamin biotin (one of the B vitamins); that is, the mold could synthesize all its required substances from the ingredients in the minimal medium. With that groundwork completed, the new system was ready for testing concepts and strategies.

According to Beadle he designed his experimental strategy for *Neurospora* based on the lessons from the transplantation experiments in *Drosophila*. He reasoned that if a mutant gene manifested a loss of a particular synthetic step, that mutant *Neurospora* would be unable to synthesize some essential substance and would thus fail to grow on minimal, or unsupplemented, medium. By determining which nutrient was needed for survival, a correlation could then be established between the mutant gene and the organism's failure to survive as a result of the blockage of a particular synthetic step along a metabolic pathway. From the biochemistry of the pathway, one could then match, so to speak, a specific synthetic step with a particular mutant gene. The experimental design was elegant in its simplicity; irradiate the asexual spores of the mold with x-rays to produce random mutations; then cross the irradiated spores with the appropriate mating type, isolate newly reproduced spores, grow them on a suitably supplemented medium, and test them on the unsupplemented medium. With exceptionally good luck, a few months later Beadle and Tatum isolated a first x-ray-induced *Neurospora* mutant. "I always knew they were fine bugs to work with," Beadle wrote to Lindegren in July 1941, "but I never fully appreciated all their advantages. We have one x-ray mutant that seems not to be able to make one of the B-vitamins but we haven't yet finished the analysis of this."[12]

By October 1941, in the *Proceedings of the National Academy of Sciences* Beadle and Tatum reported having isolated three mutant strains of *Neurospora*. One was unable to synthesize vitamin B_1; the second was unable to synthesize B_6; and the third could not synthesize paraaminobenzoic acid. The preliminary results indicated that *Neurospora* was an effective genetic system of analysis, and that the new methods could indeed be used to isolate mutants that were unable to carry out a particular step in a given synthesis—and thus to determine "whether one gene is ordinarily concerned with the immediate regulation of a given specific chemical reaction." That information, Beadle predicted, would reveal the mechanisms by which genes regulate development and physiological functions.[13] The theoretical import of these findings was far-reaching and the potential of the new experimental system immense. Even before its publication in the *Proceedings*, upon reading the manuscript Ephrussi immediately wrote to Beadle: "I want to congratulate both you and Tatum. I believe that these first results leave no doubt that you are entering an unexplored field of most promising possibilities." This promise, later fulfilled as the "one gene–one enzyme hypothesis," would become one of the cognitive pillars of molecular biology.[14]

It is instructive to examine the process of thought and validation of Beadle's discovery. Recalling his approach nearly two decades later, Beadle wrote: "It is sometimes thought that the *Neurospora* work was responsible for the 'one gene—one enzyme' hypothesis—the concept that genes in general have single primary functions, aside from serving an essential role in their own replication, and that in many cases this function is to direct specificities of enzymatically active proteins. The fact is that it was the other way around—the hypothesis was clearly responsible for the new approach." According to Beadle, although it had never been explicitly stated, he and Ephrussi had already developed that concept during the course of their work on eye pigment in *Drosophila*.[15]

Clearly influenced by hindsight, Beadle's recollections illuminate both the process of discovery and its historical reconstruction. Judging from his earlier papers, it is doubtful that during the mid-1930s Beadle had a clear vision of the relation between enzymes and genes. True, the controversial topic of enzymes and genes had intrigued him by the early 1930s, becoming increasingly important in his quest to understand phenotypic expression. As his investigations in *Drosophila* began to overlap with the European work on the biochemistry of development, the debates regarding the chemical mode of action of the gene in relation to enzyme action assumed greater significance and urgency. Undoubtedly, working in Morgan's laboratory, where these controversial debates were a response to the polemics of Richard Goldschmidt, Beadle had developed his thoughts on the subject.[14]

The lively discussions on the relations between genes and enzymes sparked by J. B. S. Haldane's stay at Caltech around 1933 must have added interesting perspectives. Haldane's studies on the heredity of hemophilia and color-blindness and his affiliation with biochemical genetics research at Frederick Gowland Hopkins's laboratory at Cambridge would have influenced Beadle's biochemical approach to genetics. In fact, Haldane had a profound appreciation for Archibald Garrod's work *Inborn Errors of Metabolism* (1909), where a vague version of a one-to-one relation between gene and enzyme had been implied, or so it seemed in retrospect. Haldane most likely discussed Garrod's work with Beadle,[17] which would explain why Beadle himself, despite claiming to have rediscovered Garrod's work only in 1942, knew of Garrod's work during the late 1930s and mentioned Garrod in passing as early as 1940. Yet it is clear that the full significance of the connection escaped him. Beadle's chance "rediscovery" of Garrod's supposedly "neglected" work a few years later appears to be a retrieval of forgotten information—information that during the 1930s was peripheral to Beadle's main focus. Only later would Garrod's conclusion become central to Beadle's work, when his own experimental findings converged on the hypothesis that a single gene regulated a chemical reaction, which in turn was regulated by one enzyme.[18]

It is difficult to pinpoint historically Beadle's intellectual position on the gene—enzyme problem: whether genes *were* enzymes or only *made* enzymes. Though well versed in the arguments equating gene replication with enzyme action, Beadle generally followed Morgan's reasoning that enzymes might be several stages removed from genes. Yet he also seemed to be strongly influenced by H. J. Muller's versions of the autocatalytic theory and by the analogy of genes and enzymes implied by W. M. Stanley's studies of the tobacco mosaic virus. That Beadle had been stimulated by these speculations was evidenced by his writings during the mid-1940s. In lectures and review articles, he frequently referred to the works of Troland, Muller, Goldschmidt, Haldane, Wright, and Stanley, citing their important contributions to the gene–enzyme problem. Even as late as 1947, in a lecture apparently deliberately entitled "Genes and Biological Enigmas" (after Troland's 1917 noted article "Biological Enigmas and the Theory of Enzyme Action"), Beadle acknowledged the influence of Troland's prescient ideas. Somehow Beadle managed to accommodate the notion of the autocatalytic theory of life with Morgan's and Sturtevant's views that enzymes were merely gene products.[19]

Beadle's effectiveness lay in his biologically reductive approach: not a theory-reduction of biology to chemistry but a reduction sustained within a biological framework. Unlike most of his predecessors in biochemical genetics who worked with higher organisms, Beadle (reflecting the nascent trends in molecular biology) streamlined physiological problems by working with microorganisms. Yet within this reductive framework he maintained a biological standpoint. Beadle did not underestimate the immense complexities of the lowly *Neurospora*. Precisely because the fungus could perform diverse syntheses from a handful of resources, Beadle inferred that its metabolic pathways were complex and therefore involved a great number of enzymes. Operationally, Beadle followed closely Morgan's reductive formalisms. By grounding his analyses in strict correlations between gene function and a specific mutation, Beadle reduced the need for untestable assumptions about the chemical nature of the gene in his experimental design. By sticking to his specialty and broaching the problem from the genetics end, and by building on the work of biochemists in a well-characterized, relatively simple system, Beadle circumvented some potential pitfalls of theory-laden methods. He introduced few assumptions about intermediate chemical steps in the chain linking gene and product while resisting the temptation (of which Sturtevant warned) to reduce gene expression to enzyme chemistry.[20]

Thus a critical reexamination of Beadle's discovery reveals considerably less novelty than previous accounts have portrayed. Beadle's originality as a biologist lay in his thinking beyond the organism, his conceptualization of the organism as a probe into the gene. In so doing he took bold steps entailing substantial career risks. Although *Neurospora* genetics was not a new field, Beadle's willingness to switch biological systems for the third time within a decade and experiment with novel techniques placed him in the intellectual vanguard. However, the challenges of his project placed him in a particularly precarious position during the institutional and professional turmoil of the war years.

Selling Pure Science During Wartime

Given the pressures of war, the potential commercial aspects of Beadle's new research were not disregarded. Even in their preliminary report Beadle and Tatum were quick to stress the practical significance of *Neurospora* biochemical genetics and its utility to other areas such as nutrition and pharmacology. The methods outlined, they argued in their 1941 paper, were of value as techniques for discovering additional substances of physiological significance. A complete medium could be made up with extracts of normal *Neurospora*; and if through mutations the ability to synthesize some substance were lost, it could then serve as a test strain for isolating the substance. "It may of course be a substance not previously known to be essential for the growth of any organism," they suggested. "Thus we may expect to discover additional amino acids if such exist."[21] This assertion was a bold one; undoubtedly, the increasing attention to commercial and military needs had influenced Beadle's research strategy.

At that time the United States was at the height of its "preparedness" phase. Laboratories in the life sciences were diverting their resources toward the war effort, and many war projects in pharmacology and biochemistry, notably the production of penicillin, were being coordinated with the Department of Agriculture and with commercial concerns such as the pharmaceutical firms of Merck and Company, E. R. Squibb and Sons, Sharp and Dohme, and Lederle Laboratories. In 1941, just when Beadle and Tatum were publishing their preliminary results on *Neurospora* and pointing out the work's projected practical applications, nonessential scientific expenditures were being trimmed back. Most researchers in the physical sciences had already organized their war-related projects under the auspices of the Office of Scientific Research and Development (OSRD). Investigators in the life sciences, in areas relevant to the priorities outlined by the Committee on Medical Research (CMR), were increasingly entering into government contracts, which were usually drawn for 6 to 12 months. Basic research in the life sciences was being gradually curtailed.[22]

For Beadle, however, the *Neurospora* program was just beginning. He envisioned a large-scale attack on the fundamental problem of the relation between genes and enzymes, work that was both time-consuming and expensive. The laborious task of running mutants through what he called a "nutritional mill," that is, through a systematic battery of tests for various vitamin and amino acid deficiencies, required many hands and substantial sums for materials. The projected research program called for expanded laboratory facilities and staff. As Beadle explained in his 1941 grant proposal to the American Philosophical Society, 34,000 *Neurospora* strains had been established and tested during the previous year; each strain, in turn, had resulted in several cultures; and from these hundreds of thousands of cultures, only 102 mutants had been isolated so far. Requesting additional funds, Beadle explained that it would take 20,000 tries to find a single additional mutant. Yet the value of each additional mutant, he argued, increased as the list approached completeness, because when matched with the corresponding chemical reaction, these last mutants would fill in the crucial remaining pieces of the biosynthetic puzzle.[23]

Beadle also appealed to the Rockefeller Foundation for support. In November 1941, one month after the first publication on *Neurospora*, he announced his findings to Weaver. After explaining some of the difficulties with the new experimental procedures, Beadle reported that since the initial publication he and Tatum had more than doubled the number of mutants having a known role in synthesis. Among the newer mutants, one had been found that lacked the ability to synthesize what appeared to be a new, unknown amino acid, which they had named neurosporin.[24]

Figure 19 Graduate student preparing *Neurospora* cultures. Courtesy of the Rockefeller Archive Center.

While progressing on the basic research front, Beadle was also being courted by the food and drug industries; genetics was making unexpected contributions to the science of nutrition. Aside from their theoretical significance, a number of the newly isolated mutants that were unable to synthesize either a vitamin or an amino acid had proved to be important in applied bioassay work. The growth rate of each mutant was a function of the concentration of the substance in which it was deficient. Therefore by measuring the dry weight of the mycelium (the vegetative form of the fungus) produced during a specified growth period, or by following the rate of progression of the mycelial frontier over the agar surface, one could obtain an estimate of the concentration of the specific substance in the medium. One of the advantages of *Neurospora* techniques, compared with other methods, according to Beadle, was the efficiency and specificity of response. The *Neurospora* bioassays were therefore attractive procedures for commercial houses that dealt with the manufacturing of vitamins and amino acids.[25]

Figure 20 Research Fellow Marko Zalokar plotting growth curves of *Neurospora*. Courtesy of the Rockefeller Archive Center.

From the point of view of the Rockefeller Foundation, links between basic research and commercial profit could potentially create delicate situations. Because the natural sciences division supported nonmedical research, questions of practical applications and patent rights were of only marginal concern. With no firm guidelines for commercial ties between university and industry, such matters were usually left to the discretion of individual investigators.[26] Thus while stressing the immediate practical value of the *Neurospora* mutants for food and drug testing, Beadle also solicited Weaver's advice regarding commercial involvements.

In November 1941 Beadle told Weaver that Merck and Company had expressed an interest in supporting *Neurospora* research. Because of the efficacy of the new analytical and culturing techniques and the precision of the vitamin and amino acid essay methods, the company was enthusiastic about entering into collaborative research with Beadle's laboratory. Of course his laboratory could benefit a great deal from cooperative ventures with Merck and similar concerns, Beadle admitted, but he was uneasy about linking his research program with the work of pharmaceutical houses. He thought that there were definite disadvantages in ties with commercial concerns due to the possibility of disagreements over such questions as manufacturing procedures and patent rights in relation to newly discovered substances. He would prefer to limit such entanglements.[27] Because the Natural Sciences had no clearly articulated policy regarding patents, the officers stated that they had no intrinsic objections to Beadle's entry into applied research and left the matter up to him.

A month later Beadle visited Merck and Company in New Jersey with the purpose of exploring the possibilities for cooperative projects. He learned that Merck was willing to support the entire *Neurospora* project in return for the patent rights. According to the Rockefeller Foundation's report after Beadle's New York visit in December 1941, Beadle was completely uninterested in the patent question; and, in fact, the patent policy of Stanford seemed to be opposed. The Foundation officer reported that Merck would probably be willing to supply funds and assistance without any patent rights; but in return for furnishing the chemical services they would expect to obtain information in advance of publication of any papers and thus to acquire an edge on their competitors. The Research Corporation was also interested and led Beadle to believe there were considerable chances for the success of an application to it unless there were complications due to patent problems. The Rockefeller Foundation officer noted: "Beadle states explicitly that he has not interest in patent or any personal profit for himself but, on the other hand, must find outside assistance to push his work rapidly. His first preference would be a grant from the RF which would free him of all obligations other than to work hard and publish freely his results. His second choice would be the Research Corporation and third, Merck."[28]

In 1942 Beadle received a grant from the Rockefeller Foundation, but he also entered into cooperative projects with Merck and Company and later with Sharp and Dohme and other commercial agencies. In 1943 the Research Corporation, which had close ties with the OSRD and heavily supported work on nutrition (particularly nutrition research at Wisconsin), awarded Beadle $10,000. Beyond this financial support, Beadle undoubtedly benefited from other services of the Research Corporation, which the Rockefeller Foundation (and other agencies) often used to "hold" patents that were licensed to commercial houses.[29]

The early links between *Neurospora* research and the food and drug industries not only broadened Beadle's financial and institutional base, they carried considerable weight in assessing the utilitarian value of his program. These connections were a testimony to the practical significance of his program at a time when relevance to nutrition and pharmacology counted for much. When Beadle reapplied for a Rockefeller Foundation grant for 1942, Stanford's president R. Wilbur not only praised the *Neurospora* program as "ushering a new era in genetics research," he promoted the broad range of practical applications. "The wide scope of the problems on which these researches bear," Wilbur wrote to his old friend Raymond Fosdick, "gives them an importance not only for further advancement along these (genetic) lines but also for more immediate applications in our present war emergency. The latter aspect alone would seem to justify an additional grant from the Rockefeller Foundation."[30]

Indeed, the grant appropriated by the Rockefeller Foundation in 1942 was reinforced by a grant from the American Philosophical Society and buttressed by the various benefits of collaborative projects with the food and drug industries, which by 1942 included Merck, the Fruit Product Laboratory, and the Western Regional Department of Agriculture. Clearly, Beadle attached a great deal of weight to the practical and commercial side of *Neurospora* work while simultaneously pursuing his main interest: the correlation between mutant genes and their biochemical deficiencies.[31]

This twofold approach to *Neurospora* research—the pure and the applied—and the base of support it attracted resulted in a considerable expansion of Beadle's program. His applied work was in great demand, and several laboratories began sending people to Stanford to learn the new techniques. Beadle also gathered junior faculty members, postdoctoral fellows, graduate students, and additional technicians. Of the new investigators who joined Beadle's group in 1942, Norman Horowitz and David Bonner from Caltech were particularly valuable to the development of the new biochemical genetics. Having studied during the 1930s in Morgan's interdisciplinary division, they were the first generation of American graduates trained in the new physiochemical biology. Both were proficient in genetics and biochemistry and possessed for the early 1940s a unique combination of skills to bring to *Neurospora* research. Upon inviting his Caltech friend Sterling Emerson to spend the summer of 1942 at Stanford, Beadle described the rapid growth of his group and boasted: "We have up our sleeves plans for a super gigantic *Neurospora* Institute for next summer."[32]

By summer 1942 the United States was deeply involved in the war, and science was heavily immersed in the war effort. Laboratory resources had been diverted to war-related projects, and junior laboratory personnel were being drafted into the armed forces. Senior researchers, even those who discontinued basic research were experiencing difficulty in maintaining postdoctoral fellows, graduate students, and technicians. The future of many research programs in the life sciences now became uncertain, a situation that presented special problems to the Rockefeller Foundation regarding its annual appropriations allocated to long-term grants awarded before the war. As part of a general survey the Foundation requested information from its principal supported investigators concerning the war's impact, or projected impact, on basic research—on the availability of personnel or the acquisition of materials and equipment. The Foundation expressed its preference for maintaining those basic and long-term research programs that could be sustained on a high level without conflicting with the demands of defense. However, if the quality of research had to suffer owing to the exigencies of war, it might be necessary to curtail the support.[33]

During the spring of 1942 Beadle still emphasized his primary commitment to basic research, but the pressures of military relevance had already begun to manifest. He wrote to the Rockefeller Foundation:

> Our facilities and the generous Foundation support are proving to be quite adequate for the basic aspect of the work and I feel confident that we shall continue to make satisfactory progress along these lines. It becomes increasingly evident, however, that it is desirable to apply these findings to the development of a rapid standardization technic of bioassay for various vitamins and amino acids. While it seems obvious that this type of work should not be done at the sacrifice of more fundamental work, it occurred to us that a number of our best qualified graduate students, who are actively looking for ways to be useful in the present emergency, might well undertake such applied work with a view toward making

efficient bioassay technics available for studies of vitamin and amino acid contents of various types of preserved foodstuffs.[34]

Beadle inquired about the Rockefeller Foundation's reaction to his proposal that the Nutrition Foundation, Inc., founded by 15 national manufacturers and headed by Karl T. Compton, might support four graduate fellowships at Stanford for two years. Following the approval of the Rockefeller Foundation, the Nutrition Foundation awarded Beadle a substantial sum for fellowships and equipment. This support further facilitated the applied aspects of Beadle's *Neurospora* program.[35]

Up until the summer of 1942, Beadle had been able to hold on to his people and to ensure the continuity of his program. He had relied on the argument that, considering the importance of the adequate protein nutrition during times of meat shortages, it was of great practical as well as theoretical importance to complete the amino acid mutant list. Because of the projected meat shortages, and because California's agribusiness was the nation's principal supplier of produce, Beadle had buttressed his argument by promoting the new assay techniques. The *Neurospora* assay methods for determining the vitamin content in produce and the means of creating "high vitamin" products were at a premium, he claimed. "This question of the vitamin content of dehydrated products will certainly become increasingly important in the near future from both military and civilian stand point. We feel that we should very soon know just how useful 'made-to-order' *Neurospora* mutants will be in vitamin research and control."[36] His arguments remained effective until mid-1942.

During the summer of 1942 the local Draft Board denied his requests for military deferments for his graduate students and assistants. Beadle was about to lose a couple of his men, and basic *Neurospora* research was now threatened by the demands of the war. He informed the Rockefeller Foundation about the new developments. Soliciting their cooperation, he asked that they intercede on his behalf and use their influence with the local Board and state appeal authorities. He suggested that the deferment of men who had training and skills in biochemical genetics of *Neurospora* could be justified on the ground that their contributions to the field of nutrition were likely to be much greater than any contribution they could make when starting from the ground up in direct military service.[37] The Foundation decided, however, as a matter of policy, not to exert pressure upon local boards.[38]

The Foundation's decision to avoid intervention in military matters helped accelerate the trend toward applied *Neurospora* research. Although Beadle graciously accepted the Foundation's refusal, stating that he could see "how a stand other than the one taken would be difficult to maintain as a general proposition," he also communicated his resolve to intensify the practical direction of biochemical genetics as a result of wartime pressures. "Several of us at Stanford feel," he wrote, "that it is becoming more and more obvious that the only way we are going to be able to continue scientific work is to turn our efforts more and more toward applied lines. Even so, contracts with governmental agencies would still be essential to the survival of research groups."[39] Accordingly, Beadle's team would now begin exploring the possibilities of obtaining one or more contracts in connection with the development of vitamin and amino acid assays.

A few months later, Beadle flew east to meet with the Subcommittee on Medical Nutrition of the CMR in order to investigate the possibility of using some of his laboratory facilities and techniques to study problems of nutrition related to the war effort. As a result of these meetings it was agreed that Beadle's *Neurospora* program could aid several ongoing CMR projects. Although the group's research would not be performed under government contract, it was concluded that *Neurospora* techniques and results should definitely aid the work of R. J. Williams at the University of Texas on paraaminobenzoic acid, the Harvard project on tetanus, and E. N. Ballantyne's project on gas gangrene. Beadle emphasized that he still planned to push forward in basic research, but greater weight would now be given to applied war research.[40]

The following month, November 1942, Beadle's program of biochemical genetics was classified as essential to the war effort under the CMR guidelines, though it did not receive a formal contract. Beadle immediately dispatched a letter to the Rockefeller Foundation in which he quoted with obvious pride excerpts from Richard's letter:

> This is equivalent to saying that it is my conviction, which I am confident would be shared by all other members of the committee, that the work is of sufficient fundamental importance and potential practical usefulness that it should not be interrupted in favor of other research which may seem to have more immediate practical utility in the War Effort. I can only assure you that we will endeavor to give such requests [deferments in the absence of government contract] the full influence of the Office of Scientific Research and Development in the case of any of your investigators whom you certify as essential and irreplaceable. Similarly in the case of critical materials for which high priorities are needed, we will do everything in our power to assist you.[41]

This official statement provided the necessary guarantee that Beadle's program of biochemical genetics would develop relatively unhindered. In fact, by having no formal contract, Beadle gained an advantage: He was free to pursue his work with fewer constraints on his facilities and time while receiving priority privileges equivalent to contracted research. He could also publish freely. "Naturally we are encouraged by this letter," Beadle wrote to Rockefeller officer F. B. Hanson: "We feel that we can now go ahead with our work with clear conscience."[42]

During the war years Beadle's group isolated about 80,000 single spores; of them, approximately 500 had given rise to mutant strains that were unable to carry out essential syntheses, and more than 100 mutant genes controlling vital syntheses had been detected. Most of the mutants were characterized by loss of the ability to synthesize a vitamin, an amino acid, or a nucleic acid component. Mutants for the synthesis of seven B-complex vitamins and twelve amino acids were established, and most of them had been shown to be essential for rat, dog, and human metabolism. Using the *Neurospora* mutants, Beadle's group had worked out bioassays for choline, paraaminobenzoic acid, inositol, pyridoxin, and leucin. With none of the constraints of the secrecy inherent in classified contract work, and with no obligations to industry due to patent restrictions, Beadle and his collaborators were free to publish most of their findings in the standard scientific journals. The numerous articles and reports about the culture techniques needed for mutants and about the various bioassays appeared in the *Journal of Biological Chemistry, American Naturalist, Physiological Reviews, American Journal of Botany, Journal of Bacteriology,* and *Proceedings of the National Academy of Sciences.* These publications made Beadle's research highly visible during wartime.[43]

Not all that Beadle touched turned to gold. At the end of 1942 he communicated to the Rockefeller Foundation his belief that the putative new amino acid neurosporin had been isolated, and furthermore that it promised to be of importance in the CMR's tetanus toxin project. There was a great deal of excitement and a flurry of activity in the laboratory as "Tatum and Bonner," according Beadle, were "burning the night lights trying to get the structure established and a synthesis worked out."[44] Soon afterward, however, Beadle had to retract the discovery. Disappointed, he reported to the Foundation that what for some time was thought to be the amino acid neurosporin had turned out to be an active crystalline material isolated from a casein hydrolysate; neurosporin was actually a mixture of valine, isoleucin, and leucin. Although still useful as a booster for the preparation of tetanus toxin, the new substance was of only marginal fundamental significance.[45]

The "false" amino acid was but a minor setback. The profundity of Beadle's program lay in its contributions to fundamental biological knowledge: elucidating the relation of individual genes to individual metabolic reactions and, in turn, to the specific enzymes regulating these reactions. Without exception, every biochemical pathway leading to the synthesis of a final product (either a vitamin or an amino acid) proved to be comprised of a series of biochemical reactions. In each case, a specific gene mutation blocked only a single biochemical reaction along the pathway and, by inference, depended on the deficiency of a specific enzyme.

The detail involved in analyzing the sequence of biochemical steps in a synthetic pathway was staggering, scores of *Neurospora* mutants being needed for a single step. Several of the pathways under study therefore contained gaps. Two important biochemical sequences and their corresponding mutants had been well characterized by the end of the war. Horowitz and his collaborators established that the synthesis of the amino acid arginine in *Neurosopora* proceeded through the synthesis of two precursors, ornithine and citrulline, and that each step in the sequence of reactions was under genic control (thus by inference under control of a specific enzyme). Furthermore, by working out the reaction cycle, they showed that the biochemical reactions were identical to the ones occurring in mammalian liver, and the same experimental approach was used to study the synthesis of the amino acid tryptophan. Bonner and his associates showed that anthranillic acid and indole were intermediate products in the synthetic pathway of tryptophan in *Neurospora* and that each was under specific genic control. These findings established for the first time a mechanism of tryptophan synthesis in this organism.[46]

At the end of 1944, in a lengthy report to the Rockefeller Foundation in which he described the conceptual and technical aspects of *Neurospora* research, Beadle presented two main conclusions. The first was that the synthesis of the essential constituents of living matter is under genic control, and that the requirements of higher animals for dietary supplements of vitamins and amino acids are the result of gene mutations that have occurred during the evolution of species. "Although it is going beyond our present information to suggest a mechanism of this control," Beadle cautioned in his understated style, "it appears that the primary action of the gene has to do with the synthesis of the enzymes which direct the chemical activities of the cell." The second conclusion was that there exists a one-to-one correspondence between gene and chemical reaction. The studies of *Neurospora* mutants made it possible to assign definite series of reactions to individual members in a series of nonallelic genes. As Beadle had predicted in 1941, reducing gene effects to simple chemical reactions was indeed the first step in the direction of analyzing the physiological bases of gene action.[47]

Significantly, it was only at this later stage of his work that Beadle saw the relevance of Garrod's work—the link between the metabolic disorder alkaptonuria and a recessive Mendelian character and, in turn, the absence of a specific enzyme regulating the breakdown of the amino acid tyrosine. In a letter to a colleague at Cambridge University in 1944 requesting the 1909 and 1923 editions of Garrod's book, Beadle emphasized Garrod's unacknowledged founding of biochemical genetics.

> I have recently become much interested in the life and contributions to biochemical genetics of Sir Archibald E. Garrod. . . . My interest is kindled by the fact that in spite of the significance of Garrod's work, most geneticists have very little knowledge of him or his work, Waddington too, is guilty of the sin. So far as I know, Haldane is the only writer who has done him justice. In view of this neglect of the man who I think should be regarded as the father of biochemical genetics, I have been considering the possibility of writing a brief account of him for the *Journal of Heredity*.[48]

Graciousness notwithstanding, by retrieving Garrod's "forgotten" work Beadle, of course, was engaging in legitimating his own findings; by setting the record straight he also carved a historical space for his own contributions to biochemical genetics.

It was also at this phase, during the last year of the war, that direct military demands did indeed take some toll on Beadle's research. In February 1944 he wired the Rockefeller Foundation that a representative of the War Production Board had just proposed that Beadle's group devote part of their facilities to inducing mutations in *Penicillium* in order to increase penicillin production. To do it would mean curtailing basic research activities for a while. Upon receiving the approval of the Foundation, Beadle somewhat reluctantly embarked on the organization of the new project, which had little relation to *Neurospora* work and which retarded the rate of progress of whatever basic research he had been managing to push forward. As young men were being drafted at an increased rate, he was also experiencing some difficulty in holding on to the men on his team. "I'm afraid one of the undesirable results of the war is going to be a missing generation of scientists," he wrote to the Foundation, lamenting the attrition.[49]

Nevertheless, by 1944 the conceptual foundations of the *Neurospora* projects and the disciplinary merger of biochemistry and genetics were firm. Even though war-related activities had retarded the rate of progress, some additional fundamental research was accomplished. By 1945, when the war ended, Beadle emerged as the leading authority in a new field linking physiological processes, biochemical reactions, and genetic controls. An astute discipline builder, Beadle was fully aware of the institutional innovations of his program and played up their importance. When in 1945 he was invited to deliver the Harvey Lecture, he chose to discuss "The Genetic Control of Biochemical Reactions" and to promote the conceptual and disciplinary accomplishments of his research program.

Deploring the evident lack of interaction between genetics and biochemistry, he referred to it as a "most unfortunate consequence of human limitations and the inflexible organization of our institutions of higher learning. The gene does not recognize the distinction—we should at least minimize it." Having been denied membership in the American Chemical Society for not being formally trained as a chemist, Beadle had an added gripe with academic rigidity.[50] According to the testimony of the officer of the Rockefeller Foundation who attended Beadle's Harvey Lecture at the Academy of Medicine in New York, Beadle received a great ovation. He concluded his lecture with the dramatic statement that until recently some students in the university entered a laboratory through a door on which was printed "Genetics Laboratory"; other students entered another door labeled "Biochemistry Laboratory"; but in the future, genetics and biochemistry were to be one subject.[51]

Not all researchers in the life sciences responded favorably to the innovative aspects of Beadle's research program. Beadle's criticism of the intellectual and institutional separation of genetics and biochemistry seemed to be partly a reaction to the tepid reception of his innovations. According to Beadle, when in 1945 he traveled across the United States on a series of 24 Sigma Xi lectures, he found many skeptics but few converts to the new interpretation that genes control enzymatically regulated chemical reactions. Even in 1951, he said, the believers could be counted on the fingers of one hand.[52]

Norman Horowitz recalled that, despite evidence to the contrary, many geneticists preferred to adhere to the old view that each gene was pleiotropic, that is, manifold in its action. Limiting the influence of hereditary determinants to merely regulating intermediate chemical reactions along a pathway was tantamount to dethroning the gene (this atavism explains and confirms the persistence of eugenic ideas). Some biochemists and physiologists thought that a microorganism was not representative of mammalian physiology—that the chemistry of *Neurospora* was too simple to prove a general rule. Others denounced Beadle's hypothesis on the basis that his methodology was unverifiable and unfalsifiable. Max Delbrück, for one, alleged that Beadle's conclusion was based on selection procedures ensuring that only mutations supporting the theory would be detected. Certainly the inference that a given gene controls the production of a single enzyme, was opposed. According to Horowitz, critiques published at the time were but pale shadows of the unpublished objections that were voiced during the 1940s and early 1950s at Cold Spring Harbor Symposia.[53]

Several salient features of Beadle's research were universally appreciated, however. The importance of the discovery of mutations that block the syntheses of vitamins and amino acids was generally acknowledged from the start. Beadle did succeed, at least partially, in blocking the circularity in the gene–enzyme dilemma of what genes are and what they do—whether genes are enzymes or only control reactions catalyzed by enzymes. The interdisciplinary innovations, the combination of theories and laboratory techniques from genetics and biochemistry, were certainly applauded by life scientists and by the Rockefeller Foundation.

Beadle's cognitive and disciplinary innovations, however, had been overinterpreted and misinterpreted in earlier historical accounts. Celebratory comments from scientists and historians have placed his work in the chain of events leading to the identification of nucleic acids as genetic determinants and to explanations of protein synthesis in terms of the "central dogma" (the sequence of events leading from DNA replication to RNA transcription to protein translation). Thus R. D. Hotchkiss wrote that "The whole development of modern genetics was comprised of a sequence of great steps linking the formal concepts of classical genetics with the science of matter. One such great connection between unit gene product and the enzyme had been propounded by Garrod, furthered by Wright and Scott-Moncrieff, and made experimental by Beadle and Tatum in the early 1940s."[54] R. C. Olby, outlining the autocatalytic theory of life and its promotion by Troland and Muller, claimed that "there is little indication in Beadle's papers that he was influenced by such speculative ideas."[55]

The introduction of sharp discontinuities between "wrong ideas" and "correct theories" and the emphasis on crucial experiments as necessary and sufficient conditions for scientific success have tended to obscure the subtle and gradual contributions that marked Beadle's intellectual program. These writings not only overlooked the resistance to Beadle's formulations but underestimated the continuities between Beadle and his predecessors in physiological genetics. Beadle was firmly grounded in the protein view of the gene, with no particular appreciation for nucleic acids as genetic determinants. Though familiar with the correlation between mutagenesis and the absorption of ultraviolet radiation by nucleic acids, and aware of Avery's work on the "transforming principle," Beadle viewed nucleic acids as extragenic material that transferred the mutation-producing energy to the protein gene.[56] As late as 1952 Beadle spoke of the giant protein molecules as the key to genetic and viral replication, antibody synthesis, and enzyme synthesis. His firm grounding in the protein view of the gene fit well with Pauling's program.

Earlier histories have also ignored the institutional and social settings of Beadle's work—the complex politics behind the development of his interdisciplinary science and the role of the war. This context was central to his success. Beadle was fully aware from the start of the commercial potentialities of his work and thus pursued a two-tiered approach to biochemical genetics: the pure and the applied. This factor too made him an ideal leader for Caltech's molecular biology program.

Beadle's Return to Caltech

In September 1945, when Pauling discussed the future of Caltech's molecular biology with the officers of the Rockefeller Foundation, the negotiations to bring Beadle back to the Institute were already well on their way. Offering Beadle a full professorship in May 1945, Sturtevant (head of the biology division's executive committee) informed Beadle that "The department has gone over the business and the staff has unanimously voted in favor of the scheme with considerable show of enthusiasm." Their excitement was matched by that of Millikan.[57] The letter marked the beginning of a five-month effort to once again persuade Beadle to return to Caltech. Pauling, now a member of the Institute's Board of Trustees and its liaison to the Rockefeller Foundation, would exert decisive influence on the process that culminated in Beadle's chairmanship and in drawing the blueprints for a world center for molecular biology.

It was no easy task to lure Beadle to Caltech. Despite the close personal and professional ties that Beadle had maintained with members of the biology division, he had little incentive to return to Caltech. He was now leading a thriving program of biochemical genetics at Stanford and was well aware of the problems of Caltech's biology division and the obstacles to full partnership with Pauling's group. As Sturtevant admitted to Beadle, "I don't have to tell you that the place isn't perfect." The toughest situation confronting Beadle would be the unruly expansion of commercial agricultural interests and the internal conflicts it had created in the division. Sterling Emerson could provide Beadle with intimate knowledge of departmental matters, and Sturtevant promised to do everything in his power to help supply "more dope-atmosphere, arguments, advice, etc., etc."[58] Well seasoned in the politics of interdisciplinary science and in coordinating projects involving university, industry, and government, Beadle would be exceptionally well suited for the task.

Figure 21 G. W. Beadle and A. H. Sturtevant. Courtesy of the California Institute of Technology Archives.

However, with a secure future at Stanford and a strong offer from the Wistar Institute, Beadle decided against this risky lateral move.

Pauling, however, would not let this opportunity slip by. Backed by Millikan and the Board of Trustees, and bypassing the biology division, Pauling insisted that Beadle should be offered greater incentives: He should receive a higher salary and an offer of the division's chairmanship, a position commensurate with his projected power. This appointment would enable him to redirect the division's course through changes in policy and staff and to propel the molecular biology program to the vanguard of American science. At stake was not only the future of the biology division but that of the chemistry division and of the Institute as a whole. Under Beadle's leadership of the biology division, in cooperation with Pauling's division, the Rockefeller Foundation would invest large sums in molecular biology at Caltech, thus scaling the program up closer to Pauling's visions. Although Sturtevant did not share these grand ambitions and thought that such large sums were not necessary, he did approve the scheme and seemed genuinely pleased to step down and hand the reins to Beadle.[59]

The added incentives placed the situation in a new light. Although the move would entail some risks, Stanford was not perfect either: Obstacles to developing a major molecular biology program existed there as well. Unlike Caltech, the university had not developed mechanisms of interdepartmental cooperation, a situation that could potentially limit Rockefeller support. For example, a memo circulated in 1945 strongly urged Stanford's life science faculty to develop cooperative research on proteins, enzymes, genes, and viruses in order to qualify for a large outside grant (undoubtedly from the Rockefeller Foundation).[60] Refusal or inability to cooperate discouraged Rockefeller Foundation support and could turn into institutional and career liabilities. These factors affected Beadle's program directly. He had experienced difficulty joining forces with Hubert Loring at Stanford's biochemistry department. Having done his postdoctoral research on the tobacco mosaic virus at Stanley's laboratory at the Rockefeller Institute during the 1930s, Loring exemplified that research school, stressing a purely chemical approach to biological questions. Like Stanley, Loring had little regard for biological methods; and after a short collaboration with Beadle, he failed to develop an appreciation for *Neurospora* genetics.[61]

"I must confess I have been disappointed that we have not arranged a more effective scheme for getting together with Loring," Beadle wrote in September 1945 in response to the Rockefeller Foundation's request for assessing Loring's cooperative performance. "However, several of us in Biology have felt that he just doesn't have enough of the biological point of view to see the possibilities in the *Neurospora* work. In addition it is evident that he is a strong individualist in science and therefore not easy to work with on a really cooperative basis."[62] Strong individualism had become an institutional liability and had little value in Rockefeller Foundation's project science. The Foundation urged Beadle to cultivate Loring a bit and expose him more fully "to the beauties of *Neurospora* work," eventually securing Beadle's promise to renew his efforts "to make a convert out of Loring."[63] The Foundation's support of molecular biology at Stanford hinged on strong cooperative ties with the biochemistry department.

It is likely that the obstacles to developing a cooperative venture with Stanford's unsympathetic biochemistry department contributed to Beadle's decision to return to Caltech. Further expansion of biological genetics at Stanford would be jeopardized by the tension between the biology and biochemistry departments. As Beadle put it, "Our experience has been that to get the biochemical aspects of the work done we have to have sympathetic biochemists working right in the same laboratory where biological aspects of the work are being done."[64] Beadle's experience at Caltech during the 1930s had approximated that ideal. Although major collaborative projects between the biochemistry, biophysics, physiology, and genetics groups had not yet developed, there existed a spirit and a mechanism of cooperation. Graduate students, postdoctoral fellows, and visiting scholars worked in close proximity to staff members in the various departments within the biology division. At Caltech, at least, there were no institutional obstacles in the way of those who sought interdisciplinary cooperation. Also, Caltech's graduates, such as Bonner and Horowitz, who now worked with Beadle at Stanford, were well trained in the cooperative approach. In fact, they were rather unusual in their command of interdisciplinary skills in biochemistry, developmental biology, and genetics (Bonner, Horowitz, and others from Beadle's group at Stanford would join him at Caltech). With Pauling's enthusiastic cooperation and an offer of the division's chairmanship, buttressed by Rockefeller resources, they could create a world-class molecular biology program.

"I am very happy to say the answer is 'Yes,' " Beadle wrote to Pauling on October 8, 1945. "I am sure everything will work out well and I want to thank you for the role you have played in making it possible for us to become a part of the best chemical-biology group in the world."[65] It was a great moment for the Institute. The promise of Beadle's leadership infused a new spirit of optimism in biology, chemistry, and in the Institution as a whole, "I don't mind letting you know now," wrote Millikan in his congratulatory letter,

> that although I never had what I would regard as an intimate personal acquaintance with you, on account of the judgments of all of the then members of the Biology Division I was greatly chagrined when Dr. Morgan, apparently out of sheer kindness to Stanford, let you out of the department here when you originally went to Stanford ten or fifteen years ago [sic], and while I was exercising the responsibilities of Chairman of the Executive Council I schemed repeatedly with the members of the department for getting you back.[66]

Millikan went right to the heart of the problem. Quick decisions had to be made on the long-range directions of the division, and Millikan wanted to know if Beadle was fully informed on departmental politics and financial issues. "I refer in this letter particularly, first, to the large progress of Went's [sic] in the field of plant physiology," Millikan explained. Was Beadle acquainted with Went's plans for expanding his research programs on plant growth by nearly 10-fold, Millikan queried. His description of the financial plans surrounding Went's new projects communicated strong reservations. He stressed to Beadle that:

> Obviously that kind of a program ought not to be entered into unless you, as Chairman of the Department, and all your advisors in the Department are thor-

oughly sold on this form of expansion, which of course, like all steps that involve expansion, raise [sic] the question which should be always the first consideration confronting decisions of any kind, namely, is this the sort of vital thing that can be done at this time, or should it take its place farther down in the list of new developments?[67]

Caltech's biology was never intended to play a service role to California's agribusiness. It was only by default, in lieu of effective leadership for Caltech's molecular biology program, that the plant physiology group had seized control of departmental resources and directions.

Beadle, of course, was well aware of the problems of Caltech's biology division. He was also well versed in the politics of agribusiness research and would thus be effective in managing a division in which such research played an important role. However, Beadle's primary focus would not be agriculture. Looking ahead to the scientific opportunities during the postwar era, Beadle and Pauling would design a cooperative scheme based on the giant protein molecules—genes, enzymes, viruses, antibodies—a molecular biology program of unprecedented scope.

Notes

1. Joshua Lederberg, "Edward Lowrie Tatum," *Biographical Memoirs of the National Academy of Science, 59* (1990), pp. 357–386; quote on p. 361.

2. G. W. Beadle, "Recollections," *Annual Review of Biochemistry, 43* (1974), pp. 7–8. Also on the separation between biochemistry and genetics during the 1930s and 1940s see H. Frankel-Conrat, "Protein Chemists Encounter Viruses," *Annals of the New York Academy of Sciences, 325* (1979), pp. 309–318.

3. The early history of biochemistry and nutrition at the University of Wisconsin is discussed in the context of agricultural development in America in Charles E. Rosenberg, "Science, Technology, and Economic Growth: The Case of the Agricultural Experiment Station Scientist, 1875–1914," *No Other Gods* (Baltimore: Johns Hopkins University Press, 1978), pp. 153–172. Additional sources are given in David Bearman, John Edsall, and Robert E. Kohler, *Archival Sources in Biochemistry and Molecular Biology* (Philadelphia: American Philosophical Society, 1980), pp. 10–12. References to nutrition studies and the discovery of vitamins at Wisconsin are also mentioned in E. V. McCollum, *A History of Nutrition* (Boston: Houghton Mifflin, 1957), passim; and D. L. Nelson and B. C. Soltvedt, eds., *One Hundred Years of Agricultural Chemistry and Biochemistry at Wisconsin* (Madison: Science Tech, 1989). See also Rima Apple, "Patenting University Research: Harry Steenbock and the Wisconsin Alumni Research Fund," *Isis, 80* (1989), pp. 315–394.

4. Edward L. Tatum, "A Case History in Biological Research," *Nobel Lectures in Molecular Biology* (New York: Elsevier North-Holland, 1977), pp. 67–68.

5. B. Ephrussi, "Chemistry of 'Eye Color Hormones' of *Drosophila*," *Quarterly Review of Biology, 17* (1942), p. 332. See also the discussion on the Beadle-Ephrussi project in Chapter 4, this volume.

6. CIT, Beadle Papers, Box 3.13, "The Gene and Biochemistry" (ca. 1945), pp. 1–2. At the time bacteria were thought to reproduce strictly through binary fission. The sexual cycle in bacteria was discovered only in 1946; see J. Lederberg and E. Tatum, "Novel Genotypes in Mixed Cultures of Biochemical Mutants of Bacteria," *Cold Spring Harbor Symposia on Quantitative Biology, XI* (1946), pp. 139–155.

7. G. W. Beadle, "Recollections," p. 8 (see ref. 2).

8. G. W. Beadle, *Genetics and Modern Biology* (Philadelphia: American Philosophical Society, 1963), p. 13.

9. G. W. Beadle, "Genes and Chemical Reactions in Neurospora," in D. Baltimore, ed., *Nobel Lectures in Molecular Biology* (see Note 4), p. 59. The history of *Neurospora* research is also given in G. W. Beadle, "Genetics and Metabolism in *Neurospora*," *Physiological Reviews*, *25* (1945), pp. 643–663.

10. G. W. Beadle, "Genetics and Metabolism in *Neurospora*," pp. 646–648 (see Note 9); and G. W. Beadle and S. Emerson, "Further Studies of Crossing-over in Attached X-Chromosomes of *Drosophila melanogaster*," *Genetics*, *20* (1935), pp. 192–206.

11. CIT, Beadle Papers, Box 1.41, Lindegren to Beadle (a response to Beadle's request), March 17, 1941.

12. CIT, Beadle Papers, Box 1.49, Beadle to Lindegren, July 25, 1941.

13. G. W. Beadle and E. L. Tatum, "Genetic Control of Biochemical Reactions in Neurospora," *Proceedings of the National Academy of Science USA*, *27* (1941), pp. 494–506.

14. CIT, Beadle Papers, Box 1.26, Ephrussi to Beadle, August 22, 1941.

15. G. W. Beadle, *Nobel Lectures*, pp. 60–61 (see Note 9).

16. See for example a review article by T. H. Morgan, "Genetics and the Physiology of Development," *American Naturalist*, *40* (1926), pp. 501–511; and Garland Allen, "Opposition to the Mendelian-Chromosome Theory: The Physiological and Developmental Genetics of Richard Goldschmidt," *Journal of the History of Biology*, *7* (1974), pp. 49–92.

17. Beadle's description of the "tremendously stimulating" experiences at Caltech during Morgan's tenure when several scholars, including Haldane, visited the biology division is mentioned in his "Recollections," p. 5 (see Note 2). Haldane's work in biochemical genetics in connection with Garrod's book are discussed in "John Burden Sanderson Haldane," *Biographical Memoirs of the Royal Society*, *12* (1966), p. 232. For Haldane's theories on the relation between genes, enzymes and autocatalysis, see J. B. S. Haldane, *New Paths in Genetics* (New York: Harper & Co., 1942). For Beadle's early mention of Garrod's work see G. W. Beadle and E. L. Tatum, "Experimental Control of Development and Differentiation," *American Naturalist*, *75* (1941), p. 109, a paper read on January 1, 1941 at the AAAS meeting. See also Jan Sapp, *Where the Truth Lies: Franz Moewus and the Origins of Molecular Biology* (Cambridge: Cambridge University Press, 1990), Ch. 2.

18. A similar situation of premature discovery and its neglect is analyzed with respect to Oswald Avery's work on the transforming principle, by H. V. Wyatt, "When Does Information Become Knowledge?" *Nature*, *235* (1972), pp. 86–89.

19. For example, G. W. Beadle, "The Genetic Control of Biochemical Reactions," *Harvey Lectures*, *40* (1945), pp. 179–194; CIT, Beadle Papers, Box 3.13, "Genes and the Chemistry of Organisms" (Sigma Xi Lecture, 1945), pp. 38–40; and Box 4.73, "Genes and Biological Enigmas" (ca. 1947). On the autocatalytic theory of life see Interlude I, this volume.

20. The problem of experimental inaccessibility into the chain of reactions leading from gene to product was explained in A. H. Sturtevant, "The Use of Mosaics in the Study of the Developmental Effects of Genes," *Proceedings of the International Congress of Genetics*, *1* (1932), pp. 304–307. The pitfalls when studying these intermediate steps were discussed by A. H. Sturtevant, "Physiological Aspects of Genetics," *Annual Review of Physiology*, *3* (1941), pp. 41–56.

21. G. W. Beadle and E. L. Tatum, "Genetic Control," p. 505 (see Note 13). See also, Lily E. Kay, "Selling Pure Science in Wartime: The Biochemical Genetics of G. W. Beadle," *Journal of the History of Biology*, *22* (1989), pp. 85–98.

22. For general background see J. P. Swann, *Academic Scientists and the Pharmaceutical Industry* (Cambridge: Harvard University Press, 1987), pp. 1–56; J. P. Swann and Irvin Stewart, *Organizing Scientific Research for War* (Boston: Little, Brown, 1948), Ch. 7. On life science and war mobilization see also Chapter 6, this volume.

23. CIT, Beadle Papers, Box 3.5, Beadle to Eisenhart, December 21, 1941.

24. CIT, Beadle Papers, Box 2.54, Beadle to Weaver, November 28, 1941.

25. Ibid.

26. RAC, RG 2.2, 205D, Box 7.93, Hanson to Pauling, June 8, 1942. Foundation officer Hanson explained this patent issue to Pauling in respect to the immunology project at Caltech, as that project combined the interests of the Natural Sciences and Medical Sciences Divisions of the Rockefeller Foundation.

27. RAC, RG 1.1, 205D, Box 10.141, Beadle to Weaver, November 28, 1941.

28. RAC, RG 1.1, 205D, Box 19.191, Hanson's report, December 15–18, 1941.

29. RAC, RG 1.1, 205D, Box 10.143, Beadle to Hanson, April 3, 1943. Also G. W. Beadle, "Recollections," pp. 1–13 (see Note 2).

30. RAC, RG 1.1, 205D, Box 10.141, Wilbur to Fosdick, December 19, 1941.

31. CIT, Biology Division Records, 1936–1946, Box 11.1, Beadle to Hanson, February 20, 1942.

32. CIT, Beadle Papers, Box 2.53, Beadle to Emerson, March 14, 1942.

33. RAC, RG 1.1, 205D, Box 7.85, Hanson to Pauling, January 21, 1942. When requesting a similar assessment of Pauling's project in immunochemistry, the officer explained the nature and purpose of the Foundation's general survey.

34. RAC, RG 1.1, 205D, Box 10.142, Beadle to Hanson, April 15, 1942.

35. Ibid.

36. CIT, Biology Division Records, 1936–1946, Box 11.1, Beadle to Hanson, February 20, 1942. For the rise of California's agribusiness, see W. Bean and J. J. Rawls, *California: An Interpretive History* (New York: McGraw-Hill, 1982).

37. RAC, RG 1.1, 205D, Box 10.142, Beadle to Hanson, July 3, 1942.

38. Ibid., Hanson to Beadle, July 7, 1942.

39. Ibid., Beadle to Hanson, July 13, 1942.

40. Ibid., Beadle to Hanson, September 17, 1942; and Hanson's report, October 14, 1942.

41. Ibid., Beadle to Hanson, November 6, 1942, p. 1.

42. Ibid., p. 2.

43. RAC, RG 1.1, 205D, Box 10.144, G. W. Beadle and E. L. Tatum, "Genic Control of Biochemical Reactions in *Neurospora*," pp. 1–11; and RAC, RG 1.1, 205D, Box 10.144, December 2, 1944.

44. Ibid., File 142, Beadle to Hanson, November 24, 1942.

45. Ibid., File 143, Beadle to Hanson, January 27, 1943.

46. G. W. Beadle and E. L. Tatum "Genic Control of Biochemical Reactions in *Neurospora*," pp. 10–11 (see Note 43).

47. Ibid. Beadle also credited Sewall Wright with proposing some of the evolutionary interpretations.

48. CIT, Beadle Papers, Box 1.11, Beadle to Catchside, January 17, 1944.

49. RAC, RG 1.1, 205D, Box 10.144, Beadle to Hanson, January 17, 1944.

50. G. W. Beadle, "Genetic Control of Biochemical Reactions," *Harvey Lecture, 40* (1945), p. 193; and CIT, Beadle Papers, Box 3.4, American Chemical Society to Beadle, February 15, 1943.

51. RAC, RG 1.1, 205D, Box 10.145, Hanson report, February 15, 1945.

52. G. W. Beadle, "Recollections," p. 11 (see Note 2).

53. N. H. Horowitz, "Genetics and the Synthesis of Proteins," *Annals of the New York Academy of Sciences, 325* (1979), p. 257.

54. R. D. Hotchkiss, "The Identification of Nucleic Acids as Genetic Determinants," *Annals of the New York Academy of Sciences, 325* (1979) p. 321.

55. R. C. Olby, *The Path to the Double Helix* (London: Macmillan, 1978), p. 148.

56. CIT, Beadle Papers, Box 3.13, "Genes and the Chemistry of the Organism," pp. 38–39.

57. CIT, Beadle Papers, Box 2.36, Sturtevant to Beadle, May 13, 1945.

58. Ibid., p. 3.

59. Ibid., October 13, 1945.

60. CIT, Beadle Papers, Box 1.4, Memo from E. B. Babcock to the faculty of the University of California (undated, ca. 1945).

61. On Stanley's strong influence on his younger collaborators see R. C. Olby, "The Protein Version of the Central Dogma," *Genetics, 79* (1975), p. 11.

62. CIT, Beadle Papers, Box 11.1, Beadle to Miller, August 24, 1945.

63. Ibid., September 6, 1945.

64. Ibid., August 24, 1945.

65. CIT, Chemistry Division, Box 1.1, Beadle to Pauling, October 8, 1945.

66. CIT, Millikan Papers, Box 18.22, Millikan to Beadle, November 6, 1945, p. 1.

67. Ibid., p. 2.

At A Crossroads: Shaping of Postwar Science

Rockefeller Foundation and the New World Order

"Shall we be free to 'work as we please'?" wondered the influential Rockefeller trustee Henry Allen Moe in 1944 in his triangulations of European, American, and the Foundation's interests. No one knows what Europe will be like after the war, he admitted, but it will be different and probably less favorable to the Foundation's activities. Up until the war, he reassured his colleagues, Europeans perceived America as a neutral and generous world benefactor, making Europe particularly accessible to American ideas and technologies. This perception enabled the Foundation "to foster directly or indirectly, certain 'American-born' ideas, without provoking any suspicion of ulterior motive—everything that came from America in the way of ideas was regarded as a sort of Gospel. This will no longer be the case," he warned.

America was no longer seen as neutral and was perceived to be committed to "a policy of status quo conservatism which runs against the impulsive ideologies characterizing certain phases of the present conflict." No matter how divided the Europeans were, he warned, they were united on one point: their fear that after the war "Americans will try to impose their views and 'weltanschauung' upon them; and they are equally united in their 'determination to resist it at all costs.'" This attitude would impede the Foundation's international activities, he predicted, particularly in areas related to the social sciences.[1]

He predicted that the Foundation would be constrained by a growing governmental (United States and European) presence in international projects, threatening to leave the Foundation with "second-rate things to do." Some advocated fence-sitting until order was restored, but Moe called for swift action. The Foundation should be on the spot during that "immediate period . . . so that we may find our place in the European world while that 'world' is still 'fluid,' since if we wait for things to 'harden,' it may be more difficult for us later." Think in terms of construction rather than reconstruction, he urged.[2]

The Foundation, of course, was no stranger to war work and reconstruction. Born on the eve of World War I, the Foundation's war activities antedated its peacetime programs abroad; the European reconstruction projects guided the Foundation's interests through the interwar period. Grants and fellowships in all divisions amounting to nearly $50 million had been appropriated to institutions spanning 22 countries from Scandinavia to Eastern Europe, with Great Britain, France, Belgium, and Germany (in that order) receiving about 70 percent of the total. The International Health Division supported health services and the study of specific diseases; research projects in the medical sciences division concentrated principally on mental hygiene; the natural sciences division expanded the molecular biology program; the division of social sciences invested heavily in training centers for economics and foreign relations; and the humanities sponsored a small program of projects and fellowships. In fact, these activities, in conjunction with those of the Council on Foreign Relations, helped shape American foreign policy, investing political, military, and economic aims with the support of research. World War II, however, had a near-paralyzing effect on Foundation programs in Europe. With the exception of a handful of research and aid projects, the Foundation essentially lost contact with Europe. The world became a black box.[3]

Following Moe's suggestion, the Foundation reversed this trend promptly. Ever since V-J Day, according to Raymond B. Fosdick, all five divisions had participated in the effort to restart European research and connect it with that in the rest of the world. Modest grants were appropriated to various institutions on ad hoc bases while the trustees deliberated postwar policy questions (decisions to be formulated during the autumn of 1946). The officers of the natural science division returned to Europe in April 1946 for the first time after the war to reestablish contacts and gather data. The problem of physical reconstruction was wholly beyond Rockefeller Foundation resources, they judged; and in any case, the Foundation's program and policy were still being discussed. There was "no present indication whatsoever of any intention to diminish our past and historical interests in European science," Weaver quoted from R. B. Fosdick's 1945 annual review. "On the contrary, we have every intention of opening up our program in Europe as promptly as possible." The natural science division expected to resume long- and short-term support for experimental biology in Europe during the coming year.[4]

In December 1946 the special committee launched its new policy: accommodating its work to a new world order, where anticommunism became an organizing principle behind the intensified efforts toward internationalism and cooperation dominated by Western interests. The fundamental commitment to the human sciences was to remain, though modifications of earlier programs would ensure more effective implementation of prior goals in the face of greater challenges. The political, economic, and social institutions are unstable and disorganized, the trustees lamented; in the democratic world, moral confusion and uncertainty prevail; outside that world, fundamental values essential to the well-being of mankind are denied. It was more evident than ever that our knowledge greatly exceeded our capacity to control, they noted under the atomic cloud of guilt.[5] The Berkeley cyclotron project, financed by the Rockefeller Foundation, was diverted during the war years from its putative medical applications to the production of fissionable material. "If it had to be, I am glad we were not in our ivory tower," Fosdick later rationalized. However, they would have preferred to distance themselves from the dubious honor bestowed by Ernest O. Lawrence, who unwittingly boasted that "if it hadn't been for the RF, there would have been no atomic bomb." As Fosdick saw it, the Foundation now shared the awesome prospect of writing the "world's obituary" and shared in the responsibility of controlling the future use of atomic energy. Bracing for the challenges ahead, the Foundation would intensify its efforts in the social universe.[6]

Thus policy objectives would follow Fosdick's old dictum: "The proper study of mankind is man." Guided by the "sailing directions" of the 1930s, projects would be distributed along three broad categories: human behavior—psychiatry, psychology, social anthropology, and experimental biology; national life—projects promoting political, economic, cultural, and spiritual values; and international understanding and cooperation—support of the activities of the United Nations, cross-cultural and linguistic projects, public health, and "other constructive work in foreign countries exemplifying good neighbor morals and ethics of universal brotherhood." The committee also reiterated the Foundation's 1934 decision to place even greater emphasis on the application of knowledge; to support research as an instrument of social engineering. "What has changed," they stressed, "is the time factor. There is urgency today."[7]

The divisions' projects would reflect these objectives. The natural sciences division would not support the physical sciences, except in special cases, for instance, the completion of Caltech's Mount Palomar Observatory. The division would confine itself to the biological sciences, including nutritional science, and the agricultural work in Mexico (in operation since 1941). The medical sciences would broaden their scope to add physiological and abnormal psychology to the ongoing programs in neuropsychiatry, neurology, endocrinology, and sexual behavior. The interest in the social sciences would intensify substantially. "Even if there is government support for social science," the trustees elaborated, "the political interest in this field is likely to be such that completely private research detached from political interest will be especially desirable." The Foundation's emphasis on the social sciences (intended to redress the imbalance relative to the natural sciences) aimed at providing world leadership in these fields. The humanities would concentrate on intercultural understanding at home and abroad.[8]

Although the Foundation voted for the maintenance and expansion of the primary routes to the "the well-being of mankind throughout the world," some of the road signs had changed—those related to eugenics and social control. The linguistic discontinuities between program descriptions of the early 1930s and late 1940s are particularly striking in light of the Foundation's implicit ongoing commitment to the biology of human behavior. The policy discussions of the early 1930s still reflected, through linguistic slips and semantic variants, the struggle with the stigma and promise of eugenics, race biology, and social hygiene. Having rejected the wayward Davenportianism, the Foundation nevertheless had placed the genetic control of human behavior at the center of its "psychobiology" program during the early 1930s, acknowledging that the goal of breeding a physically and mentally superior race was predicated on fundamental research in genetics, physiology, and neurophysiology. Throughout the 1930s, 1940s, and into the 1950s the Foundation supported a number of research projects in "human genetics," including the study of hereditary diseases at the Galton Laboratory (University of London) and the genetics of mental defectiveness at the University of Copenhagen. In fact, Alan Gregg's long-standing commitment to a sound eugenics had been inaugurated as the project of behavioral genetics in 1945. Unlike the 1930s, however, the policy discourse of the late 1940s reflected the emergent taboos surrounding the Holocaust, and the reports were sanitized of all rhetorical traces of eugenic goals.[9]

Gone also the rhetoric of social control. Ubiquitous in sociology discourse, social control during the 1920s and 1930s had concrete technocratic meanings for the rational management of society. Whereas policy discussions during the 1930s were literally conceptualized in terms of individual and group control—research aimed at "control through understanding"—these terms were absent from postwar discussions; neither the biological nor the social sciences were promoted as furnishing the rational bases for social control. The term largely disappeared from postwar annual reports of the Social Science Research Council as well as from mainstream sociology textbooks. It appears that the stigma of fascism and Nazism, which infused terms such as "social hygiene," "eugenics," and "race biology" with politically specific connotations virtually eliminated the usage of such expressions as "social control" and "rationalization of human behavior." The changing politics of meaning demanded alternative rhetorics. The promotion of the study of human behavior—research and applications—was articulated during the late 1940s in terms of "understanding" and "international cooperation" through a language divested of the specific tropes that had characterized nearly half a century of corporate-academic discourse.

The policy's basic structure would also remain in place. Looking back to the 1930s, the trustees quoted from Fosdick's speech to emphasize the prescient wisdom of interdivisional cooperation.

> Whether the problem was national defense or the fighting power of the armed forces or public health or making an atomic bomb, it immediately overflowed the boundaries of the natural science into the area of social relations; so that in the end there were no boundaries, no lines of separation. Instead there was a vast single problem which had to be met with all the tools of knowledge that were available. This is the direction in which the tide is moving.[10]

Fosdick, of course, was neither praising the war nor advocating government patronage of research. He deplored the disruptive effects of war work, which placed a crippling mortgage on the future. "The feverish activity of scientists in war time is essentially not scientific," he pronounced, echoing F. B. Jewett's argument that neither war nor government were good for science. Fosdick merely elevated the cooperative structure of war work, a model that had inspired the organization of

American science since World War I.[11] What did change after 1945 was the Foundation's amplification of international, especially European, interests. As Moe had predicted, the new world order, increasingly polarized by the escalation of the Cold War, posed unprecedented challenges to American political and economic goals and to the Foundation's international aspirations. A handful of officers and trustees might have been a bit uncomfortable with the force of the "Truman doctrine," but they all cheerfully embraced the Marshall Plan. Announced during mid-1947 and launched in 1948, the plan financed a massive recovery program to Western Europe, ($12.4 billion, or 1.2 percent of the United States gross national product) to stimulate production and trade, the governing logic being that a stable free-market Europe would be resistant to the virulence of communism. A softer version of containment, the Marshall Plan supplied a framework within which Foundation activities could be aligned with foreign policy.[12]

The Foundation, in fact, helped shape the course of the Marshall Plan. Under the leadership of Allen W. Dulles (brother of Rockefeller trustee and Eisenhower's Secretary of State John Foster Dulles), the Council on Foreign Relations undertook in 1948 a research project for the Economic Cooperation Administration to formulate prescriptions for postwar Europe in relation to other parts of the world. Its conservative advisory membership included Dean Acheson, Dwight Eisenhower, David C. Lilienthal (chairman of the Atomic Energy Commission), Isaiah Bowman (president of Johns Hopkins), and Columbia physicist Isidor Rabi. Running concurrently with the Marshall Plan, the Council's project addressed policy issues of industrial and agricultural production, as well as realignment of trade patterns within Western Europe and between Europe and the rest of the world. Though the Foundation officers often liked to portray their support of scientific research as politically neutral—blind to flags and borders—in fact, foreign policy assessments and prescriptions supplied guidelines to Foundation's interests abroad.[13]

The emphasis on international relations received an even sharper definition with Chester Barnard's presidency of the Rockefeller Foundation (1948–1952). Former president of Bell Telephone, Barnard had distinguished himself through his theoretical analyses of business organizations; his professional circle including Harvard's luminaries L. J. Henderson, Elton Mayo, W. B. Donham, A. N. Whitehead, and A. L. Lowell. Viewing all organizations as "cooperative systems," Barnard (later in collaboration with Herbert A. Simon) had developed academically acclaimed models of organizational equilibrium that could be extended to all social exchanges. The presidency of the United Service Organization (USO) during World War II, followed by his work on atomic energy for the State Department, permanently neutralized his academic proclivities. By joining the Rockefeller Foundation Barnard could apply his organizational model of cooperation on an international scale.[14]

Barnard encapsulated the Foundation's tripartite agenda as "Population, Communication, and Cooperation," spanning a continuum of biosocial, sexual, and environmental problems: languages, cultural anthropology, political science, the humanities, problems of ethics, morals, and human relations. "What is the bearing of endocrinology or pathology on the moral assessment of human behavior? What is the relevancy or application of ancient codes of individual behavior to modern

Figure 22 Chester
Barnard. Courtesy of the
Rockefeller Archive Center.

corporate action?" Deep linkages existed between such seemingly disparate categories, Barnard hinted. Whatever the answers, he was unequivocal about one point:
the limits of the scientific control of nature.

> Inherent in our systematic efforts to promote the welfare of mankind there may
> be an assumption that . . . by reason and science we may govern the future of
> unborn generations in ways that *we* know are right. . . . The bombastic phrase
> "control of nature" is a by-word of the literature of the day. Do we mean that
> because we have learned to navigate the tides we shall also control them? Because
> we have learned to clothe ourselves and to provide shelter we shall also control
> the winds? We have already begun the attempts to regulate local weather. Where
> do we think we shall stop—with the control of the speed of rotation of the earth,
> of its revolution about the sun? Shall we also learn to control the chain reaction
> in the sun whence comes all our life and power? Pride goeth before a fall. All
> our efforts will promote only disaster if they are not done in the humility appro-
> priate to our ignorance, never forgetting that we have not made the earth or the
> heavens above it.[15]

Given this wisdom, it is paradoxical that Barnard did not hear the dissonance
between his poignant words and the Rockefeller Foundation's agenda in biology,
where the primary justification for studying the fundamental mechanisms of soma
and psyche was the promise of intervening in the course of human behavior on a
global scale.

The intensification of international activities paralleled the scaling down of domestic projects. True to Moe's prediction, growing government intervention in

science threatened not only to push the Foundation to the margins but also, as the officers saw it, to fundamentally subvert the form and content of the American research enterprise. The challenge began with the initiative of Senator Harley M. Kilgore, who, having been inspired by wartime mobilization, formulated by 1944 a legislative program for peacetime science. The key element in his plan was a new federal agency, a National Science Foundation, under the direct control of the President.[16] Not everyone shared his enthusiasm. The opponents to Kilgore's plan in the private and public sectors spanned the spectrum, but the most effective opposition and ultimately the lasting influence came from Vannevar Bush, director of the Carnegie Institution of Washington and principal architect of the Office of Scientific Research and Development (OSRD).

Nurtured on Hoover's vision of privatized cooperation, the conservative Bush was no lover of the New Deal or a champion of government planning of science. Nevertheless, perceiving the altered future and adapting to the winds of the atomic age, Bush proposed a new deal for research in step with some of the dominant national trends.[17] His plan, *Science the Endless Frontier*, called for a federally funded science geared toward the nation's political and economic agenda but not controlled by the President and Congress. Bush too called for establishment of a National Foundation of Science but wanted it to be a government-sanctioned and supported agency controlled directly by scientists. In addition to promoting the best science, the new foundation would give priority to research areas of importance to the national goals of defense, medicine, and civilian technology. Biology and the social sciences were conspicuous by their absence. Although in some ways the targeting process followed the Rockefeller Foundation's policy under Warren Weaver (no peer review, for one), the funding mechanism was based primarily on the OSRD contract model and that of the program of the Office of Naval Research (ONR), in scope and structure the most significant federal organization for research during the early postwar years.[18]

Warren Weaver was deeply embroiled in these debates, and in various formal and informal capacities he helped shape science policy after 1945.[19] He chaired the advisory committee of the ONR and was a key member on Bush's advisory committee (the Bowman-Tate Committee). Weaver strongly objected to Bush's scheme. The subordination of biology to medicine, the privileging of physics and engineering, and the emphasis on applied science rather than pure research were merely symptoms of a deeper malaise, Weaver thought. His opposition reflected a fundamental divergence in social philosophy. Although conceding the effectiveness of cooperative war projects and government's efficient organization, Weaver trumpeted Jewett's "proof" against war-modeled government-sponsored science. His vision of postwar science and the rebuilding of experimental biology never strayed far from the cozy nest of private sector science, tenaciously gripping the vestiges of power of the preatomic era.[20]

Like his mentors and his colleagues at the Rockefeller Foundation, Weaver opposed the reorganization of American science under federal control and objected to the establishment of the National Science Foundation. Propounding his vision of a science unbridled by political or ideological constraints, Weaver dispatched a passionate letter to the *New York Times* in 1945 titled "Free Science." Written in response to the celebrated editorial "The Lesson of the Bomb," Weaver objected vehemently to the calls for concentrated mission-oriented peacetime research; science functioned best and advanced most when scientists were free of government intervention. In the short run, under the exigencies of battle, government could coordinate technological feats, he conceded, but in the long term scientific talent could not be pushed in predetermined directions.[21] Exalting the modus operandi of private sector science, Weaver celebrated the modus vivendi of

> free scientists, following the free play of their imaginations, their curiosities, their hunches, their special prejudices, their undefended likes and dislikes . . . sometimes working in austere isolation, sometimes in cooperative teams. . . . One can no more produce fundamental and truly original work by means of some grand-over-all planning scheme for science than one can produce great sonnets by hiring poets by the hour.[22]

Apparently, Weaver saw no contradiction between his conception of a free science and the directed autonomy of life scientists who, through enormous incentives to work cooperatively rather than in austere isolation, channeled their creative energies specifically along molecular lines.

Weaver's pleas for laissez-faire science were lost amid the choruses in praise of federal sponsorship. The debates over the specifics of the government's role and the form of the National Science Foundation (NSF) dragged on during the next few years.[23] When the National Foundation was finally established in March 1950, it was a diluted version of the original Magnuson bill (Bush's version of the NSF in opposition to that of Kilgore). Whereas the visionaries of the NSF had originally conceived it as the government's chief sponsor of basic research, the agency emerged as a mere element in a vast pluralistic infrastructure of postwar science, perhaps equal in significance to the receding power of the Rockefeller Foundation in national science. As Fosdick had guessed in 1945 in response to queries about the role of the Rockefeller Foundation during the postwar era, the Foundation's program became more prominent internationally, but its activity declined domestically in areas in which federal activity developed.[24]

Beyond concrete features and mechanisms, perhaps the most profound message of *Science the Endless Frontier* was embedded in the ideology-laden frontier metaphor. Half a century had elapsed since Turner's "frontier thesis" first supplied the discursive paradigm for expansionism and industrialization. The nation's physical frontiers had since been spent. The "Great West," tamed through hydroelectric power, agribusiness, and petroleum, now also accommodated a mushrooming aerospace industry. The pioneering spirit filtered through Turner's serviceable metaphor demanded more abstract frontiers, not limited by physical boundaries. Responding to Bush's maneuver for an executive request for a postwar science policy, Franklin D. Roosevelt announced:

> New frontiers of the mind are before us, and if they are pioneered with the same vision, boldness, and drive with which we have waged this war we can create a fuller and more fruitful employment and a fuller and more fruitful life.[25]

"The pioneer spirit is still vigorous within this Nation. Science offers a largely unexplored hinterland for the pioneer who has the tools for his task," proclaimed

Bush in his letter to President Truman, accompanying his masterplan for science. "The Government should foster the opening of new frontiers and this is the modern way to do it."[26] Science, now a frontier in its own right, would justify its own endless expansion.

Designing "Big Science": Caltech's "Magnificent Plan"

In December 1945 Linus Pauling (in collaboration with Beadle) submitted to the Rockefeller Foundation a plan for molecular biology that was remarkable in its scope, structure, and language. Deploying metaphors of exploration of unknown terrains, the 25-page grant proposal charted "the great problems of biology." It did so by placing heavy emphasis on group projects organized around scientific technologies, on instrumentation as a driving force of knowledge and a dominant conceptual framework for research, and on protein chemistry as the central paradigm of the new biology. With a price tag of $6 million, the new design called for two new buildings and an annual budget of $400,000 spread over a 15-year period, the most comprehensive and costly plan in the life sciences ever proposed to the Foundation.[27]

It was somewhat of gamble, given the many anticipated changes internationally, nationally, and locally; Pauling's extravagant visions had to be executed with skill and precision. Beadle (still at Stanford) was "anxious to see how we come out in the gamble with the Rockefeller Foundation" and was in favor of expediting the grant application "for chemical biology or whatever it is to be called." He had not thought of all the projects to be included but agreed with Pauling that the main danger was lack of imagination.[28] Pauling had discussed strategies with Weaver. Regarding the question of timing, the application should be on hand by the first of the year so it receives immediate consideration, Weaver suggested. The price tag was another matter. Somewhat surprised, Weaver managed to keep himself from dropping the telephone, according to Pauling, but communicated a strong interest. Prodded by Pauling, Beadle dispatched within a few days tentative plans of "how . . . Biology might be able to spend a quarter of a million a year." He admitted that he found it staggering at first but realized "that with a bit of effort one can get used to the idea." Seriously, though, he warned Pauling that a program of such magnitude would need firm direction on all levels.[29]

At Caltech a president had not yet been appointed, and the administrative and fiscal policies were still unclear, but Pauling assured Weaver that the Chairman of the Board and the Executive Council had already endorsed the plan. "We believe that the science of biology is just entering into a period of great fundamental progress, similar to that through which physics and chemistry have passed during the last 35 years," Pauling assessed from the upper rungs of the Comtean ladder, "and we believe that the California Institute of Technology will be able to make a very significant contribution to this progress during the coming decades."[30]

With rhetoric sparked by the euphoria of victory and by the seemingly endless opportunities for growth, Pauling inaugurated his technology-based utopianism by staking the welfare of humanity on progress of fundamental science. His prose

Figure 23   Linus Pauling and George Beadle. Courtesy of the California Institute of Technology Archives.

resonated with the national temper and the Foundation's sense of mission. Comparing the project in molecular biology to the 200-inch telescope at Mount Palomar (recently funded by the Rockefeller Foundation), Pauling recycled Fosdick's images of the telescope as a "mighty symbol, a token of man's hunger for knowledge, an emblem of the undiscourageable search for truth.[31] "The 200-inch telescope at Mount Palomar will extend our region of knowledge of the great space about us far beyond the island universes a million light years away," Pauling amplified Fosdick's rhetoric. "We hope to extend similarly our understanding of the biological structures of molecular dimensions, and to provide a firm basis for later work in biology and for progress in medicine."[32] Like astrophysics, the organization of biological research would consist of project teams working in close collaboration; postwar molecular biology was being transformed into "big science"—a multiunit research enterprise centered around sophisticated technology. There was, however, a fundamental difference between the molecular biology project and astrophysics, Pauling noted. No single great instrument analogous to the 200-inch telescope could probe the problems of biology; many tools were needed. In accord with the Foundation's program, the emphasis would lie in focusing techniques of physics and chemistry on the central problems of biology.

    Pauling went to great lengths to outline these central problems. The fundamental problems of biology and their solutions, according to him, were best understood in terms of the dimensional characterization of life. Forty years ago, Pauling explained, the dark forest of the unknown stretched from somewhat less than 10^{-4} cm—the limit of the visible microscope—back indefinitely into the region of smaller dimensions. The extraordinary recent advances in physics and chemistry (notably Pauling's own work on the electronic structure of molecules) had now begun illuminating the dark molecular landscape. Physicists, chemists, and biologists were exploring the region from 10^{-4} to 10^{-6} cm with the electron microscope, and the region from 10^{-7} to 10^{-12} cm with x-ray and electron diffraction methods. Physicists penetrated the atomic nuclei, below the 10^{-12} cm range. These activities, however, bypassed a crucial locus. It was Pauling's conviction that the answers to the most basic problems of biology—the nature of growth; gene and cell replication; enzyme action; physiological activity of hemoglobin, drugs, hormones, and vitamins; and neurological functions—were hidden in the folds of the giant protein molecules, in the remaining unknown region of the dimensional forest, in the strip between 10^{-7} and 10^{-6} cm.[33]

The dimensional characterization of life was not new to Pauling. Researchers since the nineteenth century have viewed life as a continuum from molecules to higher organisms, wondering where the demarcation line between the inanimate and animate should be drawn. For example, colloid chemists had credited molecular aggregates around 10^{-6} cm—the world of neglected dimensions—with fundamental life properties. Some, such as Wendell Stanley, regarded giant virus molecules as the "twilight zone of life."[34] No biology program, though, had ever privileged the lowest order of magnitude of structure as the exclusive domain of explaining life, health, and disease. It is of special historical significance and illustrative of the Rockefeller Foundation's pervasive influence that Pauling, a physical chemist and an outsider to the dominant intellectual traditions of biology—traditions that he neither understood nor respected—could boldly enter a new field and define what was interesting, important, and worth doing in biology. By reducing biology to the narrow strip between 10^{-6} and 10^{-7} cm, Pauling succeeded in promoting a lasting trend of an interventionist technology-based biology.

As Pauling described it, the new biology was the combinatorial and convergent effect of many technologies. The exploration in the new terrain could be accomplished only by combining the versatile tools of the physical sciences. Each approach—x-rays, ultracentrifuges, light-scattering techniques, biochemical assays, isotope tracers, or the electron microscope—contributed only one piece of the biomolecular puzzle. It was the sum of these techniques, through successive approximations, that could yield insights into nature's animate secrets. The crucial feature here was that this molecular representation of nature was not merely aided by various interventionist techniques. In Pauling's technological vision of life, biological representations were predicated exclusively on intervening.

The strip between 10^{-6} and 10^{-7} consisted of invisible entities that could be "visualized" only through several modes of manipulation. In fact, Pauling stressed that the search for new and powerful methods would be one of the aims of the program. He outlined 15 group projects: x-ray studies of proteins, chromatography, molecular weight and shape of protein molecules, electron microscopy, protein chemistry, enzyme chemistry, immunochemistry, nucleic acid chemistry, serological genetics and embryology, chemical genetics, virus studies, microbiology, general physiology, metabolism, and biophysics. Some of these areas were already productive, but others were in need of development.[35]

The x-ray studies of proteins occupied a primary place in the plan. Accurate determinations of structure had been made for only two amino acids (glycine and alanine) and two peptides (diketopiperazine and glycylglycine); this work been done at Caltech mostly prior to the war. The group intended to forge ahead and obtain similar information on all the naturally occurring amino acids, several peptides, and the prosthetic groups of proteins. With the availability of the sophisticated x-ray spectrographs, cameras, and the IBM punchcard technology for speedy calculations of Fourier analysis, Corey and his team could complete structural determinations of amino acid crystals at an accelerated rate of about one crystal per year.[36]

The crystallographic investigations would be complemented by biochemical studies of amino acids and proteins. This project aimed mainly at determining the order of amino acid residues in polypeptide chains of various proteins, relying heavily on chromatographic analysis. Chromatography—the technique of separating organic mixtures in solution by their differential rates of migration in glass columns filled with different absorbing agents—was invented in Russia in 1906 but was relatively neglected until the late 1930s. Its development was greatly accelerated during the war years. In fact, Hungarian plant biochemist Leszlo Zechmeister, who joined the Crellin Laboratory staff in 1940, was one of the pioneers in the revival of chromatography.[37] Though as a foreigner Zechmeister could not engage in war

Figure 24 Punch-card instrument for mathematical calculations for x-ray and electron diffraction determination of molecular structure, ca. 1947. Courtesy of the Rockefeller Archive Center.

research, his chromatographic techniques were developed and refined by his Caltech associate, Walter A. Schroeder. Working under OSRD contracts, Schroeder developed a scheme of quantitative chromatographic analysis for systems containing minute amounts of closely related substances; he would lead the group on protein composition.

In addition to chromatography, the project would also utilize the newest methods of fractional distillation, recent technologies of radioisotope tracers, and mass-spectrographic analyses. A different but complementary vantage point on protein structure and composition would come from studies of molecular weight, shape, charge, and related physical properties of proteins. Under the direction of the physical chemist and crystallographer Richard M. Badger, results from ultracentrifuge work, measurements of birefringence of flow, scattering of light, and electrophoretic data would be coordinated with the findings of the crystallographers in Corey's groups.[38]

Electron microscope studies were already in progress. Though a German invention, the electron microscope was developed in the United States in 1940 by RCA at their headquarters in Camden, New Jersey. Within a couple of years, their electron microscopist, Caltech graduate Thomas F. Anderson, was engaged at RCA in collaborative work including studies of the tobacco mosaic virus with W. M. Stanley and bacteriophage work with Max Delbrück and Salvador Luria. Caltech's

Figure 25   Representations of molecular structure of the amino acid L-threonine. Crellin Laboratory, 1947. Courtesy of the Rockefeller Archive Center.

Figure 26 Chromatographic analysis at the Crellin Laboratory, ca. 1947. Courtesy of the Rockefeller Archive Center.

chairman of the division of physics, mathematics, and electrical engineering, W. V. Houston, had constructed an electron microscope before the war and was now carrying out studies with the RCA model to increase its resolving power to 20 Ångstroms. At that resolution it would be possible to obtain visual information about such structures as viruses and genes. Pauling, who had had his eye on that powerful instrument for a few years, was now determined to exploit it for biological work.[39]

Carl Niemann's group spanned the widest range of studies; Niemann was considered to be chemistry's rising star. He would direct the work in organic chemistry of proteins, isolating and characterizing simple naturally occurring peptides and determining their composition and activities. His group would also study the mode of action of polysaccharides in bacteria and viruses and the role of phospholipids in neurophysiological functions. Closely tied to immunochemistry on the one hand and to protein chemistry on the other, Niemann's project would also complement the proposed study of enzyme chemistry, still a "paper project." Caltech had no biochemists who could perform the kind of sophisticated physical enzymology that Pauling had in mind. Borsook's old-fashioned metabolic analyses, tied mostly to nutrition, were inadequate for the new project. Given the central role of enzymes in physiology and genetics, Pauling intended to hire first-rate enzymologists as soon as possible.[40]

Figure 27   Equipment for the isolation and characterization of bioorganic substances. Crellen Laboratory, ca. 1947. Courtesy of the Rockefeller Archive Center.

The immunochemistry project under Campbell would generally continue along
the research path carved during the war, including the controversial problem of
manufacturing artificial antibodies. It would focus also on serological reactions in
relation to bacterial toxins, allergies, and similar afflictions. As before the war,
these studies would be loosely related to serological genetics and embryology in
an effort to gain insight into the question of biological and genetic specificity, for
example, Emerson's attempts at genetic modification by using antibodies.[41]

Although proteins were Caltech's primary focus, by 1945 Pauling and Beadle
acknowledged the potential significance of nucleic acids, especially in relation to
the gene problem. The gene was now generally regarded as a nucleoprotein, and
with the impact of Oswald Avery's 1944 discovery of the "transforming principle"
it was increasingly clear that chemical and structural studies of nucleic acids could
illuminate the problem of reproduction. This area too was still a "paper project"
and required assembling a group headed by a nucleic acids specialist, a rare breed
in United States. Ironically, Columbia's biochemist Erwin Chargaff, who, inspired
by Avery's discovery, had just embarked on chromatographic analyses of nucleic
acids, was anxious to join Caltech's program. Pauling's curt reply that there was
no suitable opening in chemistry or biology at the Institute effectively closed that
line of inquiry; neither the project nor the candidate received priority consideration.[42] A nucleic acids team was not assembled until the 1950s.

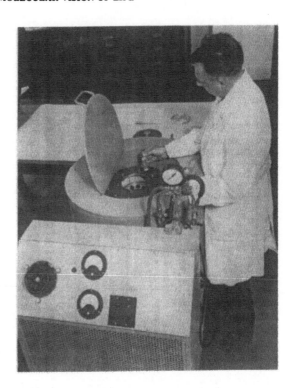

Figure 28 Crellen Laboratory's high-speed centrifuge (20,000 rpm) for molecular studies of proteins. Courtesy of the Rockefeller Archive Center.

Beadle's section of the proposal detailed the work in chemical genetics in *Neurospora*. Emphasizing the importance of combining biological and organic chemistry methods, chemical genetics was portrayed as a bridge between these two previously insular areas and a source of techniques for metabolic research. Beadle also stressed the urgency of virus studies, a point hardly requiring justification at a time when the polio epidemic was exacting a heavy toll on a terrified population. Viruses were not merely a medical topic, he emphasized. These microorganisms were crucial to understanding the gene problem and should be investigated by various physico-chemical techniques, including the electron microscope. For this central project Beadle had in mind Max Delbrück at Vanderbilt. As a closely related field, Beadle hoped to develop microbiology in order to investigate the general properties of viruses, bacteria, algae, and protozoa.[43]

General physiology needed attention as well. Utrecht physiologists C. Wiersma and J. van Harreveld, imported by Morgan a decade earlier, were plodding along their pedestrian path in neurophysiology, studying electrical conductivity in nerves. The biology division, however, lacked a strong research program in general physiology or animal physiology to supply broader interpretations to the molecular studies on proteins, nucleic acids, vitamins, and hormones and to place Borsook's work on intermediary metabolism within a more biological framework.[44]

The last area, biophysics, though thematically least coherent, was in a sense the division's lynchpin. Originating with Morgan's appointee Robert Emerson, biophysics at Caltech during the 1930s meant mostly photosynthesis research. In anticipation of Delbrück's appointment, biophysics would now also include the application of theoretical physics to phage. Being well connected with the international physics community, Delbrück was expected to make biology atractive to Caltech physicists, thereby serving a dual purpose as biophysicist (a term he despised) and virologist. A plethora of techniques were subsumed under the general rubric of biophysics—electron microscopy, x-ray crystallography, ultracentrifugation, light-scattering methods, radioactive isotope studies, mass-spectrography, and Geiger counting—cutting-edge technologies that would become synonymous with Caltech's prowess in molecular biology.[45]

Biophysics represented the extreme case of technology-driven research. If one thinks back to biology as it was practiced only a few decades earlier, one is struck by Pauling's cognitive construction of biology as a relay system of technologies having a high degree of specificity with respect to particular problems. This design, of course, captured the very essence of the molecular biology program as envisioned by Weaver. It would be a mistake, however, to advance the structuralist argument of the Foundation's imposing its policies on biological research. It would be equally simplistic to adopt a functionalist view that Pauling was merely exercising his grantsmanship skills—deploying the appropriate rhetoric to obtain funds—as Pauling trumpeted his interventionist conceptions of life on many other occasions unrelated to fund-raising. Something more profound was at work: a cognitive and social resonance. The Foundation's technocratic vision of social engineering and its representational strategies were articulated on the discursive level of program and policies; the scientist's technocratic vision of life was represented at the bench. The primacy of Caltech on the Rockefeller Foundation's roster reflected these deeply shared interests and convergent social and scientific ideologies.

The Rockefeller Foundation promised to offer significant support. The question, according to Weaver, was how much and for how long. The Foundation was in a state of transition, its future work fraught with political and fiscal uncertainties. The debates over federal sponsorship of science signaled a greater government role in American science. A corresponding decline of influence of the private sector on science policy loomed ominously on the horizon, throwing the entire future of the Rockefeller Foundation into turmoil; long-range commitments were definitely unrealistic at that time.[46] Given the global situation and America's emergent foreign policy (soon to culminate in the Marshall Plan), the clearest and most pressing need, according to Weaver, was to provide emergency relief to the war-damaged European scientific community. Ironically, just at that time, the Foundation's office was flooded with requests from American scientists "pleading that their grants be at least doubled in order to take care of expanding opportunities and increased costs." The Foundation's officers were supportive of Caltech's program and indeed viewed it as a priority for the long-term future, Weaver assured Max Mason. However, in view of the current turbulence, Weaver regretfully informed his mentor, they were in no position to think of funding in the immediate future any large fraction of the program proposed by Pauling and Beadle.[47]

During that transition period—at Caltech and the Rockefeller Foundation—the champions of the molecular biology plan lobbied vigorously on several fronts.

As Weaver assured Mason, and as he further elaborated to John D. Rockefeller III on May 1946, the program of physicochemical biology was central to the Foundation's postwar goals; the welfare of mankind hinged on the understanding of man, which in turn was grounded in fundamental biological research. Caltech's proposal, Weaver repeatedly argued to the trustees, stood at the top of any major program of the natural sciences division, being the most important plan for joining biological and chemical forces. Without a doubt, it was a "magnificent plan," Weaver wrote to Beadle, underscoring his hope and willingness to help put it into effect, even if it were on a more limited basis.[48]

As always, local support was key to securing grants. Mason, acting as liaison between the Foundation, the Institute, and Pasadena's business community, was fully aware of the Foundation's predicament; he was already at work, persuading the Institute's trustees to support the plan. "We are using every opportunity here to have our trustees become really enthusiastic over this field and make it a major, if not the major effort at the Institute," Mason promised Weaver. "We will keep on this course with a view to seeing what support in the community can be obtained when the time is ripe. I hope this will not be too far distant, for both Pauling and Beadle are at the peak of their productive capacity and we should use this to the utmost."[49]

Their productive capacities were not measured strictly by laboratory research. Beadle and Pauling were energetically tapping other sources of support for the new biology. They approached the National Foundation of Infantile Paralysis (NFIP), requesting 50 percent of the total sum on the grant, $3 million, over a period of about two decades. The NFIP was a likely generous source. Founded in 1938 by F. D. Roosevelt specifically to eradicate the dreaded polio epidemic, the NFIP had been remarkably successful in raising funds for research related to viruses and proteins. According to Weaver, it had also acquired a reputation of being an agency high on funds and short on ideas. The NFIP did make a substantial contribution to Caltech in January 1947, a five-year grant for $300,000, though a mere 10 percent of the requested amount. Apparently the fact that they considered Pauling's work on artificial antibodies to be nonsense and objected to the lack of viable virology research at Caltech at the time of application, contributed to their reticence to make a greater commitment.[50]

While in New York in March 1946 to negotiate with the NFIP, Pauling also met with Weaver to impress upon him the gravity of the impending emergency at Caltech. Beadle's Rockefeller grant and Pauling's own grant for immunology would expire in a few months. Many of the staff members had been on OSRD salaries, which would expire during the summer, and the original $70,000 per annum support level had now dropped to half. The immediate outcome of the emergency, Pauling reiterated a couple of weeks later, was that there were no funds for the program on the structure of proteins. Could they at least obtain a one-year grant to continue these crucial investigations? A month later, a $50,000 grant for 1946–1947 was awarded to Pauling and Beadle for joint researches in chemistry and biology. Although only a short-term infusion, the funds kept Caltech's molecular biology program at a steady rate of growth during the transition phase.[51]

In the meantime the period of administrative vagueness at Caltech was drawing to a close. On May 1946 physicist Lee Alvin DuBridge, Mason's protégé, assumed the Institute's presidency. By most accounts it was a splendid appointment, the choosers' first choice. At an administratively tender age of 45, DuBridge had already garnered an impressive array of titles to his name: former chairman at the University of Rochester Physics Department, dean of the Faculty of Arts and Sciences, and director of MIT's celebrated Radiation Laboratory (1940–1946). Almost single-handedly, beginning with 48 employees working in a space of 4050 square feet on radar—a "new toy with possibilities"—DuBridge had built up the "Rad Lab" into the largest single-purpose scientific research plan in history, larger even than the atomic bomb project. He had been a reasonably successful experimental physicist but distinguished himself mainly by his enormous capabilities in teaching, organizing, and managing. Having led war projects and large scientific teams, DuBridge was experienced in fostering cooperative relations between academe, industry, and the military.[52]

It would be difficult to imagine a candidate better connected with Caltech's establishment or better suited to their goals. Born in the Midwest in 1902, DuBridge, like his colleague Weaver, studied with Max Mason at the University of Wisconsin, receiving his doctorate in physics in 1926. As recipient of the Rockefeller Foundation's National Research Council Fellowship, DuBridge had worked at Caltech with Millikan during 1926–1928, thereafter climbing steadily through the ranks to reach by the 1930s one of the top places among second-generation physicists in America. An associate editor of *Review of Scientific Instruments* and president of the Rochester Optical Society, DuBridge had always shown a predilection for the technological aspects of science, a particularly important asset for an institute of science and technology.[53]

Figure 29 Lee A. DuBridge, 1948. Courtesy of the California Institute of Technology Archives.

According to DuBridge, it was Mason's influence, more than any other factor, that persuaded him to accept that position. "It was your persuasiveness that brought me to the Institute in the first place, and it was your support during the early years which was so valuable to me," DuBridge recalled five years later on the occasion of Mason's retirement. "For all that you have done for me since I first sat in your class in Madison 28 years ago, I am most sincerely grateful."[54] Weaver too was elated,

> Some time ago, and after extensive conversations with Max Mason, Jim Page, and others, I wrote them a letter in which I listed individuals which I thought they ought to consider. Looking back at the letter I find that I had you down as the No. 1 candidate, and the other men mentioned were candidates 20 to 25, inclusive, there being nobody between you and No. 20.

Weaver seemed particularly delighted with the new professional affiliation. From the days of his apprenticeship with Millikan during the early 1920s, Weaver was still technically on Caltech's faculty roster, he reminded DuBridge, "Thus I welcome you as my new boss!"[55] As James McKeen Cattell had observed more than two decades earlier, Caltech and the Rockefeller Foundation often formed incestuous relationships.

Within a month, at Weaver's request, DuBridge had canvassed Caltech's financial profile and particularly the situation at the Gates and Crellin Laboratories. Once again, Pauling's immunology was a credibility gap in the molecular biology program, and DuBridge had to assure Weaver that just as soon as "the money is made available, Pauling will secure a competent immunologist, and I shall try to see to it that he does." DuBridge also gathered some embarrassing information regarding improper allocations of funds. "I can only assure you," he wrote to Weaver, "that I am anxious, indeed, to see this entire matter straightened out, . . . I shall try to see that if this new grant goes through that we start with a clean slate and make it perfectly clear just where the money is going."[56]

The money flow would be substantially heavier than previously. According to DuBridge, Caltech needed about $20 million within the immediate future for additional buildings and endowment. He fully appreciated the Foundation's state of flux and its inability to commit at this time the vast sum specified in the molecular biology grant, but he urged the Foundation to support Pauling's program. "All the information I have suggests that it is not only eminently worthwhile but might be one of the most important enterprises in the country."[57]

Caltech's proposal was never actually turned down, but it would be nearly two more years before the Rockefeller Foundation's trustees agreed to commit vast long-term resources to Caltech's program. Though more limited in scope, their commitment would eventually be encouraged by the remarkable expansion of the molecular biology program over a period of two years and its diverse resource base. In addition to another one-year grant from the Rockefeller Foundation, the grant from NFIP, the Gosney Research Fund, and the Eversole Lecture Fund, life science at Caltech was supported by 1948 through a wide network of grants and contracts from government, pharmaceutical companies, and agricultural industries, as well as by gifts and endowments from Southern California's business community. A substantial fraction of these resources was channeled into molecular biology.

Support for the biology division came from the American Cancer Society, the Loomies Institute for Scientific Research, F. S. Markham, The Markle Foundation, Merck and Company, Charles E. Merril, The Nutrition Foundation, Pioneer Hi-Bred Corn Company, Purdue University Corn Research Fund, The Sugar Research Foundation, Sunset Magazine, United States Public Health Service, the United States Rubber Company, Williams Waterman Fund for the Combat of Dietary Diseases, and the Research Corporation. Research in protein synthesis and radiation genetics was being supported through contracts with the United States Atomic Energy Commission, operating through the Office of Naval Research; and research in corn cytogenetics through the United States Department of Agriculture. It was evident that Beadle's fund-raising skills and the strong links he had forged with commercial concerns and government agencies while at Stanford were now a boon to Caltech's biology division and to the molecular biology program. Clearly his managerial style was an effective complement to that of Pauling.[58]

DuBridge proudly reported to Weaver in 1948 the swelling budgets of chemistry and biology under his administration. Chemistry's budget climbed from $222,000 in 1941–1942 to $502,000 in 1947–1948. More significantly, the budget of biology had risen from $134,500 in 1941–1942 to $403,000 in 1947–1948.[59] Undergraduate enrollment grew from 640 just before the war to 800, and the graduate school now had 600 students. Beadle was campaigning to triple the number in the biology undergraduate body and to double the graduate student body in biology.[60]

Caltech had already taken large strides to realize the "magnificent plan." Two senior professorial appointments had been made. John G. Kirkwood, a prominent protein chemist and an expert on electrophoresis from Cornell would lead the research on physical chemistry of proteins; and Max Delbrück from Vanderbilt, the maverick leader of phage genetics, would head virology and play a leading role in biophysics and electron microscopy. Norman H. Horowitz, a graduate of Caltech's biology division who had worked with Beadle at Stanford, was now appointed associate professor of chemical genetics to study the relation between genes and enzymes in *Neurospora*. Ray D. Owen, an immunologist from the University of Wisconsin and former Gosney Fellow, was appointed associate professor to head the team in serological genetics.[61]

A spirit of great optimism enveloped Caltech. The bonds between science, government, and industry promised to accelerate the production of knowledge, stimulating technologies that would penetrate life's deepest mysteries and control man's biological destiny. The exploration of the endless scientific frontier coincided with the dawn of an era of seemingly unlimited opportunities and resources. "I am very optimistic about the way things are going," Beadle wrote to Weaver in May 1947. "Lee DuBridge is doing a wonderful job and the effects are very evident. Morale throughout the Institute has taken a marked upward turn during the past year."[62]

When DuBridge resubmitted the grand plan for molecular biology in February 1948 to the Rockefeller Foundation, he could point to the striking changes that had taken place since 1946. Many of the activities originally planned were already in effect; some had gathered considerable momentum. In keeping with Foundation policies, the grant would stimulate rather than initiate growth. "It can now be said that the program proposed to the Foundation is no longer largely a paper one, but is a program actually under way, at least in rudimentary form, but which needs only substantial strengthening and expansion in certain critical areas." Despite the diverse resource base, the requested amount did not change substantially. Due to rising costs and salaries, and because of constraints inherent in some of the other short-term grants, DuBridge explained, the support requested from the Foundation came to $350,000 per year for 15 years.[63]

No harm in trying. They were skilled maneuvers indeed, but the chance for a financial investment of that magnitude during the turbulent period at the Rockefeller Foundation was negligible. Such a financial commitment was completely out of line; and, according to R. B. Fosdick, 15 years was too long a period. Weaver fought heroically for his most important project. In a confidential letter to DuBridge on the outcome of his meeting with the trustees, Weaver reported that he had hoped it would be feasible to propose a $1 million project for 10 years, but to no avail. In characteristically hyperbolic Foundation language he explained that "the somewhat terrifying atmosphere of the moment, and the fact that many of our Trustees have such responsibilities and such knowledge as makes them peculiarly sensitive to the present uncertainty, has led to the conclusion that it would be definitely unwise to propose a grant as long as this or for as much money. . . . I remain very hopeful that they will make a substantial contribution, and in any event, I can assure you that the officers will present the item with all the persuasiveness they can command, for we feel very strongly that it is absolutely first-rate."[64]

Given the circumstances, it was indeed a great coup for Weaver and Caltech to have persuaded the trustees to appropriate a $700,000 grant in 1948 for a period of seven years—still one of the largest research grants ever to be received by the Institute. It was great cause for celebration, and DuBridge saw to it that it received ample publicity. Sporting a different service role, in step with postwar trends, the program deployed a rhetoric of legitimation quite different from the discourse of the 1930s. Though not affiliated with medical schools, Caltech's biology was now portrayed as advancing medical knowledge. Under DuBridge's guidance the news release (presumably drafted by Beadle) described the long-term grant in molecular biology as an attack on the fundamental problems of medicine: molecular structure of proteins and nucleic acids, self-replication of viruses and genes, and problems of cellular differentiation.[65]

Gone was the rhetoric of biological improvements of the race, which still trailed behind the grant for biology and chemistry just a decade earlier. Gone also was Caltech's connection to the old eugenics. With the death of E. S. Gosney in 1942, the trustees of the Human Betterment Foundation (including Millikan and a few Caltech trustees) agreed that the Foundation's interests would best be served by transferring its activities to Caltech. In October 1943 an agreement was drawn up, dissolving the Human Betterment Foundation as such and turning over its assets to the Institute. The Institute, in turn, agreed to use these resources "and the proceeds thereof to establish the Gosney Research Fund, the income from which will be devoted in perpetuity to the promotion of research into the biological bases of human qualities and for making known the results of research for the public interest." Linguistically watered down from its eugenic potency, the Gosney Fund would support postdoctoral fellowships in "those branches of biological science basic to our understanding of human welfare."[66]

The *New York Times* too refurbished its prose. The announcement emphasized the medical connection: "$700,000 to Trace Polio and Cancer—Rockefeller Grant is Made for California Institute Research in Molecular Biology." The article stressed that the future progress of medical science rested on the new knowledge in chemistry and biology and that modern medicine could no longer be content with cures that were not based on fundamental scientific understanding. "Our program is not aimed directly at a cure of human disease," DuBridge was quoted as saying, "nevertheless the knowledge which is sought is basic to the development of future treatment of some of mankind's most serious ailments."[67]

It was not the only instance where the Caltech–Rockefeller axis exerted considerable power over its image-making in the public media. Soon after the award of the Rockefeller grant, George W. Gray, America's most prominent science writer and staff member of the Rockefeller Foundation, visited Caltech to write a story for *Scientific American*. No stranger to that network, Gray had visited Caltech before to prepare Confidential Monthly Reports on Caltech for the Foundation's trustees. The cozy collaborative relations resulted in his article entitled "Beadle and Pauling," placing the program in the vanguard of life science research. Dramatically illustrated with pictures describing the protein template idea and the formation of antibodies (a central project in the program), Gray's article recounted the origins and scope of Caltech's molecular biology. Extolling the virtues of the molecular approach to life, he pointed to the "striking example of present-day partnership of chemistry and biology—a union which had been solemnized at the Institute in a large new joint project of its chemical and biological divisions."[68]

A keen and seasoned observer of science, Gray understood the problems of service role and the tensions at Caltech between basic research and medicine. The chemists at the Institute, Gray reported, were not consciously after therapeutic agents or clinical application. Theirs was a quest for fundamental knowledge, a systematic search into the behavior of the body's giant molecules, such as genes, antibodies, viruses, hormones, biological pigments, and related structures. "Science is still far from completely analyzing these biological agents," Beadle was quoted as saying, "but the investigations tend to show that the molecular form known as protein is the key structure. Apparently most of the bodies that we are studying in our program are either simple proteins or conjugated proteins." When the functions and structures of these giant molecules were elucidated, Beadle went on, the solution of practical problems in medicine would follow inevitably. The utopian vision hinged on reciprocity between representing and intervening. "Fortunately medical men are able to use biologically active molecules without knowing very much about them; but they crave control of processes and results which fundamental knowledge would give. Along this road, the scientists believe, lies the unmasking of stubborn mysteries; the elucidation of cancer, of aging, of the divine spark itself."[69]

Notes

1. RAC, RG1.2, 700, Box 1110, H. A. Moe, memoranda, March 13, 1944, pp. 1-2; and June 5, 1944, p. 1.

2. Ibid., June 5, 1944, p. 3; and March 13, 1944, p. 3.

3. RAC, RG1.2, 700, Box 1110; T. Appleget, "The Rockefeller Foundation in Europe," survey 1913-1946; RG1.2, 100S, Box 57.440, Council on Foreign Relations, Aid to Europe, January 21, 1949.

4. RAC, RG3, 915, Box 1.4; Warren Weaver's report, April 23, 1946.

5. RAC, RG3.2, 900, Box 28.155, Report of the Special Committee on Policy and Program, 1946, p. 2.

6. RAC, RG1.1, 205D, Box 13.191; memos from R. B. Fosdick, September 12 and September 20, 1945.

7. RAC, RG3.2, 900, Box 28.155, Report of the Special Committee on Policy and Program, 1946; quote on p. 7.

8. Ibid., quote on p. 12.

9. For the Foundation's support of projects related to eugenics see, for example, the Rockefeller Foundation Annual Report, 1946, pp. 117-118 and 1948, pp. 137-141. Daniel J. Kevles has characterized the changing interests in human heredity and reproduction during the 1930s as "reform eugenics," *In the Name of Eugenics* (Berkeley: University of California Press, 1985), Ch. 11. Others have identified longer and sharper lines of continuity into the 1950s; see, for example, Diane B. Paul, "The Rockefeller Foundation and the Origins of Behavior Genetics," pp. 262–283, and Garland E. Allen, "Old Wine in New Bottles: From Eugenics to Population Control in the Work of Raymond Pearl," pp. 231-262, in Keith Benson, Jane Maienschein and Ronald Raigner, eds., *The Expansion of American Biology* (New Brunswick, NJ: Rutgers University Press, 1991). See also Barry Mehler and G. E. Allen, "Sources in the Study of Eugenics," *The Mendel Newsletter, 14* (1977).

10. RAC, RG3.2, 900, Box 28.155, Report of the Special Committee on Policy and Program, 1946, p. 14.

11. Rockefeller Foundation Annual Report, 1946, pp. 24-25; Frank B. Jewett, "The Promise of Technology," *Science, 99* (1944), pp. 1-6.

12. Melvyn Leffler, "The American Concept of National Security and the Beginning of the Cold War, 1945-1948," *American Historical Review, 89* (1984), pp. 346-381; Imanuel Wexler, *The Marshall Plan Revisited* (Westport, CT: Greenwood Press, 1983).

13. RAC, RG1.2, 100S, Box 57.440, Council on Foreign Relation, 1948-1953, passim. Consider, for example, the correlation between foreign policy interests and the Foundation's programs. The support of physics in Brazil: RAC, RG3, 915, Box 4.35, Warren Weaver, August 31, 1948, and Weaver to Chester I. Barnard, January 4, 1950. Public health projects in Crete: RAC, RG3, 900, Box 17.127, Organization, R. B. Fosdick, March 31, 1948; and RG1.1, 749, Box 2.16. In Greece: D. E. Wright to G. K. Strode, July 19, 1948; and RG1.1, 749, Box 1.4, Health, Organizational Study of the Greek Ministry of Hygiene, August 12, 1949. For implementation of the Marshall Plan see Chester I. Barnard, "Central European Rehabilitation," Rockefeller Foundation Annual Report, 1948, pp. 45-48; and Doris Zallen, "The Rockefeller Foundation and French Research," *Cahiers pour L'histoire du CNRS, No. 5* (1989), pp. 35-58. For support related to the rehabilitation of Japan see Warren Weaver Diary, RAC, RG12.1, Box 68, March 17, 1947, pp. 30-31. For postwar reconstruction of Japanese science see James R. Bartholomew, *The Formation of Science in Japan* (New Haven: Yale University Press, 1989), Epilogue, and Kazuo Kawai, *Japan's American Interlude* (Chicago: University of Chicago Press, 1960).

14. F. J. Roethlisberger, "Barnard, Chester I." *International Encyclopedia of Social Sciences*, Vol. 2 (New York: Macmillan, 1968), pp. 12-13.

15. Rockefeller Foundation Annual Report, 1948, "President's Review," pp. 12, 22, and 24.

16. Daniel J. Kevles, *The Physicists* (New York: Vintage Books, 1977 edition), p. 344; Daniel J. Kevles,"The National Science Foundation and the Debate over Postwar Research Policy, 1942–1945, *Isis, 68* (1977), pp. 5-26; and David F. Noble, *Forces of Production* (New York: Oxford University Press, 1986), Ch. 1.

17. Nathan Reingold, "Vannevar Bush's New Deal for Research, or the Triumph of the Old Order," *Historical Studies in the Physical Sciences, 17* (1987), pp. 299–344.

18. Vannevar Bush, *Science the Endless Frontier* (Washington, DC: National Science Foundation, reprinted 1980); Daniel J. Kevles, *The Physicists*, pp. 347–348 (see Note 16); and Nathan Reingold, "Vannevar Bush's New Deal for Research," p. 42 (see Note 17).

19. Nathan Reingold, "Choosing the Future: The U.S. Research Community: 1944–1946," paper prepared for the conference, "Science and the Federal Patron," National Museum of American History, Smithsonian Institution, September 1989.

20. RAC, RG3, 915, Box 2.13, "Natural Sciences," Warren Weaver, p. 32.

21. Warren Weaver, "Free Science," in his *Science and Imagination* (New York: Basic Books, 1967), pp. 10–14 (slightly modified from Weaver's "Occasional Pamphlet," No. 3, November 1945, Society for Freedom in Science).

22. Ibid., pp. 12–13.

23. Daniel J. Kevles, *The Physicists*, pp. 357–358 (see Note 16).

24. Nathan Reingold, "Choosing the Future," p. 17 (see Note 19).

25. For the discussion on Turner's "frontier thesis," see Chapter 2, this volume. F. D. Roosevelt's quote (November 17, 1944) in Vannevar Bush, *Science the Endless Frontier*, p. iv (see Note 18). For the background to the letter see Daniel J. Kevles, *The Physicists*, p. 347 (see Note 16).

26. Vannevar Bush, *Science the Endless Frontier*, p. v (see Note 18).

27. RAC, RG1.1, 205D, Box 4.23, grant application, Pauling to Weaver, December 4, 1945.

28. CIT, Chemistry Division, Box 1.1, Beadle to Pauling, October 28, 1945.

29. Ibid., Pauling to Beadle, November 6, 1945; and Beadle to Pauling, November 10, 1945.

30. RAC, RG1.1, 205D, Box 4.23, grant application, Pauling to Weaver, December 4, 1945 "The Possibilities for Progress in the Fields of Biology and Biological Chemistry," p. 1.

31. R. B. Fosdick, Rockefeller Foundation Annual Report (1945), p. 7.

32. RAC, RG1.1, 205D, Box 4.23, grant application, Pauling to Weaver, December 4, 1945, p. 7.

33. Ibid., p. 2.

34. Robert Olby, "Structural and Dynamical Explanations in the World of Neglected Dimensions," in T. J. Horder, J. A. Witkowski, and C. C. Wylie, eds., *A History of Embryology* (Cambridge: Cambridge University Press, 1983), pp. 275–308; and Lily E. Kay, "W. M. Stanley's Crystallization of the Tobacco Mosaic Virus, 1930–1940," *Isis, 77* (1986), pp. 450–472.

35. RAC, RG1.1, 205D, Box 4.23, grant application, Pauling to Weaver, December 4, 1945, Ibid., p. 3. For a more extended discussion see Ian Hacking, *Representing and Intervening* (Cambridge: Cambridge University Press, 1983), especially pp. 130–146; Evelyn Fox Keller, "Physics and the Emergence of Molecular Biology," *Journal of the History of Biology, 23* (1990), pp. 389–409; and Lily E. Kay, "Life as Technology: Representing, Intervening, and Molecularizing," paper presented at the Boston Colloquium for the Philosophy of Science, April 16, 1991.

36. RAC, RG1.1, 205D, Box 4.23, grant application, Pauling to Weaver, December 4, 1945, Ibid., "Some Detailed Statements about the Work to be Done," p. 1.

37. Ibid., p. 3. Historical aspects of chromatography, Zechmeister's contributions, and Schroder's innovations are described in the proceedings of the conference on chromatography, *Annals of the New York Academy of Sciences, 48* (1948), pp. 141–325.

38. RAC, RG1.1, 205D, Box 4.23, "Detailed Statements," pp. 4–5.

39. Ibid., pp. 5–6. For Anderson's account of his early career in electron microscopy see his "Electro Microscopy of Phages," in J. Cairns, G. S. Stent, J. D. Watson, eds., *Phage and the Origins of Molecular Biology* (New York: Cold Spring Harbor Laboratory of Quantitative Biology, 1966), pp. 63–78.

40. RAC, RG1.1, 205D, Box 4.23, "Detailed Statement," pp. 7–10.

41. Ibid., pp. 11–13.

42. Ibid., pp. 10–11; APS, Chargaff Papers, Box 4.2 (O,P,R), Chargaff to Pauling, August 14, 1945 and Pauling to Chargaff, September 17, 1945.

43. RAC, RG1.1, 205D, Box 4.23, "Detailed Statement," pp. 16–21.

44. Ibid., pp. 21–24.

45. Ibid., pp. 24–25.

46. RAC, RG1.1, 205D, Box 4.23, Weaver's interview with Pauling, March 22, 1946.

47. Ibid., Weaver to Mason, January 13, 1946.

48. Ibid., Weaver to Beadle, January 28, 1946.

49. Ibid., Mason to Weaver, January 21, 1946.

50. Ibid., Weaver's report, March 25, 1946, and Weaver's interview with T. Rivers, April 8, 1946.

51. Ibid., Weaver's interview with Pauling, March 22, 1946.

52. *Current Biography* (New York: W. Wilson Company, 1948), pp. 161–163.

53. Ibid.

54. A statement prepared by Lee DuBridge shortly after Mason's death. Quoted in Warren Weaver, "Max Mason, 1877–1961," Biographical Memoirs of the *National Academy of Sciences, 37* (1964), p. 225; and CIT, DuBridge Papers, Box 126.1, DuBridge to Mason, November 6, 1951.

55. CIT, DuBridge Papers, Box 112.1, Weaver to DuBridge, April 25, 1946.

56. RAC, RG1.1, 205D, Box 4.23, DuBridge to Weaver, June 15, 1945.

57. Ibid.

58. CIT, "Annual Report to the Board of Trustees," *Bulletin of the California Institute of Technology, 58* (1947–1948), pp. 52–60.

59. RAC, RG1.1, 205D, Box 4.25, DuBridge to Weaver, February 23, 1948.

60. RAC, RG1.1, 205D, Box 4.24, MC's interview at Caltech, February 20, 21, 26, 1947.

61. Ibid., Beadle to Weaver, May 24, 1947.

62. Ibid.

63. RAC, RG1.1, 205D, Box 4.25, DuBridge to Weaver, February 23, 1948.

64. Ibid., Weaver to DuBridge, March 22, 1948.

65. Ibid., draft attached to DuBridge's letter to Weaver, April 14, 1948.

66. George W. Beadle, "The Gosney Research Fund," *Engineering and Science* (May 1947), p. 26.

67. *New York Times*, April 29, 1948, 9:5.

68. CIT, Chemistry Division, Box 2.1, Gray to Pauling, August 3, 1946; Box 2.3, Gray to Beadle, June 22, 1948. Also G. W. Gray, "Beadle and Pauling," *Scientific American, 180*, No. 5 (1949), p. 16.

69. G. W. Gray, "Beadle and Pauling," p. 21 (see Note 68).

CHAPTER 8

Molecular Empire (1946–1953)

Life in a Black Box: Rise of Delbrück's Phage School

During the decade after World War II, through the partnership of Beadle and Pauling and the implementations of the "magnificent plan," Caltech became the leading edge of the nascent discipline of molecular biology. Virus research occupied a strategic position within these plans: It defined fundamental researches in genetics, microbiology, and biophysics; and by legitimizing the claims to medical utility it was well positioned to attract private and federal support. Thus one of Beadle's first steps upon assuming the chairmanship of the biology division in 1946 was to offer the German physicist Max Delbrück, now at Vanderbilt University, a professorship in biology. "You know our plans for the future development of physical and chemical biology at the Institute," Beadle wrote. "We need you as a key member of a team that will further elaborate them as well as carry them out."[1]

At 40 years of age, Delbrück's impact as a man of action, a charismatic leader of an expanding phage school, was matched by his reputation as a "think-man," whose theoretical grasp of problems in physics, chemistry, and biology endowed him with an unusual ability to integrate knowledge across scientific boundaries, an approach of particular importance to Caltech's interdisciplinary agenda. A cultivated eccentric, similar to Morgan, Delbrück's iconoclastic individualism was well balanced by his exceptional talent for group cooperation, a prized asset at Caltech. As a Rockefeller officer had observed, since his early career in Europe Delbrück had consistently developed "group interest" in science. "All this makes it evident that Delbrück is not only cooperative, enthusiastic and broadly interested in science, but he also has the drive and personality which makes such attributes effective."[2]

*   *   *

Seven years had elapsed since Delbrück's departure from Caltech in 1940. During that time, despite his primary duties as associate professor of physics at Vanderbilt, he had managed to build with the aid of the Rockefeller Foundation a remarkably attractive program in viral genetics. His work was supplying provocative insights into what he had termed "the riddle of life"—the subcellular mechanisms of replication and mutation—and he was attracting physicists and chemists to molecular biology. Although Vanderbilt had been sympathetic to Delbrück's endeavors in biology, he found little time and only meager resources for his phage work during the academic year. It was Cold Spring Harbor, with its summer symposia and cross-disciplinary research in quantitative biology and the energetic cooperation of geneticist Milislav Demerec, that enabled Delbrück to maintain contact with the community of geneticists and to establish his career during the years 1940–1947.[3]

The 1941 Cold Spring Harbor Symposium on genes and chromosomes, chaired by H. J. Muller, was particularly important for Delbrück's career; there he made his phage debut. An invitation to present a paper at that prestigious meeting signaled professional recognition that placed Delbrück visibly on the map of genetics research. Ironically it was probably his worst and least representative paper, though historically a significant one. Like most of the symposium's offerings on the structure and material properties of chromosomes, it dealt with protein chemistry. Like most of the participants, Delbrück expressed the dominant view that self-replication was analogous to enzymatic autocatalytic reaction, and that the active hereditary component of chromosomes was an enzyme or enzyme-like protein. Although Delbrück deprecated biochemistry (his nascent program blended mathematical physics and the formalism of Morgan's genetics), he was clearly intrigued with the material basis of replication. His symposium paper, "A theory of autocatalytic synthesis of polypeptides and its application to the problem of chromosome reproduction," was a heroic attempt to postulate a chemical mechanism of reproduction based on the autocatalytic theory of life. "My own feeling toward its contents are divided," he confessed to Linus Pauling a few months later. "I could not persuade myself either into believing or into not believing that the idea contained some truth, but I thought the idea was too good to be discarded right away."[4] Perhaps not right away, but he would discard it later with little tolerance for those who had promulgated the autocatalytic theory before him.

That summer of 1941, Delbrück also formed the social nucleus of the phage school by joining forces with Salvador Luria, who became his life-long collaborator and with whom he and Alfred Hershey would share the 1969 Nobel Prize.[5] Symposium organizer Demerec, reporting the success of that summer's events, described Delbrück's and Luria's contributions with special enthusiasm. In a series of experiments, utilizing two strains of phage, these researchers had begun to generate "unexpectedly good results," findings that offered a perplexing glimpse into the replication process inside the cellular "black box."

As early as 1941 the "unexpectedly good results" revealed unforeseen complexities in Delbrück's simple "black box" approach to the gene problem. The strategy for penetrating the cell by using a combination of two viruses (mixed infection) backfired, and the bacteria persistently behaved as if they were infected by only one virus, characterized by its plaque shape and burst size. This exclusion phenomenon provided the first clue that intracellular events, which involved virus and bacterium, represented mechanisms far more tangled than had been previously envisioned. Additional complications ensued the next summer, when collaborative experiments indicated that bacteria undergoing spontaneous mutations were resistant to phage attack. That too implied that replication mechanisms of phage were intertwined with those of bacteria in complicated ways.[6]

In 1943, with the aid of an electron microscope, Delbrück was able to "see" bacteriophage for the first time. The micrographs, which were prepared in collaboration with Caltech graduate and biophysicist T. F. Anderson, revealed tadpole-shaped or sperm-like organisms with texturally distinct head and tails, varying with different phage strains. As Delbrück had to admit, "while no harm is done by calling viruses molecules, such terminology should not prejudice our view regarding the biological status of viruses, which has yet to be elucidated."[7] The infection mechanisms too had yet to be explained, as, paradoxically, the micrographs indicated that the infecting phage itself never entered the bacterial cell.

This observation, together with the phenomenon of mutual exclusion, strongly suggested to Delbrück the common analogy between phage penetration of bacteria and the interaction of sperm and egg. He observed that penetration of the first virus made the cell membrane impermeable to other virus particles, just as fertilization of an egg by one spermatozoon made the egg membrane impermeable to other spermatozoa. This analogy, conferring a high degree of specificity between virus proteins and bacteria, complemented well the studies on protein specificity between antibodies and antigen. It lent support to the view that the specificity of proteins involved in genetic replication was in some ways related to the specificity involved in the formation of antibodies, a mechanism explicated by Pauling's protein template theory and further elaborated by Caltech embryologist Albert Tyler. In fact, some of the immunological studies of phage conducted for several years by Alfred D. Hershey (Delbrück's close collaborator), at the Carnegie Institution of Washington, were based on just these concepts of biological specificity.[8]

The unforseen setback to the "black box" model and the incongruities of phage replication hardly discouraged Delbrück. On the contrary, realizing the magnitude of the problem and the expanding scope of his project, he began to mobilize collaborators and institutional resources. He had only a couple of postdoctoral fellows at Vanderbilt in 1943; but Luria, who had just moved to Indiana University, would assemble a phage group within a short time. Mark Adams at New York University was beginning to conduct phage experiments, and T. F. Anderson would continue the electron microscope studies of phage. Hershey was already pursuing collaborative phage experiments with bacteriologists and immunologists at Washington University, and Demerec of Cold Spring Harbor, who himself had just begun experimenting with phage, was delighted to spread the new scientific gospel during summer meetings.

By 1944, Delbrück's third summer at Cold Spring Harbor, the network of phage workers under Delbrück's leadership decided to standardize the phage systems, as well as to establish the Phage Information Service (modeled after the *Drosophila*

Information Service). Unbridled individualism was curbed. Instead of each investigator developing his own collection of phages and host bacteria, Delbrück insisted that all phage workers concentrate their efforts on a set of seven phages, named T_1 through T_7 and defined according to morphological, serological, and growth properties. He also insisted that experimental conditions be strictly defined and standardized. With mounting complexities and a rapidly growing body of information, these laboratory conventions facilitated comparison between studies and improved the dissemination of knowledge between phage workers under Delbrück's coordination. These practices also helped consolidate his disciplinary power.[9]

Delbrück readily admitted to gross underestimation of the phage problem and to his mobilization campaign. With the characteristic blend of playful arrogance and reluctant humility, he impressed upon his audience at his 1945 Harvey Lecture that during the eight years of phage research he had not come any closer to solving the problem of genetic replication. He conceded having made a "slight mistake": Instead of a few months it would take a few decades and the help of a few dozen people willing to join him in solving the riddle of life. To optimize the enlistment process, Delbrück organized in 1945 an annual phage course at Cold Spring Harbor, thus setting off a rapid chain reaction of recruitment. Even after his return to Caltech in 1947, when his laboratory hosted frequent phage conferences and teemed with undergraduate and graduate students, postdoctoral fellows, and visiting professors, Delbrück upheld the Cold Spring Harbor tradition, continuing the course well into the 1960s. Phage workers by then numbered in the hundreds.[10]

Designed primarily for those who either expected to work with phage or wished to obtain a better understanding of molecular genetics, the course attracted scientists interested in borderland problems of biology, physics, and chemistry. In addition to having the students master simple laboratory procedures, Delbrück laid great stress on the theoretical foundations of phage work, requiring fairly sophisticated mathematical preparations, at least from the standpoint of biologists. The prerequisites for participating in the phage course (facility in calculus and differential equations and thorough familiarity with exponential functions, statistics, and numerical analysis) were strictly enforced with admission tests that obviously favored those trained in the exact sciences. Successful "graduates" of the course often stayed on for several weeks, conducting experiments directed by Delbrück and submitting their results to his unyielding scrutiny.[11]

Phage work under Delbrück was especially hospitable to physicists. Through his Vanderbilt position and his close relationship with Niels Bohr, he had remained in contact with the physics community. In fact, immediately after the war, Bohr visited Delbrück at Cold Spring Harbor and communicated an enthusiastic response to Delbrück's program in the international physics community. Erwin Schrödinger's little book, "What is Life?" extolling Delbrück's physicomathematical approach to genetics, also circulated among physicists; it too contributed to Delbrück's fame. The book stirred up intellectual excitement for the life sciences and validated biology as a respectable line of inquiry for physicists. With the onset of the atomic era and the profound moral dilemma facing nuclear physicists in the wake of the Manhattan project, biology research under the charismatic leadership of a physicist opened up alternatives that must have indeed seemed like a "scientific playground" for serious children.[12] Although Delbrück's mission in molecular biology was motivated by non-reductionistic, non-mechanistic approach to animate phenomena, most of his phage disciples and the migrant physicists were unconcerned with these noble goals. The minimalist phage work was accessible, and yielded rapid results.

The intellectual migration was welcomed by the many biologists who shared faith in the primacy of the molecular knowledge of life. The presence of physicists at Cold Spring Harbor enhanced the prestige of biology, whose academic status, especially during the war years, lagged behind that of the physical sciences. With obvious elation, organizer Milislav Demerec reported in 1946 that Delbrück's phage course "brought to Cold Spring Harbor, among others, the nuclear physicist Leo Szilard, Aaron Novick from the University of Chicago, P. Morrison from Cornell, [and] R. B. Roberts from the Department of Terrestrial Magnetism of Carnegie Institute of Washington, with whom we are glad to discuss our problem."[13] Delbrück's dual citizenship in the communities of physics and biology was an asset that matched the cooperation ideal as envisioned by the Rockefeller Foundation, Morgan, and Caltech's Executive Council.

In addition to social and disciplinary expansion, the phage school broadened its cognitive scope. As reflected in the 1946 Cold Spring Harbor symposium, "Heredity and Variation in Microorganisms," viruses were brought into the realm of Mendelian genetics. This new perspective, in turn, began to tie together previously unrelated observations, eventually striking one of the fatal blows to the enzyme theory of life. At that symposium Delbrück reported his latest findings on induced mutations in bacterial viruses, and Hershey announced his surprising discovery of spontaneous mutations in phage. Both papers demonstrated that bacterial viruses exhibited a remarkably complicated genetic behavior, and that they underwent mutations principally during their intracellular existence—during the least studied, darkest aspect of the "black box," the phase that later came to be called the "eclipse."[14]

Studying two classes of mutations in T_2H phage—one affecting host range, the other affecting plaque type—Hershey showed that each of these phenotypic characters was determined by several loci of independent mutation. The genetic complex determining plaque types apparently contained a number of independently mutating genetic factors that exerted their influence through a single physiological mechanism, lysis inhibition. Contemporary methods did not enable researchers to determine the structural relation between these genetic factors and alleles, crossing-over units, and linkage groups of conventional genetics. Hershey was optimistic, however, that a purely genetic analysis modeled after *Drosophila* studies would soon yield valuable clues to the genetic structure of viruses and therefore to the organization of the gene itself.[15]

His optimism was amply reinforced by Delbrück's 1946 findings that simultaneous infection of a bacterium by two related phage strains (mixed infection) resulted in an apparent segregation of genetic factors. The phenomenon of "mutual exclusion" in mixed infections as reported by Delbrück and Luria four years earlier still applied to unrelated phages (i.e., unrelated morphologically, serologically, with respect to host range, and so on). However, for two related viruses, in this case a wild type and its mutant, the mixed infection gave rise to novel types of virus progeny, characterized as mutants of the infecting types. Though theoretical interpretations of these findings were of course premature, Delbrück and Hershey advanced the bold hypothesis that these novel viral biotypes in some respect looked more like genetic transfers or even exchanges of genetic materials.[16]

The time was ripe and the audience prepared for considering such bold speculations. The *Neurospora* studies of Beadle and Tatum during the early 1940s had shown rather convincingly that fungi underwent mutations that were amenable to genetic analyses, just like *Drosophila*. In 1946, Joshua Lederberg in collaboration with Tatum reported genetic recombination in bacteria, or what they termed sexual mating. These studies were complemented by the researches at the Pasteur Institute by André Lwoff and Jacques Monod on biochemical genetics in bacteria. These various observations underscored the realization that a microorganism was neither a bag of enzymes nor an amorphous protoplasmic glob, propagating itself exclusively through binary fission, but a sophisticated reproductive system possessing transmittable specific genetic factors that determined phenotypes.[17] The participants at the 1946 Cold Spring Harbor symposium were also more receptive than during years past to the possibility that the genetic factors were nucleic acids. Researchers were citing Oswald T. Avery's landmark 1944 studies on chemical transformation in pneumococcal bacteria, in which he concluded that the material responsible for transforming nonvirulent bacteria into virulent types was DNA. Some began to incorporate his findings into their own explanatory frameworks.

Avery had reported having isolated a nonproteinaceous (and noncarbohydrate) substance having these transforming properties as far back as 1934; but intellectual, psychological, and institutional factors coalesced to produce nearly a decade's delay in his "going public." Avery's retiring personality was poorly suited to handle some of the aggressive objections to his interpretations (led primarily by Mirsky at the Rockefeller Institute), thus contributing to his reluctance to publish his findings before 1944. Even in 1944 the initial reaction was rather muted. To assign biological specificity to nucleic acids seemed incompatible with the prevailing view of the tetranucleotide structure of DNA, the postulated simple structure in which nucleotides were supposed to be arranged in a fixed sequence. The argument against Avery's results was that it was the traces of protein trapped in the nucleic acid solution that were the causative agents of bacterial transformation. It seemed also counterintuitive before 1946, at the height of the autocatalytic protein theory of replication, to speak of sexual reproduction in bacteria, let alone ascribe to bacteria a mode of replication based on nucleic acid chemistry. On many counts, Avery's work before 1946 seemed to have only marginal relevance to mainstream problems of reproduction and genetic specificity.[18]

Delbrück, who was familiar with Avery's unpublished work through an association at Vanderbilt with Roy Avery (O. T. Avery's brother), did not seem at first to recognize the implications of Avery's studies in bacteria for his own investigations in bacteriophage genetics or for Luria's concurrent work on bacterial mutations. In 1944, however, he was willing to entertain the notion that Avery could be right. "He [Avery] believes that it [the transforming principle] is ribonucleic acid," Delbrück wrote to Hershey. "Mirsky seemed to think that the evidence is no good, but I don't know. . . . It seems to me that genetics is definitely loosening up and maybe we will live to see the day when we know something about inheritance in bacteria, even though the poor things have no sex."[19] Hershey, although finding Avery's paper scientifically satisfying, had to confess in 1944 that, "Of course one is disturbed by the conviction that the gene is a trace substance par excellence, but if this is wrong it's time to find out."[20]

A couple of years later, as the 1946 Cold Spring Harbor meeting attested, the intellectual climate was far more receptive to Avery's work and to the new challenges of microbial genetics. Venturing his own interpretations of mutations in phage, Hershey suggested that the transformation of pneumococcal types, the recently observed fibroma-myxsoma transformation, the induced mutations with respect to lysis inhibition in bacterial viruses, and the directed mutations with respect to host range could all be explained for the first time through a common mechanism, one involving a new biological phenomenon of general importance.[21] In response to these developments, Delbrück now acknowledged that the phage system was exciting "principally by the blow it deals to our fond hopes of analyzing a simple situation."[22]

During the next five years Hershey and Delbrück would pursue a two-tier approach. Developing the work on phage recombination, on the one hand, they would show that one could construct a genetic map of phage analogous to *Drosophila* maps, with mutant loci arranged in a linear order based on the frequencies of recombinant progeny for various mutated characters. This work formed the basis for later reconstruction of a fine-structure map of the phage genome, which in turn redefined the classical concept of the gene by showing that units of recombination, mutation, and function were operationally separate. On the other hand, in tandem with the genetics analyses, they would also design biochemical experiments to furnish decisive evidence for the primacy of DNA during replication and mutation in phage.[23]

The research conducted between 1940 and 1946 placed Delbrück at the vanguard of genetics and microbiology. At the same time his uniqueness as a physicist freelancing in biology—an image he actively cultivated—intrigued and attracted many institutions eager to develop molecular biology and benefit from Rockefeller support. The case of the University of Illinois was emblematic. The biology faculty was strongly interested in Delbrück, but they sought the advice of the Rockefeller Foundation, as they did not know what to make of Delbruck's remark that "he had never read any biology." Inasmuch as they wanted him as a biophysicist "to leaven the whole University lump of physics and bio- and physical chemistry as applied to biological problems," his remark caused a "little unfavorable comment." The chairman's assessment, however, exemplified the consensus at Illinois, Caltech, and the other interested institutions.

> Personally, I do not believe, for the purposes of Delbrück's present objectives, that he needs to know anything about the bulk of biology, not even much about bacteriology. . . . To my way of thinking it is perfectly incidental that he is working with bacteria and bacteriophage. His objective is an analysis of the nature of the gene—like Beadle working with molds, Sonnerborn with paramecium. . . . To "leaven this lump" of researches in basic life processes here at Illinois, it is necessary that our man be cooperative. He should know no departmental bound-

aries. He should be tolerant of earnest workers equipped with slower working neopallia than his own. I know that he is exacting and perhaps a bit Prussian; but if the right spirit is there, that's all that matters.[24]

As this communication demonstrates, Delbrück's playful but nuanced approach to biology was largely misunderstood by most scientists and administrators, who saw him as biophysicist applying highly specialized tools to specific problems. We can also see the close involvement of the Rockefeller Foundation network in academic decisions, the capillary workings of the knowledge-power nexus.

At the time of Caltech's overtures, Delbrück had "turned down numerous attractive U.S. offers," though he still sought the advice of the Rockefeller Foundation, "to which he feels great obligation because of fellowship and research aid given to him at Vanderbilt" regarding an attractive professorship of biophysics at the University of Manchester.[25] The strong offer, the cross-disciplinary structure, and a sense of nostalgia tilted the balance in favor of Caltech.

After nearly a decade and a half of freelancing in biology, Delbrück could at last set up a permanent phage center for research and training at Caltech. "I am very happy about this," he wrote to his mentor Niels Bohr in 1947, "because it signals the completion of my metamorphosis into a biologist, and because I believe that Caltech in the coming years will be to biology what Manchester was to physics in the 1910s."[26] As a key team member of the molecular biology project Delbrück was expected to lead a double life at Caltech, "that of general think-man (because I know physics, and because I put my foot into various theoretical problems) and as a phagologist," a dual role he relished.[27]

Key Team Member: Delbrück and the Phage Cult

When Delbrück joined the Caltech staff in January 1947, one of the first things he did was to offer the directorship of his new phage laboratory to biochemist Mark Adams, thus implicitly acknowledging the importance of biochemical tools in phage research. This was a radical step for Delbrück, who deprecated biochemistry, and he approached the issue cautiously. He regarded Adams as the only competent man in the phage biochemistry field, convinced that "other biochemists have caused, and are still causing, more confusion than enlightenment."[28] There was no doubt whom he meant. He considered H. J. Northrop of the Rockefeller Institute and his assistant A. P. Krueger to have muddled the phage field with their enzyme autocatalysis theories; and he regarded Stanley's work on plant viruses to be principally technique-driven and theoretically impoverished. Stanley's main problem was that he had too many centrifuges, Delbrück was fond of saying. Even though he had been initially inspired by these researchers, he preserved little tolerance for them; they in turn did not readily accommodate the upheavals he caused in the virus field.[29]

Their lack of theoretical rigor, Delbrück thought, was mainly due to the strong ties that still linked biochemistry to medicine. When Adams, declining Delbrück's offer, compared Caltech with the Rockefeller Institute, Delbrück was quick to

Figure 30 Max Delbrück conducting an informal lunchtime seminar in virology, ca. 1948. Courtesy of the Rockefeller Archive Center.

point out that although the comparison was a good one there were important qualifications. "The facilities at Caltech are not quite as good as at the Institute; on the other hand, the aim at Cal Tech is fundamental biology, quite uncontaminated by the MD spirit. There is nobody at Cal Tech who kills 100,000 mice per annum for futile nonsense. Cooperation is real, and so is the community of interest."[30] Delbrück seemed to have overlooked Pauling's futile attempts to produce artificial antibodies in mice and his repeated attempts to establish closer ties with the medical community.

Delbrück's aversion to biochemistry as an appendage of medicine and his dislike of medical research reflected the constant tension between fundamental biology at Caltech as envisioned by Morgan and the recognition that a link with biomedical research provides strong leverage when promoting public support and fund-raising. Although Delbrück was dismissive of the historical reality, bacteriophage had been from the outset of keen interest to physicians and medical bacteriologists—hence Delbrück's 1945 lecture for the Harvey Society. Delbrück would also soon realize that Beadle was hoping to establish collaborative phage work with the nearby Huntington Hospital, where "we plan to look into the possibility of a closer tie up," and even establish an entire laboratory devoted to medical research.[31] With a 15-year plan and a $6 million budget that played up the benefits of molecular medicine, it was only reasonable to cultivate some ties with the medical community.

Delbrück's disdain for biochemistry was based mostly on ignorance. He objected on several grounds: the social component—contamination by the "MD spirit"; the philosophical premise—the mechanistic approach to biology; and, from an aesthetic point, its cumbersome laboratory techniques and its lack of theoretical elegance. Conjuring up images of an alchemical laboratory, Delbrück's impressions were a throwback to the beginning of the century, when primitive equipment and crude procedures frequently resulted in chemical artifacts and dubious theories. Obviously biochemistry's early reputation in Europe, *"Tier Chemie ist Schmear Chemie,"* still lingered in some circles.

He soon found out upon returning to Caltech that things had changed a great deal: Precision instruments and sophisticated gadgets from the physical sciences generated novel and reliable data, potentially complementing the theoretical work in phage genetics. Though he had criticized Stanley's technological enterprise, Delbrück was visibly impressed with Caltech's technological prowess. In a typically playful letter to Hershey, inviting him for a year's collaboration, Delbrück described his "snooping around in chemistry and physics" and finding that they not only had wonderful centrifuges, electron microscopes, "Tiseliuses," and so forth, but they had competent people in charge of these gadgets who were anxious to do something sensible with the instruments—and with whom it would be fun to explore possibilities. Such explorations would of course reinforce Delbrück's role as a "unifying principle" between the divisions. "Very good students from physics and chemistry come and say: Show us some, we also want to try our hands at the riddle of life, etc.," Delbrück told Hershey. "I could of course sit in my shell and refuse to be tempted and go on using only Petri dishes and pipettes, but the fact is that I am tempted."[32] The temptation to play was reinforced by serious incentives. With indirect mathematical analyses yielding only partial knowledge of subcellular events, it was increasingly difficult to resist the temptation to tackle the riddle of life by penetrating the "black box" directly by physicochemical means.

One of the most powerful, yet least disruptive, new physicochemical tools in molecular biology was the radioisotope tracer. Having emerged during the years immediately after the war, radioactive isotopes were rapidly proving to be indispensable to the study of cellular mechanisms in general and to phage in particular. In August 1946 the Isotope Division o the Oak Ridge National Laboratory (ORNL) first made reactor-produced radioisotopes available. Naturally occurring radioactive substances had been used since 1913 to "tag" atoms in order to follow the course of chemical reactions. Biologists had explored the potential of this method as early as the 1920s, even though the handful of natural radioactive substances were suited for only a narrow range of biological investigations. Although the availability of artificial cyclotron-produced isotopes after 1934 did broaden the scope of radioisotope research in physiology, high costs and meager quantities prevented wide use of the technology. After 1945, with the establishment of several national laboratories and as the uses of nuclear energy and radioactivity expanded and diversified, isotopes became more readily available, though they were still tightly controlled by the government in 1946.[33]

Additional constraint retarded the use of radioisotopes in biological research in 1946. Because most biochemical studies required that tracer atoms possess highly specific chemical forms, sophisticated procedures of chemical reprocessing were invariably needed for preparing the isotopes for scientific research. At the end of the war, in anticipation of a trend in "nucleonics" toward cooperation between science, industry, and the military, the Atomic Energy Act of 1946 (the McMahon Act) was passed, loosening government monopoly over radioisotopes. A greater variety of reprocessed tracers were made available through the Oak Ridge laboratory and the National Bureau of Standards to authorized users in medical institutions, and for approved university research.[34] The availability of these "made-to-order" molecular tracers played a major role in the studies that eventually determined unequivocally the functional role of nucleic acids and proteins in the phage life cycle.

By 1947, radioisotopes were being used in phage studies. At the 1947 Cold Spring Harbor Symposium "Nucleic Acids and Nucleoproteins," Seymour Cohen, a biochemist at the University of Pennsylvania Medical School (formerly with Stanley at the Rockefeller Institute), had completed a series of studies on the synthesis of bacterial viruses by following the turnover of phosphorus in DNA. This work, in which radioisotopes (^{32}P) were used in phage work for the first time, further undermined the autocatalytic theory of replication and the notion that enzymatic bacteriophage precursors were normally present in bacteria and were merely triggered by infection. By exposing bacterial cultures to radioactive phosphorus (^{32}P) before infection, and by analyzing the phage progeny for their relative content of radioisotope, Cohen showed that most of the DNA in phage particles therefore could not be derived from preexisting bacterial precursors. Cohen also demonstrated that there was no increase in DNA during the eight to ten minutes after infection, and that DNA synthesis occurred after the latency period, results that were soon confirmed by Delbrück's collaborator, A. H. Doerman.[35]

That same year Luria, following up on mixed infections experiments, discovered that an ultraviolet-inactivated phage particle, though unable to reproduce itself, was far from being physiologically inert. It might kill a host cell, interfere with multiplication of other phages, and even regain its ability to reproduce after exposure to visible light. If two or more such ultraviolet-inactivated phage particles were adsorbed to the same host bacterium, Luria observed, they tended to cooperate.

Luria explained the phenomenon "multiplicity reactivation," as he called it, in terms of genetic exchange of undamaged parts between irradiated phage particles. An extension of Delbrück's and Hershey's interpretations of recombination, Luria's explanation implied that upon entering the bacterial cell the phage broke down into several independently reproduced subunits, which later, during the intracellular phase, were assembled into mature virus. A number of experiments ensued to follow up on the multiplicity reactivation phenomenon. The investigations of phages damaged by x-rays and their patterns of recombination became the main focus of James Watson's doctoral dissertation in Indiana under Luria and Delbrück, which in turn prepared Watson for studying the general relations between structure and function in viruses. The accumulated inferences regarding the intracellular picture left no choice but to peer behind the cover—by lysing the cell prematurely and analyzing its material content.[36]

A. H. Doerman, who had completed his doctorate at Stanford in biochemical genetics of *Neurospora* under Beadle, and his postdoctoral training at Vanderbilt under Delbrück, was entrusted with the "crucial experiment" of premature lysis.

His experiment (1948) led to a surprising finding; no infective phages whatsoever were found in any of the infected bacteria lysed artificially within the first 10 minutes following infection. Infectivity associated with the original parental phage was lost at the outset of the reproductive process. After the initial 10 minutes, however, an ever-increasing number of infective phages emerged on the intracellular scene, until the litter of progeny was produced during the normal lysis period, just as Cohen's findings had indicated. The latent period, when no infective plaque-forming phage particles appeared, became known as the "eclipse," and the noninfective immature virus was called the "vegetative phage."[37]

Doerman's results carried additional significance. They provided an impetus for a redefinition of the phage life cycle that illuminated the murky problem of lysogeny, thus contributing to further integration of bacteriology, virology, and genetics. Lysogeny, the hereditary property of some bacteria to produce phage without infection, had hitherto been an unexplained phenomenon, poised on the crossroads of virology and bacteriology. Lysogeny had been ridiculed in some circles and accepted in others, where it appeared to furnish evidence for the existence of a cellular precursor to an autocatalytic virus molecule.[38]

Delbrück and his powerful "phage church," as bacteriologist André Lwoff referred to it, had steadfastly rejected lysogeny through the 1940s as a nonphenomenon. They ignored the evidence in support of lysogeny even though it had already been well documented at the Pasteur Institute during the early 1920s that certain bacterial strains permanently "carried" phage and could cause lysis of other susceptible bacterial strains. Opposing interpretations that the "carrier" was a mere contaminant seemed to Delbrück equally reasonable, especially as the bacterial strains used in his typical phage experiments happened to be nonlysogenic. Pasteur Institute bacteriologist Elie Wollman, who arrived in Pasadena in 1948 as a Rockefeller Fellow, recalled his feelings of intimidation upon finding at Caltech's bibliographic index a 1937 paper on lysogeny by his parents labeled "Nonsense." In that paper the Wollmans proposed that intracellular phage may be in a noninfective phase, as they failed to recover plaque-forming samples from disrupted bacteria.[39]

As late as 1946, N. W. Pirie presented a paper at the Cold Spring Harbor meeting suggesting that at least part of the intracellular virus was present in a form that was complexed with host cell material—a form later identified as lysogenic provirus—but the paper generated little interest. With Doermann's elucidation of the "eclipse" and the "vegetative phage," the idea that bacteria could act as carriers of latent noninfective phage, or provirus, supplied strong evidence in favor of lysogeny. Lysogeny studies could now be coordinated with the findings of Delbrück, Hershey, and Luria regarding segregation and recombination of viral genetic material during the vegetative phase. In fact, in 1950 Lwoff demonstrated that similar mechanisms existed in lysogenic bacteria, thus adding important clues for explaining the mechanisms of replication and mutation in bacteria.[40]

The knowledge of the phage life cycle in the bacterial cell was extended to other animal and human viruses. In 1950 Renato Dulbecco, having just completed his postdoctoral research with Luria, accepted a permanent offer in Delbrück's laboratory. Delbrück suggested that he apply some of the methods of phage research to animal viruses. Until that time viruses could not be studied in their cellular environment, as they could not be grown in tissue culture. Dulbecco, with the expert collaboration of cell biologist Marguerite Vogt, was the first to develop a method for growing animal cells that produced viruses in culture dishes. Within the next two years the two researchers had established a reliable plaque assay for viruses similar to the one in phage, thereby bringing animal viruses into the realm of molecular genetics. This important work, which won Dulbecco the Nobel Prize (while bypassing acknowledgment of Vogt), generated a flurry of new studies. These studies, in turn, extended virus research to areas of tumor growth and cancer and to viruses involved in neurological pathologies such as poliomyelitis. Caltech thus emerged by the early 1950s as a shrine of virus research, with Delbrück the high priest of molecular virology.[41]

Indeed, Delbrück was much more than a leader of a research program or even an interdisciplinary liaison; he was a scientific cult figure. Lwoff's encounter with the "phage church" and Wollman's sense of intimidation at Caltech were by no means isolated accounts of Delbrück's autocratic rule and sharp tongue. A brash young Turk in biology, Delbrück impressed his followers by his affiliation with illustrious physicists and dazzled his audience with his broad knowledge and wit. As a master of practical jokes, he enchanted disciples and colleagues with pranks and angelic smiles. There was a kind of intriguing contradiction to Delbrück's charisma, his totalitarianism and eccentricity being neutralized by his humanism and communal style. Few could, or chose to, resist his charm and authority. After only one year at Caltech, Beadle, reporting to Weaver on the new appointments in biology, exclaimed incredulously: "Max has a biophysics seminar that has theoretical chemists and physicists not only attending but actually presenting papers!"[42]

From his confrontation during the late 1930s with the bacteriophage medical establishment through his standardization of phage work and the founding of phage courses during the mid-1940s, and now in his role as "unifying principle" at Caltech, Delbrück displayed remarkable originality and organizational talent. His success as founder and leader of a research school stemmed only in part from his ability to challenge, inspire, and teach. To a large extent, Delbrück's scientific popularity derived from a powerful social model that had shaped Delbruck's early career: Niels Bohr and his "Copenhagen spirit." He adopted this model after returning to Caltech but tailored it to the local culture.[43]

Phage work in Pasadena, like physics in Copenhagen, was much more than an intellectual pursuit. While working on the cutting edge of science in a well-funded field, Delbrück's colleagues and disciples participated in an unusual social experience. Delbrück was quick to adapt the traditions at Bohr's Institute in Copenhagen—social cohesiveness, collective intellectual criticism, jocular behavior, outdoor activities—to the Southern California milieu. Delbrück's garden in Pasadena became a favorite meeting place for colleagues, students, and visitors, and his house an intellectual and cultural sanctuary. He also drew on Caltech's unique tradition of weekend camping trips, a social activity cultivated by Noyes during the 1920s. Having been converted to that tradition during the late 1930s after joining Went's and Dobzhansky's frequent expeditions, which combined social and field activities, Delbrück now exploited and amplified that experience, shrouding it with the mystique and prestige of cult initiation; several renowned researchers would participate in the exotic activities during these outings.

With the self-assurance of an absolute monarch, Delbrück often proclaimed Wednesday and Thursday as weekend to avoid crowds and highway traffic on camping trips to the desert. There, by a crackling campfire under a star-sprinkled sky, Delbrück would lead lively discussions and raucous celebrations. Hikes in the wilderness and mountain climbing expeditions often spawned friendships, promoting camaraderie and loyalties that were reflected in collaborative laboratory projects. With the exception of Luria, who would not come to Caltech unless guaranteed immunity from camping, the blend of merrymaking, rarefied intellectual atmosphere, and romance of the wild were potent enticements for recruits and veterans alike.[44]

The sense of mission, peer acceptance, and exotic adventures under the leadership of a revered taskmaster had all the manifestations of a cult experience. Young recruits and assistants would cheerfully boast about being locked up for three days at Caltech's Marine Biology station at Corona Del Mar and ordered to write up their experiments, which were typed by Manny, Delbrück's wife. In keeping with the "Copenhagen spirit" of open criticism, brutal intellectual honesty, and cooperation, drafts would be mercilessly dissected with Delbrück's logical scalpel and the ideas refined through his perennial skepticism. As James Watson's revealing title "Growing Up in the Phage Group" suggests and as the numerous accounts by phage disciples attest, the process of enculturation was central to the project. Phage work was no place for loners; cultivation of the cooperative ethos was the social cement that held together phage genetics at Cold Spring Harbor and Pasadena.[45]

Protein Victory, Pure and Applied

The weight of accumulated evidence during the late 1940s increasingly pointed to nucleic acids as a crucial component of replication and mutation in phage. This awareness, however, did not substantially attenuate the thrust of Caltech's protein-centered program. During these years and until 1953, Beadle and Pauling frequently stressed that the central problems of molecular biology, and in turn the basis for a rigorous approach to medicine, were either those of simple or conjugated proteins. The chemical key to replication and biological specificity was the nucleoprotein template in which the nucleic acids were an extragenic component. That important component, they conjectured, somehow imparted energy in the physicochemical processes involved in the replication and mutation of proteins, processes outlined by Pauling and Delbrück in 1940 and further elaborated in Pauling's template theory of antibody formation. Nucleic acid research remained an undeveloped area at Caltech until 1953, while protein research flourished.

No project could have been more effective in swiftly confirming the claim that the study of proteins was the key to physiology, and that the etiology of disease was to be sought on the molecular level, than Pauling's project on sickle cell anemia. Published in 1949 under the provocative heading "Sickle-Cell Anemia, a Molecular

Disease," the work and the wide publicity it received appeared to validate the view that normal and abnormal physiological functions were basically molecular puzzles.[46] It reinforced the argument that the crucial problems of biology resided principally within the folds of the giant protein molecules—the submicroscopic dimensions between 10^{-6} and 10^{-7} cm—and that the explorations of this dimensional forest were best executed by physicists and chemists equipped with powerful molecular probes. Pauling would repeatedly deploy the potent term he coined, extrapolating the explanatory range of molecular disease to nearly all of medicine and human behavior.

Research on sickle cell anemia was not new; the disorder had been studied at several medical schools since its discovery in 1910. By the 1940s it had been well established that the sickling behavior was reversible, depending on the pressure of oxygen and the degree of acidity or alkalinity of the blood. It was known also that sickling followed the cell's release of oxygen and that arterial blood of anemic patients rarely contained sickled cells. There was no explanation, however, as to why cellular deformity took place and why it was lethal.[47] Because sickle cell anemia occurred almost exclusively among blacks, it was reasonable to assume it was genetically based. Several researchers had shown during the 1930s that the tendency to sickle was inherited according to Mendelian law and that a dominant gene was involved. However, because until 1948 no distinction had been made between sickle cell anemia and sicklemia—a milder form of the disease in which only a fraction of the cells are deformed—it was difficult to sort out the hereditary mechanisms.[48]

Pauling first became interested in sickle cell anemia in February 1945, when dining at the Century Club in New York with other members of the Committee on Medical Research, the group that later contributed the Bush Report, "Science, the Endless Frontier." A member of the group had described some aspects of his work on sickle cell anemia; and, according to Pauling, as soon as he heard that red blood cells of diseased patients were deformed, or sickled, in the venous circulation but resumed their normal shape in the arterial circulation it occurred to him that the fault might lie with the hemoglobin molecule rather than with the cell as a whole.[49] His own work during the 1930s on the difference between venous and arterial blood had already exposed him to some of the physicochemical properties of the hemoglobin molecule. It seemed reasonable to Pauling to trace the problem to the level of hemoglobin, as it comprised the only part of the cell concerned with the attachment of oxygen and because cells were deformed after being relieved of oxygen. A year later Pauling enlisted two young local medical researchers, Harvey A. Itano and Ibert C. Wells, and a junior physical chemist from Yale, Seymour J. Singer, to work on the sickle cell anemia project. In January 1948, soon after launching the project, Pauling left for England on a guest lectureship, returning a year later to a brilliantly executed project.

Tackling the sickle cell anemia project, the Caltech team examined blood samples from 30 patients: 15 with sickle cell anemia, eight with sicklemia, and seven normal adults. The researchers extracted the hemoglobin from the erythrocytes of each group and embarked on a battery of chemical and physical studies, looking for differences in the three types of hemoglobin. Hoping to explain the divergence in shape between normal and sickle cell hemoglobin, the team first ran a series of ultracentrifugation studies to measure changes in molecular weight and size of the three types of hemoglobin. The results revealed no difference. They then turned to the study of the electrochemical properties of the hemoglobin molecules, subjecting the three molecular species to analysis in their "homemade" Tiselius electrophoresis apparatus (which measured differential rates of migration of charged molecules in an electric field). They exposed the three species of hemoglobin to varying magnitudes of electrical force in solutions of various degrees of acidity and alkalinity until a point was reached where the effect was striking. At that point (pH 6.9) the normal hemoglobin migrated to the positive electrode, whereas the sickle cell hemoglobin migrated to the negative electrode; the hemoglobin sample from patients with the milder form of the disease behaved as a mixture of normal and sickle cell hemoglobin molecules in roughly equal proportion. The boundary between health and disease was marked by a small difference in electrical charge; hence the two species differed either in amino acid composition or arrangement.[50]

Just as the project neared completion, geneticist James V. Neel from Ann Arbor showed that sickle cell anemia was a manifestation of a homozygous condition; that is, both parental genes contributed to the offspring's sickle cell trait. Sicklemia then was a result of heterozygous condition with only one parental gene contributing to the sickle cell trait. In fact, Neel published his results in the same issue of *Science* in which Pauling's article appeared.[51]

Pauling arrived at the same genetic interpretation based on the physicochemical data. He concluded, in agreement with Neel, that the manufacture of abnormal sickle cell hemoglobin is controlled by a gene that when present in double dose, causes the red blood cells of the individual to contain only the abnormal hemoglobin. Following up on the hemoglobin work after returning to Caltech, Pauling reasoned that the nature of the charge of the anemic hemoglobin indicated its high chemical affinity. Enlisting his template, or "instructive theory" of protein synthesis, he visualized the sickling process as a formation of self-complementary structures, where "one end of the molecule is able to form a bond with the opposite end of another hemoglobin molecule. Under these circumstances the molecules clamp onto one another to undergo a sort of pseudocrystallization, which then twists the red blood cells out of shape."[52] Here was the first demonstration of how molecular architecture depended on the genetic protein blueprint—the first example of what Pauling called a "molecular disease."

With Pauling's own enthusiastic promotion, in both scientific circles and the popular media, the work was regarded as a spectacular achievement, its significance reaching far beyond explaining this particular medical syndrome. Intended to prove a general rule, the finding appeared to validate the molecular vision of life. George Gray saw to it that the research received lavish coverage. Combining Pauling's generous explanations with his own material prepared for the Rockefeller Trustees Report, Gray spun out a richly illustrated article on sickle cell anemia for *Scientific American* recounting the discovery process. Undoubtedly inspired by Pauling's technological vision, Gray speculated on the possibility of engineering molecules that would lock permanently to the defective hemoglobin to prevent it from misbehaving; he concluded that one of the most significant lessons of these studies "is that life is basically an affair of molecules. It is significant that the detective who tracked sickle cell anemia down to its lair was a physical chemist, and the chief instrument of his research was that versatile tool of physical chemistry—the electrophoresis apparatus" (on which Gray published an article in *Scientific American*).[53]

Equally significant was the relatively tepid reception of competing explanations and alternative frameworks of defining life, health, and disease. During the 1950s it was recognized that in its heterozygous form the gene responsible for the mutant sickle cell hemoglobin also conferred resistance to malaria. Despite being a genetic deficiency, it was also an adaptive advantage for black Africans living in malarial regions. It is quite instructive that even though Pauling was actively involved in these studies these findings and their implications did not benefit from enthusiastic promotions. Evolutionary explanations complicated the definitions of genetic deficiency and fitness, and diluted the impact of Pauling's neat conception of molecular disease.[54]

By the time the electrophoretic differences between normal and anemic hemoglobin were discovered (suggesting a difference in amino acid composition) a great deal of knowledge had accrued at the Gates and Crellin Laboratories on the physical structure of proteins and the properties of individual amino acids. Because hemoglobin, like all proteins, consisted of chains of amino acids, W. A. Schroeder of the Crellin Laboratory performed extensive chromatographic analyses, which showed in 1950 that both normal and diseased hemoglobin contained the same 17 amino acids. Further analysis, however, suggested that there was a minute difference in the relative amounts of the amino acid components. It would be another six years before the difference would be tracked down by researchers in other laboratories to a replacement of a single amino acid.[55]

The concept of molecular disease did not merely carry cognitive power, it was the linchpin in Pauling's long-standing quest to develop medical research at Caltech (as originally envisioned by Noyes). He liked to stress, in scientific forums and the popular media, that the progress of medicine hinged on molecular knowledge, an approach that would transform clinical practice into an exact science. As the Caltech group saw it, the spectacular success of the sickle cell anemia project supplied tangible evidence for these claims, prompting yet another scheme for developing medical research at Caltech. In January 1950, Beadle and Pauling submitted a proposal to the Rockefeller Foundation for establishing a Laboratory of Medical Chemistry at Caltech. "The time is now ripe for establishing a permanent laboratory devoted to the field of what might be called "medical chemistry," they proclaimed.[56] As Noyes had outlined in his proposals during the 1920s and 1930s, and according to Pauling's attempts to persuade the trustees during the intervening years, the prime purpose of such a laboratory would be to bring together the knowledge of biologists, chemists, physicists, and those with a medical interest and focus this knowledge on the basic studies of biological systems with the particular aim of understanding their chemical processes.

"It is doubtful whether any place in the country is more suitable for such a venture than the California Institute of Technology, where there is such a long tradition of cooperation between biologists and chemists," Pauling and Beadle argued.

The proposed Laboratory of Medical Chemistry would bring together these two groups and would add to their numbers, and would provide specific incentives to bring into these groups additional men who have had experience in the field of medicine itself. One of its great goals would be to train young doctors of medicine in the field of medical chemistry by actual participation in medical-chemical research for a period of years. These men would then return to medical research centers and spread the influence of this laboratory throughout the country.[57]

The plan was to construct a laboratory adjacent to Crellin and connected with it, ultimately linking it with the Kerckhoff Laboratories. It would not have special staff but, instead, would be used by the staff of the division of chemistry and chemical engineering.

The Rockefeller Foundation officers were skeptical about the new venture. After a site visit and meetings with Pauling, Beadle, and DuBridge, the Foundation officer apparently assessed that Caltech could not maintain such a large operation, that "They are very conscious of a somewhat precarious financial position much too dependent upon government projects."[58] Weaver wondered if the new venture was not merely a new terminology for the existing biology-chemistry program, which was already amply funded by the Foundation. However, "If this was an expansion into a new field," as he suspected, "I am *very* doubtful just because they need to consolidate, not expand." He seriously questioned an enterprise of "medical chemistry" in a setting where there was no natural medical contact and resident medical conferences.[59] Alan Gregg from the medical sciences division concurred that such a program would be better placed in or near a good medical school or teaching hospital.[60]

Beadle and Pauling did not give up easily. Together with Charles Newton, assistant to DuBridge, they flew East; and in addition to meetings at the Rockefeller Foundation, they approached nearly all potential patrons of their proposed venture in the New York area: Josiah Macy Foundation, Dazian Foundation, Samuel S. Fund, Donner Foundation, The Kresge Foundation, Carnegie Corporation, National Foundation, NIH, Milbank Memorial Fund, Alfred P. Sloan Foundation, The Commonwealth Fund, Life Insurance Medical Research Fund, Metropolitan Life Insurance, Eli Lilly, Lederle Laboratories, Abbott Laboratory. According to Rockefeller accounts, however, they managed to extract some promises of modest support for only a project or two. "Perhaps the most important thing they have learned is the desirability of dropping the name Laboratory of Medical Chemistry, since almost universally it has proved to be a disturbing description which confuses them and does not really represent the intent of the group at Cal Tech."[61]

The New York experience took the wind out of the sails of Caltech's crew. "I must say," Beadle confessed to Weaver, "that our several days' pounding the streets of New York was really liberal education for all three of us. One conclusion certainly became clear—that was that the term 'medical chemistry' used in our proposal is generally misleading,"[62] a rather narrow conclusion. The term medical chemistry was symptomatic of a deeper problem: the danger of substantially expanding the molecular biology program beyond its intended scope. After giving the matter more serious thought, especially to the warning from the Rockefeller Foundation about the danger being too large, he conceded to Rockefeller officer Gerald R. Pomerat: "[A]fter growing a couple of new stomach ulcers over contract negotiations with AEC [Atomic Energy Commission], I've about come to the conclusion that maybe we ought to formulate some careful plans for some judicious contraction and consolidation on one or two fronts."[63] The reverberations from the short-lived campaign for a medical laboratory soon died down, and Pauling refocused his energies on the central project of protein structure.

Considerable advances, quantitative and qualitative, had been made on the protein problem since Pauling first outlined the various strategies of attack to Weaver in the 1946 grant proposal. The work in organic chemistry and biochemistry of proteins, especially Schroeder's chromatographic analysis, had established the amino acid content in various proteins. Kirkwood's group, using mainly the Tiselius electrophoresis apparatus, had gathered data on charge distribution of these proteins and in turn on the general properties of globular proteins. The x-ray crystallographers and computers under Corey and Pauling, in coordination with Badger's ultracentrifuge and spectroscopic methods, had determined by 1949 the physical properties and crystal structures of many amino acids. They also worked out the three-dimensional geometry involved in the arrangement of amino acids in several large components of proteins, dipeptides, and tripeptides.[64]

The steady accumulation of these small and large jigsaw pieces enabled Pauling literally to begin fitting them together toward the construction of the physical model of alpha-keratin, a protein structure he had begun pondering during the late 1930s. He relied heavily on the building of molecular models. Even before leaving for England in 1948 he had been building molecular models, a method aimed at testing various molecular structures of proteins based on exact knowledge of the structure of amino acids and the smaller peptides. He attempted to build up the protein piecemeal, from the ground up, and then check the structure by x-ray methods.[65]

Pauling's approach departed radically from the conventional mode of point-by-point crystallographic analysis of full-size fibrous proteins. As British crystallographer J. D. Bernal once put it, the conventional method was analogous to the reconstruction of one of Christopher Wren's cathedrals in London from a collection of rubble after the blitz. The combination of new methods of organic chemistry and model building, however, rendered the task of molecular reconstruction far less confusing, as though a house was cracked up in parts—into basement, bedrooms, kitchen, and attic, rather than into individual bricks and boards.[66] Pauling's model-building approach was novel to both crystallography and biological research. It became crucial to the investigations of protein structure, allowing precise visualization of the molecular arrangements and interactions hitherto hidden. Wooden and plastic balls of all colors were designed and made at the laboratories and shops of the chemistry division, their scales and shapes represented such atoms as carbon, oxygen, and nitrogen as they exist in proteins. They could be added and subtracted at will, thereby bringing some order to the process of building by trial and error without a clear blueprint.[67]

The model-building approach became Pauling's hallmark, winning him the title "atomic architect." Molecular architecture, indeed, captured the essence of the scientific process and products of Pauling's program. Molecular models generated several possible pictures of the world, encompassing observations and calculations; those possessing the highest degree of convergence became a representation of biological reality. Molecular architecture was an epistemology, a technology, and a metaphor that he propagated among scientists and the lay public. As America's premier chemist and the recipient of numerous awards and appointments, Pauling was frequently photographed displaying some molecular edifice. "Dr. Pauling keeps cardboard and wooden models of the molecules he builds close by him in his office," wrote the author of a 1949 article in *Science Illustrated* entitled "Atomic Architect."[68] A smiling Pauling, pointing to replicas resembling kindergarten toys, predicted that within 20 years medicine could become an exact science, if only medical researchers became atomic architects.

He was in pursuit of the mechanisms that explained and controlled life in its most fundamental form: protein. "Proteins hold the key to the whole subject of the molecular basis of biological reactions," he was quoted as saying in 1949.[69] Proteins would unlock the mysteries of growth; the replication of viruses, genes, and cells; the action of enzymes, hormones, and vitamins. The unraveling of protein structure was not merely a theoretical interest. He proclaimed his hope to create life in the laboratory, by designing and building self-reproducing protein molecules. "And if anyone within the next 25 years becomes master of this second creation, it very probably will be Pasadena's wizard of atomic architecture," the article concluded.[70]

Here was a clearly articulated vision of science as doing, science as technology affecting change in the world. Intervention in this context was not merely an instrumentality of representing but also an instrumentality of design and control. There is little doubt that Pauling envisioned the region between 10^{-6} and 10^{-7} as the basic level of explanation as well as the fundamental level of intervention: The same technologies that would be used to represent these macromolecular phenomena would also be also mobilized to manipulate matter and control processes, to effect biological change, to create artificial life.

Pauling neared the solution to the alpha-keratin structure while he was still in England in 1948, lecturing on proteins and biological specificity. He fell ill during that period and, deprived of his laboratory gear, he worked on the structure with only paper, pencil, and a ruler. He constructed various hypothetical models that incorporated what had been established with respect to bond angles, rotations, amino acid geometries, hydrogen bonds, and the folding of simple peptides. The planarity of the amide bond (a point he had established by 1935 and that was further elaborated by Corey) helped determine the sites of rotation of the polypeptide chain. The work on the role of hydrogen bonding in the three-dimensional configuration of proteins (1936) aided in the proper placement of these bonds along the chain. To these structural constraints, Pauling added a few other assumptions regarding symmetry and amino acid equivalence. The results were two helical structures; one was the alpha helix, in which the pitch of the turn occurred every 3.7 amino acid residues.[71]

The alpha helix, with its nonintegral repeat, represented a radical departure from previous protein structures. According to J. D. Bernal, "The stroke of genius on the part of Pauling was to abandon the idea of integral repeats along a helix and to substitute a helix of peptides with an irrational and therefore, not exactly repeating structure."[72] However, this extreme divergence from the fixed idea that a helix was possible in crystals only with a helicity of two-, three-, four-, and sixfold symmetry worried Pauling. Rather than announcing his discovery in 1949, he resolved to wait until more data could support such a counterintuitive claim.[73]

Model-building played a central role in confirming the aperiodic structure of the helix. Based on data generated by his collaborators, Pauling had determined mathematically that 36 combinations were theoretically equally possible when constructing a polypeptide chain. In order to test whether these theoretical structures were possible in nature, Pauling built several of these molecular models. Attempting to place the atomic billiard balls in various positions, he found that 32 of his theoretical structures were impossible in nature because the balls did not fit properly. Further study showed that two of the four remaining structures represented forms of the spiral he had already worked out in England.[74]

Pauling's bold approach and novel model building method did not convince everyone, especially as Pauling had been wrong in the past. The flurry of activity in his laboratory did not impress Rockefeller officer W. F. Loomies during a site visit to Caltech in February 1951. Loomies was shown the various spiral structures Pauling believed accounted for chemical and physical data but that differed considerably from the structures postulated by all the leading crystallographers in England. Loomies had serious reservations about Pauling. "He certainly is imaginative, daring, and brilliant, but he has gone off the deep end in some cases (such as the 'artificial antibody' story) and his many stimulating pictures, models, etc., may be largely figments of his own imagination rather than lasting and sound science. Like R. J. Williams, he has no further worlds to conquer in straight science, so why not shoot at the moon."[75]

A month later Pauling dispatched a letter to Weaver announcing the structure of the alpha helix. He recounted his experience at Oxford three years earlier, his discovery of a configuration of a promising polypeptide chain, then the discovery of the second helical form, and the period of evidence-gathering. He thought that "one of these two helical structures, which we predict with every atom located to couple of hundredth of an Ångstrom, is represented in contracted muscle, ordinary hair, many other fibrous proteins, and in hemoglobin, myoglobin, and other globular proteins." He also had figured out the structure of collagen and gelatin, which involved the interaction of three such helical chains. "I feel," Pauling concluded, "that in a sense this represents the solution of the problem of the structure of proteins."[76] The alpha-keratin structure was announced a month later in the *Proceedings of the National Academy of Sciences*, followed up by seven detailed papers in the May issue of that journal.[77]

It was a major victory for Pauling, Caltech, and the Rockefeller Foundation, which had supported these efforts for nearly two decades, a giant step toward the solution of what had been defined for decades as the central problem in the life sciences. From the perspective of the protein view of life, which had equated protein with biological destiny, the solution of protein structure seemed to represent the primary aspect of the solution to heredity. Caltech's news release of the alpha helix story highlighted this very point, reminding the public that the new structure was part of the protoplasm of all living things, a substance Thomas Huxley had called "the physical basis of life."[78] One unresolved issue, of course, remained: There was no hint how Pauling's new structure could be capable of self-replication.

Warren Weaver was elated with Caltech's breakthrough. Inspired, he wrote up a lively and lengthy account of the protein problem and its solution for circulation at the Foundation; one reader remarked that Weaver had almost outdone George Gray. Unquestionably, it was also a great coup for Weaver, who when persuading the Rockefeller trustees to invest immense sums in the molecular biology project at Caltech at times had to defend Pauling's work. Weaver's most important long-term project had finally yielded substantial and tangible returns on investment, justifying the Foundation's confidence and Weaver's judgment.[79]

Weaver would be retiring in eight years and was already in the process of taking stock, examining the broad contours of the natural sciences territory. He was now thinking "in terms of consolidating some of the gains which have been made in the NS [natural sciences] program in the last 20 years. This would involve thinking about the program as a whole, and deciding what small number of situations may deserve consideration, over the next eight years, for some really definitive treatment." Weaver concluded that in all likelihood the results of such a consideration would place Caltech high on the list of priorities, in which case he would recommend that the Foundation contribute "sometime during the next 18 months, a major sum (say $2.0 million or $2.5 million) to C.I.T. for research in biology and in those aspects of chemistry which are intimately related to biological problems."[80]

Thus when DuBridge submitted to Weaver in July 1953 a major grant proposal for Caltech's long-term needs in molecular biology, he was already quite confident of the outcome. Caltech's impressive annual reports for chemistry and biology, accompanied by references to George Gray's confidential monthly report (January 1949) buttressed the proposal. "We believe," DuBridge wrote Weaver, "that what we have done over a period of years in the field of chemical biology . . . as a basis for judging us, is worth more to you and your Foundation associates than any promises or predictions we can make for the future."[81]

The $1.5 million grant to Caltech in December 1953 was indeed a great show of confidence. "We haven't yet come down to earth after hearing the wonderful news of the new grant." Beadle told Weaver. "When we do, we'll start seriously worrying about the matching problem and about how we can do a good enough job in the coming years to justify such confidence on the part of you and the Foundation."[82] There was much work to be done. Among the raucous accolades and laudatory orations of Caltech's contributions to molecular biology, a casual statement in the grant resolution implied that the answer to the protein problem did not solve the central problems of biology, and that nucleic acid may be the hereditary determinant: "Many signs indicate that the golden age of biology is only now beginning."[83] Caltech had led the research to the threshold of another era, during which a molecular biology based on DNA genetics would assume cognitive and institutional dominance.

Figure 31 R. A. Millikan and L. A. DuBridge in front of Throop Hall, ca. 1950. Courtesy of the California Institute of Technology Archives.

Notes

1. CIT, Delbrück Papers, Box 2.10, Beadle to Delbrück, December 19, 1946.
2. RAC, RG1.1, 205D, Box 164.2015, Tisdale to Slack, September 8, 1939.
3. On Delbrück's scientific ascent see Lily E. Kay "Conceptual Models and Analytical Tools: The Biology of Physicist Max Delbrück," *Journal of the History of Biology, 18* (1985), pp. 207–247, and Ernst P. Fischer and Carol Lipson, *Thinking About Science: Max Delbrück and the Origins of Molecular Biology* (New York: Norton, 1988).
4. OSU, Pauling Papers, Box 6.1, Delbrück to Pauling, November 16, 1941; and *Cold Spring Harbor Symposia on Quantitative Biology, ix* (1941), passim. For Delbrück's contribution see "A Theory of Autocatalytic Synthesis of Polypeptides and Its Application to the Problem of Chromosome Reproduction," pp. 122–126, in that volume.
5. On Luria's account of these collaborations see Salvador E. Luria, *A Slot Machine, A Broken Test Tube: An Autobiography* (New York: Harper & Row, 1984), Chs. 3–5.
6. M. Delbrück and S. E. Luria, "Interference between Bacterial Viruses. I. Interference between Two Bacterial Viruses Acting on the Same Host, and the Mechanism of Virus Growth," *Archives of Biochemistry, 1* (1942), pp. 111–141; S. E. Luria and M. Delbrück, "Mutations of Bacteria from Virus Sensitivity to Virus Resistance," *Genetics, 28* (1943), pp. 491–551.
7. S. E. Luria, M. Delbrück, and T. F. Anderson, "Electron Microscopy of Phages," in J. Cairns, G. S. Stent, J. A. Watson, eds., *Phage and the Origins of Molecular Biology* (New York: Cold Spring Harbor Laboratory of Quantitative Biology, 1966), pp. 63–68.

8. M. Delbrück, "Interference Between Bacterial Viruses. III. The Mutual Exclusion Effect and the Depressor Effect," *Journal of Bacteriology*, *50* (1945), pp. 166–167; a discussion on the "penetration hypothesis."

9. T. F. Anderson, "Electron Microscopy of Phage," *Phage and the Origins of Molecular Biology*, p. 73 (see Note 7).

10. G. S. Stent, "Waiting for the Paradox," in *Phage and the Origins of Molecular Biology*, p. 60 (see ref. 7). The purpose and content of the phage course are outlined in "The Max Delbrück Laboratory Dedication Ceremony" (New York: Cold Spring Harbor Laboratory, 1981).

11. L. E. Kay, "Conceptual Models," p. 241 (see Note 3).

12. CIT, Delbrück Papers, Box 31.29, Bohr to Delbrück, March 28, 1946. The reception of Erwin Schrödinger's "What is Life?" (New York: Macmillan, 1944) has been discussed by R. C. Olby, "Schrödinger's Problem: What is Life?" *Journal of the History of Biology*, *4* (1971), pp. 119–148; and E. J. Yoxen, "Where Does Schrödinger's 'What is Life?' Belong in the History of Molecular Biology," *History of Science*, *17* (1979), pp. 17–52.

13. M. Demerec, "Annual Report," *Carnegie Institutions of Washington Year Book*, 1946–1947 (Baltimore: Lord Baltimore Press, 1947), p. 127.

14. M. Delbrück and W. T. Bailey Jr., "Induced Mutations in Bacterial Viruses," pp. 33–37, and A. D. Hershey, "Spontaneous Mutations in Bacterial Viruses," pp. 66–77, both in *Cold Spring Harbor Symposia on Quantitative Biology*, *XI* (1946).

15. A. D. Hershey, "Spontaneous Mutations," pp. 74–75 (see Note 14).

16. M. Delbrück and W. T. Bailey Jr., "Induced Mutations," pp. 36–37 (see Note 14).

17. J. Lederberg and E. L. Tatum, "Novel Genotypes in Mixed Cultures of Biochemical Mutants of Bacteria," *Cold Spring Harbor Symposia on Quantitative Biology*, *XI* (1946), pp. 139–155.

18. RAC, RU 439, Box 6, Reports of the Rockefeller Institute: O. T. Avery and C. M. MacLeod, "Studies on the Specific Types of Pneumococcus," Vol. 23 (1934–1935), pp. 177–183. See also Maclyn McCarty, *The Transforming Principle* (New York: W. W. Norton, 1985); and O. T. Avery, C. M. McLeod, and M. McCarty, "Studies on the Chemical Transformation of Pneumococcal Types," *Journal of Experimental Medicine*, *79* (1944), pp. 137–158. Avery had worked on pneumococcal transformation for about eight years before publishing that classical paper, and then his findings did not have a strong impact on geneticists and biochemists. The historical problem—why Avery waited so long to announce his results, and why their reception was so muted—has been debated by several scholars, notably by G. S. Stent, "Prematurity and Uniqueness in Scientific Discovery," *Scientific American*, *227* (1972), pp. 84–93; H. V. Wyatt, "When does Information Become Knowledge?" *Nature*, *239* (1972), p. 234; and R. J. Dubos, *The Professor, the Institute, and DNA* (New York: Rockefeller University Press, 1976). R. C. Olby in *The Path to the Double Helix* (London: Macmillan, 1974) and H. F. Judson in *The Eighth Day of Creation* (New York: Simon & Schuster, 1979) also discussed this issue, listing extramural factors such as the institutional bias of the Rockefeller's "enzyme camp," Alfred E. Mirsky's (a powerful protein chemist at the Institute) animosity and overt opposition to Avery, and Avery's own psychological impediments. Unfortunately, Mirsky's records at the Rockefeller Archive Center have not shed light on this issue.

19. CIT, Delbrück Papers, Box 10.35, Delbrück to Hershey, February 25, 1944, p. 2.

20. CIT, Delbrück Papers, Hershey to Delbrück, March 6, 1944, p. 2.

21. A. D. Hershey, "Spontaneous Mutations in Bacterial Viruses," p. 75 (see Note 14).

22. M. Delbrück, "A Physicist Looks at Biology," in *Phage and the Origins of Molecular Biology*, p. 14 (see Note 7).

23. G. S. Stent, "Introduction," *Papers in Bacterial Viruses* (Boston: Little, Brown, 1960).

24. RAC, RG1.1, 200D, Box 164.2016, Hartman to Hanson, October 4, 1945.

25. Ibid., H. M. Miller's report, September 25, 1946.

26. CIT, Delbrück Papers, Box 1.7, Delbrück to Bohr, January 11, 1947. As Robert H. Kargon described in *Science in Victorian Manchester* (Baltimore: Johns Hopkins University Press, 1977), Manchester (like Pasadena) had been a provincial backwater at the turn of the century, an unlikely place for the pursuit of science. By the 1920s, however, Manchester had become the nursery of modern physical sciences, where Niels Bohr, in Ernest Rutherford's laboratory, forged his scientific career.

27. CIT, Delbrück Papers, Box 1.7, Delbrück to Adams, January 7, 1947.

28. CIT, Delbrück Papers, Box 8.2, Delbrück to Fraser, February 17, 1948.

29. T. F. Anderson, *Phage and the Origins of Molecular Biology*, pp. 65–69 (see Note 7). See also correspondence between Delbrück and J. H. Northrop from 1941 to 1945 (CIT, Delbrück Papers, Box 16.32); also personal communication, J. Bonner to L. E. Kay, January 22, 1990.

30. CIT, Delbrück Papers, Box 1.7, Delbrück to Adams, January 7, 1947.

31. CIT, Delbrück Papers, Box 2.10, Beadle to Delbrück, May 26, 1947.

32. CIT, Delbrück Papers, Box 11.1, Delbrück to Hershey, December 5, 1947; and Delbrück's own account of his somewhat irrational resistance to biochemistry in CIT, Oral History, Delbrück, p. 2.

33. G. Hevesy, "Historical Sketch of the Biological Application of Tracer Elements," *Cold Spring Harbor Symposia on Quantitative Biology*, XIII (1948), pp. 129–150. See also Hilde Levi, *Georg de Hevesy: Life and Work* (Rhodos Press, 1985); and Robert E. Kohler, *Partners in Science: Foundations and Natural Scientists, 1900–1945* (Chicago: University of Chicago Press, 1991), pp. 379–381.

34. C. A. Ziegler, "Looking Glass Houses: A Study of the Process of Fissioning in an Innovative Science-based Firm," Ph.D. dissertation, Brandeis University, 1982, pp. 98–110: a brief prehistory of commercial isotopes.

35. S. S. Cohen, "The Synthesis of Bacterial Viruses in Infected Cells," *Cold Spring Harbor Symposia on Quantitative Biology*, XIII (1947), pp. 35–49.

36. M. H. Adams, *Bacteriophages* (New York: Interscience, 1959), Chs. 6 and 11.

37. D. H. Doermann, "Lysis and Lysis Inhibition with *E. coli* Bacteriophage," *Journal of Bacteriology*, 55 (1948), pp. 257–276; D. H. Doermann, "Intra-cellular Phage Growth as Studied by Premature Lysis," *Federation Proceedings*, 10 (1951), pp. 591–594; and G. S. Stent and R. Calendar, *Molecular Genetics* (San Francisco: W. H. Freeman, 1978), pp. 301–305.

38. D. H. Doermann, "The Eclipse in the Bacteriophage Lifecycle," *Phage and the Origins of Molecular Biology*, p. 79 (see Note 7).

39. Elie L. Wollman, "Bacterial Conjugation," *Phage and the Origins of Molecular Biology*, pp. 216–225 (see Note 7). On the history of lysogeny see Charles Galperin, "Les bactériophages, la lysogénie, et san determinisme génétique," *History and Philosophy of the Life Sciences*, 9 (1988), pp. 175–224.

40. André Lwoff, "The Prophage and I," *Phage and the Origins of Molecular Biology*, pp. 88–99 (see Note 7).

41. R. Dulbecco and M. Vogt, "Some Problems of Animal Virology as Studied by the Plaque Technique," *Cold Spring Harbor Symposia on Quantitative Biology*, 18 (1953), pp. 273–290; R. Dulbecco, "The Plaque Technique and the Development of Quantitative Animal Virology," *Phage and the Origins of Molecular Biology*, pp. 287–291 (see Note 7), and Daniel J. Kevles, "Renato Dulbecco and the New Animal Virology: Medicine, Methods, and Molecules." (Paper presented at the Mellon Workshop "Building Molecular Biology," MIT, April 1992).

42. RAC, RG1.1, 205D, Box 4.24, Beadle to Weaver, December 9, 1947.

43. See also L. E. Kay "The Secret of Life: Niels Bohr's Influence on Delbrück's Biology Program," *Rivista di Storia della Scienza*, 2 (1985), pp. 487–510.

44. Testimonies of impressions and experiences of numerous participants are given in *Phage and the Origins of Molecular Biology*, passim. See also Fischer, *Thinking About Science*, passim (see Note 3).

45. J. Watson, "Growing up in the Phage Group," *Phage and the Origins of Molecular Biology*, pp. 239–245 (see Note 7).

46. L. C. Pauling, H. A. Itano, S. J. Singer, and I. C. Wells, "Sickle-Cell Anemia, a Molecular Disease," *Science*, 110, No. 2 (1949), pp. 543–548; L. C. Pauling, "The Hemoglobin Molecule in Health and Disease," *Proceedings of the American Philosophical Society*, 96, No. 5 (1952), pp. 556–565; G. W. Gray, "Sickle-Cell Anemia," *Scientific American*, 185, No. 2 (1951); and a transcript of a talk on sickle cell anemia by L. C. Pauling, ca. 1950, CIT.

47. G. W. Gray, "Sickle-Cell Anemia," p. 56 (see Note 46).

48. L. C. Pauling et al., "Sickle-Cell Anemia, a Molecular Disease," p. 547 (see Note 46.).

49. CIT, Pauling Papers, transcript of talk on sickle cell anemia by Pauling, ca. 1950.

50. See Note 46.

51. J. V. Neel, "Inheritance of Sickle-Cell Anemia," *Science*, 110, No. 1 (1949), pp. 64–66.

52. L. C. Pauling, "The Hemoglobin Molecule," p. 562 (see Note 46).

53. CIT, Chemistry Division Papers, Box 2.4, Pauling–Gray correspondence, November-December 1950. See also G. W. Gray, "Sickle-Cell Anemia," pp. 56–57, 59 (see Note 46).

54. See for example review article by E. Zuckerkandl and L. Pauling, "Evolutionary Divergence and Convergence in Proteins," in V. Bryson and H. J. Vogel, eds., *Evolving Genes and Proteins* (Orlando, FL: Academic Press, 1965). On the debates on evolutionary fitness see John Beatty, "Weighing the Risks: Stalemate in the Classical-Balance Controversy," *Journal of the History of Biology, 20* (1987), pp. 289–320.

55. G. W. Gray, "Sickle-Cell Anemia," p. 57 (see Note 46).

56. RAC, RG1.1, 205D, Box 4.26, "Proposal for the Establishment of a Laboratory of Medical Chemistry at the California Institute," January 27, 1950, p. 2.

57. Ibid.

58. RAC, RG1.1, 205D, Box 4.26, Report of C.F.K., February 20, 1950.

59. Ibid., Weaver's hand-written comments on the above report.

60. Ibid., Gregg's hand-written comments on the same report.

61. Ibid., entry in officer G.R.P.'s diary, March 17, 1950.

62. Ibid., Beadle to Weaver, March 23, 1950.

63. Ibid., Beadle to Pomerat, March 23, 1950.

64. G. W. Gray, "Pauling and Beadle," *Scientific American, 180, No. 5.* (1949), pp. 19–20; L. C. Pauling, "Fifty Years of Progress in Structural Chemistry and Molecular Biology," *Daedalus, 99* (1970), pp. 1003–1004; R. C. Olby, *The Path to the Double Helix,* pp. 279–280 (see Note 18); and J. D. Bernal, "The Patterns of Linus Pauling's Work in Relation to Molecular Biology," in A. Rich and N. Davidson, eds., *Structural Chemistry and Molecular Biology* (San Francisco: W. H. Freeman, 1968), pp. 370–379.

65. J. D. Bernal, "The Patterns of Linus Pauling's Work," p. 372 (see Note 64).

66. "Signs of Life," *Electronic Medical Digest* (Summer 1949), pp. 35–36.

67. RAC, RG1.1, 205D, Box 4.27, News release, September 4, 1951, p. 4. See also R. C. Olby, *The Path to the Double Helix,* pp. 287–289 (see Note 18): and R. B. Corey and L. Pauling, "Molecular Models of Amino Acids, Peptides, and Proteins," *Review of Scientific Instruments, 24* (1953), pp. 621–627.

68. Linus Pauling, "Atomic Architect," *Science Illustrated* (January 1949), p. 39.

69. "Signs of Life," p. 35 (see Note 66).

70. "Atomic Architect," p. 40 (see Note 68).

71. R. C. Olby, *The Path to the Double Helix,* pp. 280–281 (see Note 18).

72. J. D. Bernal, "The Patterns of Linus Pauling's Work," p. 373 (see Note 64).

73. R. C. Olby, *The Path to the Double Helix,* pp. 281–282 (see Note 18).

74. RAC, RG1.1, 205D, Box 4.27, News release, September 4, 1951.

75. Ibid., "WFL's Diary," February 1951.

76. Ibid., Pauling to Weaver, March 8, 1951.

77. L. Pauling, R. B. Corey, and H. R. Branson, "The Structure of Proteins: Two Hydrogen-Bonded Helical Configurations of Polypeptide Chain," *Proceedings of the National Academy of Science USA, 37* (1951), pp. 205–211; and L. Pauling and R. B. Corey, "Atomic Coordinates and Structure Factors for Two Helical Configurations of Polypeptide Chains," in the same volume, pp. 235–285.

78. RAC, RG1.1, 205D, Box 4.27, News release, September 4, 1951.

79. Ibid., "Why Should Joe Care about Protein Structure Studies?" by Warren Weaver, September 1951.

80. Ibid., Weaver's Diary, April 27, 1951.

81. RAC, RG1.1, 205D, Box 4.28, DuBridge to Weaver, July 28, 1953.

82. Ibid., Beadle to Weaver, December 10, 1953.

83. RAC, RG1.1, 205D, Box 4.22, Grant to the California Institute of Technology, December 1, 1953, p. 2.

Paradigm Lost? From Nucleoproteins to DNA

When the Rockefeller Foundation appropriated $1.5 million for Caltech's molecular biology program in December 1953, the rumblings of a scientific revolution were already audible. There were good reasons to suspect that the solution of the protein problem was not the right answer to the "riddle of life." Life now appeared to be controlled by another giant molecule. A self-replicating DNA spiral held out the promise for controlling and creating life; a new golden age of biology was dawning, the Rockefeller Foundation forecasted.

The signals were visible for several years. As early as 1950, as Pauling directed his energy to the alpha helix, several phage workers were shifting their focus to nucleic acids. The mounting details on phage replication and its life cycle strongly suggested that the virus could no longer be regarded as a form of protoplasm or even a nucleoprotein. By 1950 a virus particle came to be viewed by most phage researchers as a dual structure of protein and DNA; its nuclear and cytoplasmic functions were perceived as distinct, pointing to DNA as the genetic material. In fact, these changing perceptions were the impetus behind Delbrück's and Luria's decision to send their promising disciple James D. Watson to Europe for postdoctoral studies of the chemistry of nucleic acids.

By 1950 electron microscopist T. F. Anderson had embarked on experiments aimed at differentiating between the structural components of phage. His new micrographs showed that osmotic shock ruptured the phage to produce empty-headed "ghosts," and that phages attached themselves to bacteria by their tails, forming a dynamically unstable union with the host cell surface. He also found that violent agitation of mixed suspensions of phage and bacteria in a Waring blender prevented tail attachment, and therefore no infection occurred.[1]

Assimilating these and other clues, enzymologist Roger Herriott, Northrop's protégé (then working on phage), sent his interpretations to Hershey: "I've been thinking—and perhaps you have too—that the virus may act like a little hypodermic needle full of transforming principles; that the virus as such never enters the cell; that only the tail contacts the host and perhaps enzymatically cuts a small hole through the outer membrane and then the nucleic acid of the virus head flows into the cell."[2] Even Northrop, the main protagonist of the enzyme theory of bacteriophage, was moved to contemplate the possibility that "the nucleic acid may be the essential autocatalytic part of the molecule, as in the case of the transforming principle of the pneumococcus . . . and the protein portion may be necessary only to allow entrance to the host cell."[3]

Linking these bits of evidence and conjecture, Alfred Hershey and Martha Chase conducted a radioisotope study in 1951 that became one of the classical experiments in molecular biology, providing one of the most compelling arguments for the functional independence of protein and DNA components in phage and for the role of DNA as the genetic determinant. Using the radioactive phosphorus label ^{32}P to follow the trail of phage DNA, and radioactive sulphur (^{35}S) to trace the fate of viral proteins, they devised a simple procedure that enabled them to measure how much ^{32}P-labeled nucleic acid or ^{35}S-labeled protein was present after infection in the various fractions of bacteria and phage. Exposures to a Geiger-Muller radiation counter indicated that practically all the viral ^{35}S-labeled protein remained at the surface of the infected bacterium, and that most of the ^{32}P-labeled viral DNA entered the cell at the beginning of the intracellular phase. Once the attachment of the protein tail to the surface was completed and the DNA injected into the host, the empty protein heads appeared to have no further function in the intracellular reproductive process. However, during the early eclipse period, the DNA deprived of its protein shell could not carry out the infective functions of a maturing phage.[4]

Hershey's decisive results had a strong impact on Watson, then a postdoctoral fellow in Europe. Groomed for success by Delbrück and Luria, the 22-year-old Watson in collaboration with British physicist Francis Crick in Cambridge mounted in 1951 a concentrated attack on the problem of DNA structure. During the two-year period that culminated in their determination of the double-helix structure of DNA in 1953, Watson remained in close contact with Delbrück, informing him of the details of his progress. Delbrück, in turn, communicated the developments to Pauling's group, by then also in the race to the solution of DNA structure.

In May 1952, when he reported his initial attempts at finding a helical structure for DNA to Delbrück, Watson was already working at a frantic pace. Strongly influenced by the epistemology and methodology of Pauling's molecular architecture and by the structure of the alpha helix, Watson and Crick were constructing large-scale molecular models of DNA.[5] By spring 1953, they announced the double-helix structure of DNA: two antiparallel chains comprised of a sugar-phosphate backbone held together on the inside by the complementary hydrogen binding of purines to pyrimidines. An understated comment at the end of the note to *Nature* disclosed that: "It has not escaped our notice that the specific pairing we have postulated immediately suggests a possible copying mechanism of the genetic material."[6] Watson's biological thinking along functionalist lines, combined with Crick's structuralist approach, had begun to unravel the structure and function of the gene.

Delbrück was enamored with the new DNA structure. During the following two weeks he discussed the model with several phage workers and with Pauling's group and then listed in a letter to Watson possible objections to the model, arguments in its support, and some logical corollaries. He arranged for Watson's unveiling of the new structure at the 1953 Cold Spring Harbor Symposium on viruses and predicted that "if your structure is true, and if its suggestions concerning the nature of replication have any validity at all, then all hell will break loose, and theoretical biology will enter into a most tumultuous phase."[7] According to one enthusiastic participant, the Symposium was dominated by discussions of the implication of the Watson-Crick DNA structure. No one who listened to Watson's lecture, even if they did not fully agree, needed much imagination to grasp its immense significance.[8]

Outside the Cold Spring Harbor virus community the reactions to the Watson-Crick model of DNA and to its implications for heredity ranged from exultation, to mild interest, to skepticism.[9] Not everyone was convinced. "This picture, which is inspired by certain suggestions recently put forward by the English biochemists J. D. Watson and F. H. C. Crick," argued the noted biochemist Linderstrom-Lang (voicing the sentiment of other protein chemists),

> entirely neglects the question of the part played in protein synthesis by nucleic acids. . . . The fact that nucleic acids are genetic regulators, that they are present in cells whenever protein synthesis occurs, and that they are important as transforming factors and as initiators in virus formation does not necessarily mean that they participate intimately in the synthesis of the primary peptide chain. They seem to be far too unspecific for this process.[10]

He believed that it was the mechanisms of protein synthesis that posed the greatest challenge to modern biochemistry and to the study of life.

At Caltech, Beadle and Pauling cautiously accepted the significance of the new developments. "Recent investigations have verified the opinion that nucleic acids play as important a part in biology as proteins, especially in our study of the structure of the gene," they conceded in their 1953 report to the Rockefeller Foundation. "Both chemical and biological studies make it clear that, contrary to our earlier beliefs, nucleic acids are of many kinds, presumably differing both in purine and pyrimidine composition and in sequence of these bases." Pauling and Corey also proposed a molecular structure for DNA, and they noted, pointing to their share in the action. "There is now widespread agreement that the Watson-Crick structure is substantially correct," thus forecasting advances that "will be of the most profound significance in biology and the disciplines of medicine and agriculture." Naturally they expected Caltech to assume a leading role in these developments; Watson's move to Caltech in 1953 (though he accepted Harvard's offer a year later) reflected that resolve.[11]

The shift of the molecular vision of life from the protein paradigm toward the emerging paradigm of DNA signaled a great reshuffling of ideas and techniques. The new phase, prophesied Delbrück, would only partially involve structural and analytical chemistry. The more important part would consist in attempts to take a fresh look at many of the problems in genetics and cytology that had come to dead ends during the previous 40 years, the period during which the protoplasmic theory of life stalled in explaining the physicochemical mechanisms of replication and mutation.[12] Equally significant, when the precise mechanisms by which nucleic acids exerted their putative power as the chemical blueprints of life were elucidated, molecular biology would claim greater cognitive authority and technological potential when addressing the unresolved problems of biological deterioration and rational social planning.

During the following decade researchers would direct their efforts principally at answering what have been defined as the two central questions of molecular biology: First, how does the DNA manage to reproduce itself autocatalytically to generate an exact copy of every gene, or, more specifically, how is parental DNA replicated to yield two DNA molecules of identical composition? Second, how does the DNA control heterocatalytically the synthesis of a polypeptide to provide a cell with its essential apparatus? A decisive answer to the first question came in 1957, when Matthew Meselson and Franklin W. Stahl, working in Pasadena under Delbrück's direction, tagged the DNA of *Escherichia coli* bacteria with a heavy nitrogen isotope. Tracing the distribution of the parental heavy nitrogen·through the cycle of bacterial replication, they showed that the bacterial DNA followed the semiconservative mode predicted by the Watson-Crick model, each helical chain synthesizing the chain complementary to it, thereby conserving half of itself.[13]

Biochemist Arthur Kornberg at Stanford approached the problem from a radically different vantage point: enzymology. He set out to test his bold theory that the replication of polynucleotide chains had to be catalyzed by an enzyme. Effecting successful synthesis of a polynucleotide by using *E. coli* extracts, a DNA template, and DNA's four building blocks, he christened the newly discovered enzyme as "DNA polymerase." The manner in which the polynucleotide synthesis was catalyzed by the *E. coli* DNA polymerase indicated that, as implied by the Watson-Crick mechanism, DNA acted directly as a template for the copolymerization of its replicas, without mediation of other substances, as some had suggested.[14] Though still limited theoretically and technically encumbered, the successful execution of an in vitro synthesis of DNA marked one of the primary links in the chain connecting molecular biology with genetic engineering.

The representation of gene action in terms of a genetic code forged additional links in that chain. One of the most compelling ideas in science to burst onto center stage immediately following elucidation of the DNA structure, the notion of a genetic code was developed as the answer to the problem of heterocatalysis: how only four bases of DNA—adenine, thymine, guanine, and cytosine—could specify the assembly of 20 amino acids into the myriad proteins present even in as simple an organism as a bacterium. Erwin Schrodinger's suggestion of a code script for the gene had intrigued scientists since the mid-1940s, and the idea crystallized during the summer of 1953 that there had to exist some type of code relating the base sequences in polynucleotides to amino acid sequences in polypeptides. It was proposed (and experiments designed to test this hypothesis confirmed it over the next few years) that the heterocatalytic function could be represented as a two-stage process: (1) the DNA template's transcription into messenger RNA; and (2) the translation stage: After carrying the coded information to the cytoplasm, the nucleotide sequences were translated into polypeptide chains of predetermined primary structure.[15]

Researches on the primary structure of protein had been in progress for decades, of course, and a particularly important breakthrough coincided with the search for the code's heterocatalytic message. At that time British biochemist Frederick Sanger was in the final stretch of his biochemical marathon, painstakingly determining the complete sequence of amino acids in insulin, a relatively short polypeptide. By 1955, when the reconstruction was completed, it elucidated more than the composition of insulin. His techniques could now be extended to determination of the primary structure of the longer and compositionally more complex polypeptide chains that characterized enzymes. In principle the biochemical translation could now be matched with the genetic message.[16]

Simple permutational arithmetic dictated that a triplet constituted the smallest set of four bases possessing coding capacity for at least 20 amino acids. The first scientist to establish a formal scheme for a genetic code was the physicist-cosmologist George Gamow, proposing in 1954 an overlapping code. Though beset by internal contradictions, his scheme inspired several researchers, among them Sydney Brenner and Francis Crick, to develop alternative paper solutions for reading the message written in a triplet code (codons). None of these elegant and clever representations unlocked the code.[17] The key to the lock was discovered by chance in 1961 by Marshall Nirenberg and Heinrich J. Matthaei at the NIH. Using an artificially synthesized RNA consisting only of uracil monomers (polyU), they managed to produce a polypeptide comprised solely of phenylalanine—a triplet U coded for the amino acid phenylalanine. The code could now be deciphered with chemical probes. Within a short time 64 of 124 codons—those consisting of repeating units (AAA, CCC, and so on)—were accounted for. The decoding experiments (some of the important ones conducted by Severo Ochoa and Marianne Grunberg-Manago) thus continued by trial and error, relying on artificial random RNA fragments, until in 1964 Gobind Khorana managed the chemical feat of synthesizing a heterogeneous messenger RNA. In addition to giving rise to three polypeptides, these experiments provided powerful insights into the organization of the code and its reading order. By 1965 the representation of heterocatalysis in terms of a genetic code had been completed: The transcriptional specifications for all the amino acids were on paper, and in theory molecular biology had the cognitive apparatus for direct genetic intervention and for designing artificial life.[18]

Were these researches in any sense mission-oriented? That is, were they carried along with the momentum generated since the 1930s by the goal-directed agenda of the Rockefeller Foundation? Were there ideological continuities? The complexities of the political and institutional configurations of postwar science, exacerbated by the paucity of studies on the subject, hinder a clear assessment of the social forces behind the various research programs in molecular biology after the mid-1950s. However, preliminary glances at the historical landscape do reveal some remarkable lines of continuity. To be sure, the Rockefeller Foundation sponsored only a fraction of these researches. The Foundation's investment in molecular biology had declined substantially during the 1950s; with Weaver's retirement in 1958, an era ended. Within the widely decentralized postwar life science, molecular biology enjoyed a pluralistic mode of patronage, spanning government, military, industry, and private foundations. The molecular approach to life and its attendant technocratic approach to health, disease, and human relations, previously concentrated within a few centers of power, diffused across diverse institutional contexts. The scientific ideology of social control, so neatly articulated and localized by the ruling academic-business elite of the 1920s, was fragmented and lodged within pockets of power dispersed throughout the multiple administrative structures of postwar science.

For example, during the early 1950s, as the Rockefeller Foundation's domestic interests in human relations declined, the Ford Foundation entered the field. With both Rockefeller Foundation officers and Caltech scientists playing a significant advisory role, the Ford Foundation launched an enormous program in the behavioral sciences aimed at promoting research and applications in areas of political, social, and individual behavior. Introducing the new term "behavioral sciences," the program director stressed that problems of human relations would not be approached from a vantage point of political science or economic but from that of psychology, anthropology, sociology, and related fields. The discourse on personality needs and social maladjustment echoed earlier calls for action. By 1957, five years after launching the program, the Ford Foundation had granted nearly $24 million to behavioral science research, about $13 million to the mental health program, and $1.6 million to population studies.[19]

Pauling received nearly $1 million in grants from the Ford Foundation for biochemical studies of mental deficiency. As a member of the Hixon Fund committee at Caltech, his interest in the biological basis of human behavior dated back to the 1940s, but by the late 1950s these involvements had acquired a sharper focus through the concept of molecular disease and the cascade of discoveries related to the mechanisms of DNA replication, transcription, translation, and the genetic code. Like other leading practitioners of molecular biology at that time, Pauling's intrigue with the triangle of heredity, intelligence, and social planning assumed more precise technocratic meanings.[20]

In a 1958 television broadcast entitled "The Next Hundred Years," Pauling described his vision of scientific utopia, attained through detailed knowledge of the molecular structure of humans.[21] The study of sickle cell anemia, he stated, set a precedent for that kind of approach. Recounting the biochemical and genetic aspects of the discovery of that first molecular disease, "discovered in our laboratory," Pauling postulated that there were "thousands, tens of thousands of molecular diseases." Like other physiological abnormalities, Pauling believed that mental deficiencies were genetically determined molecular abnormalities. His vision of the nearing Golden Age was a move away from mere palliative action: biology turning molecular, medicine maturing into an exact science, and social planning becoming rational. Like some of his peers, Pauling saw the deterioration of the human race as the most compelling challenge for the new biology. "It will not be enough just to develop ways of treating the hereditary defects," he said. "We shall have to find some way to purify the pool of human germ plasm so that there will not be so many seriously defective children born. . . . We are going to have to institute birth control, population control."[22] Pauling's interventionist concepts of social control, which had previously resonated with those of the Rockefeller Foundation, now buttressed the Ford Foundation's program he had helped shape.

Outlining the approach to the studies on the biochemical bases of mental deficiency, Pauling's early papers in psychobiology explained that the new program was just a natural progression of the studies of the previous two decades. The progress in molecular biology, he noted, had related mainly to somatic and genetic aspects of physiology, rather than to psychic aspects. "We may now have reached the time," he proclaimed, "when a successful molecular attack on psychobiology, including the nature of encephalonic mechanisms, consciousness, memory, narcosis, sedation, and similar phenomena can be initiated."[23] Indeed by the mid-1960s, Pauling had coined a new term and annunciated his concept of "orthomolecular psychiatry," proposing a treatment of mental diseases that involve providing the optimal molecular environment of mind, including the introduction of nucleic acid into the cells to correct genetic abnormalities.[24]

Pauling was not alone in his utopian visions of rational control based on the new biology. At a meeting sponsored by the Ciba Foundation in 1963, about two score distinguished scholars, primarily biologists, gathered to speculate on "man and his future," in light of the revolutionary discoveries in molecular biology.[25] Unleashing a scientific imagination uninhibited by the kind of public alarm and social accountability that soon came to characterize the 1960s era, these researchers pondered the scope of a new eugenics. Based on an implicit assumption of the deterioration of the human race, H. J. Muller advocated measures that would enhance evolutionary selection through molecular genetics. "It is more economical in the end to have developmental and physiological improvements of the organism placed on a genetic basis, where practicable, than to have to institute them in every generation anew by elaborate treatments of the soma," he argued. Deploring the social naivete and offensive reactionary attitude of the old eugenics, Muller promoted a new eugenics free of class and race prejudices and based on biological and social merit.[26]

Joshua Lederberg, though not in full agreement with Muller, predicted that in "no more than a decade" the molecular knowledge of microbes would be applied to the human genome. In light of these rapid developments, "Why bother now with somatic selection, so slow in its impact?" Investing a fraction of the effort, we could soon learn how to manipulate chromosomes ploidy, homozygosis, gametic selection, full diagnosis of heterozygotes, to accomplish in one or two generations of eugenic practice what would now take ten or one hundred." As he saw it, "the ultimate application of molecular biology would be the direct control of nucleotide sequences in human chromosomes, coupled with recognition, selection and integration of the desired genes, of which the existing population furnishes a considerable variety. These notions of a future eugenics are, I think, the popular view of the distant role of molecular biology in human evolution."[27]

In a utopian universe constructed by an intellectual elite, the most prized feature of genetics resided in the control of intelligence. "Surely the same culture that has uniquely acquired the power of global annihilation must generate the largest quota of intellectual and social insight to secure its own survival," Lederberg argued. To dramatize the relation of mental science to molecular biology, he promoted a developmental engineering (euphenics) aimed at controlling hormonal function, brain size, and intelligence. When these technologies had been refined, he predicted, they would supply a "catalogue of biochemically well-defined parameters for responses now describable only in vague functional terms. Then we shall more confidently design genotypically programmed reactions, in place of evolutionary pressures, and search for further innovations."[28]

These ideas were further explored by the symposium participants in the discussion session. Francis Crick was generally in agreement with Muller and Lederberg, and with the eugenic goals and technological potentials of molecular biology; his reservations centered primarily on the feasibility of implementation of such biological-social technologies. Others had deep, principled objections. "Dr. Lederberg, what makes you think that we could make ourselves less likely to blow ourselves up by a genetic increase in intelligence?" challenged Alex Comfort, who pointed out that personality problems had a far greater causal role in human affairs than I.Q. To Lederberg's reply that personality too was under genetic control, Comfort pointed out that, although that may be so, training and upbringing were far more direct social means than breeding. There were also reminders of the Nazi experience, which "should warn us against evaluating our plans for the race solely in terms of technical feasibility."[29] Jacob Bronowski, representing the humanistic position of science, was appalled by the different arguments, by Muller's, Lederberg's, and Sir J. Huxley's positions, and he demanded quantitative evidence for the premise that the human population was deteriorating genetically. "What problem are we trying to solve?" he insisted. "What genes are we trying to boost?"[30]

Such voices of dissent did not dissuade other scientists from their enthusiasm for a new eugenics. By the late 1960s the potential of social control through molecular biology was shared by its leading practitioners. Reflecting on the new biology, Pauling suggested a "yellow star" policy of eugenic prophylaxis.

> There should be tattooed on the forehead of every young person a symbol showing possession of the sickle-cell gene or whatever other similar gene. . . . It is my opinion that legislation along this line, compulsory testing for defective gene before marriage, and some form of semi-public display of this possession, should be adopted.[31]

Robert Sinsheimer at Caltech rejoiced in the new powerful technologies for perfecting the remarkable product of two billion years of evolution.

> The old eugenics was limited to a numerical enhancement of the best of our existing gene pool. The new eugenics would permit in principle the conversion of all the unfit to the highest genetic level.[32]

These technocratic ideologies of controlling life and behavior, as well as the opposition to these visions, foreshadowed the heated scientific and public debates over recombinant DNA technologies of the 1970s and over the human genome project of the 1980s. True, most of the Ciba Symposium's participants later retracted some of their speculative forecasts of a brave new molecular world, confessing to their social and scientific naivete. Nevertheless, the Symposium's proceedings stand out as an instructive historical lesson: The compartmentalization of intellectual puzzles cannot escape the mutually reinforcing powers of the scientific and social imagination.[33] These attitudes and the other sweeping expressions of faith in technologies of selective breeding attest to the durability and resilience of the quest for science-based social control.

Thus the perceived role of molecular biology as a rational basis for human behavior and social planning outlived the Rockefeller Foundation's program, but it underwent modification within the changing ecology of biological knowledge during the postwar era. New structures of patronage generated different coalitions, with novel implications for life science. The displacement of proteins by DNA as the genetic determinants certainly stands out as a paradigm shift, a principal discontinuity in biological theory and practice. Yet the lines of continuity are also striking. The eugenic goals, which had informed the design of the molecular biology program and had been attenuated by the lessons of the Holocaust, revived by the late 1950s. Dredged from the linguistic quagmire of social control, a new eugenics, empowered by representations of life supplied by the new biology, came to rest in safety on the high ground of medical discourse and latter-day rhetoric of population control.

More importantly, the authority of the underlying epistemological stand did survive the paradigm shift and change in patronage. The premise that the soma and psyche are essentially the outcome of genetically determined activity of macromolecules, and that these mechanisms of upward causation should be the principal basis for intervening in higher-order life processes, has acquired even greater intellectual vigor and social legitimacy.

Notes

1. T. F. Anderson, "Destruction of Bacterial Viruses by Osmotic Shock," *Journal of Applied Physics*, 21 (1950), p. 70.

2. Quoted in A. D. Hershey's "The Injection of DNA into Cells by Phage," in J. Cairns, G. S. Stent, J. D. Watson, eds., *Phage and the Origins of Molecular Biology* (New York: Cold Spring Harbor Laboratory of Quantitative Biology, 1966), p. 102. Hershey, however, was not entirely convinced until 1953.

3. Ibid.

4. A. D. Hershey and M. Chase, "Independent Function of Viral Proteins and Nucleic Acid in Growth of Bacteriophage," *Journal of General Physiology*, 36 (1951), pp. 39–56.

5. CIT, Delbrück Papers, Box 27.20, Watson to Delbrück, May 20, 1952. See also James D. Watson, *The Double Helix* (New York: Mentor Books, 1968); Robert C. Olby, *The Path to the Double Helix* (London: Macmillan, 1974), Section 4–5; and Horace F. Judson, *The Eighth Day of Creation* (New York: Simon & Schuster, 1979), Part I.

6. J. D. Watson and F. H. C. Crick, "A Structure for Desoxyribose Nucleic Acids," *Nature*, *171* (1953), pp. 737–738.

7. CIT, Delbrück Papers, Box 23.20, Delbrück to Watson, April 14, 1953.

8. Gunther S. Stent and Richard Calendar, *Molecular Genetics: An Introductory Narrative* (San Francisco: W. H. Freeman, 1978), p. 219.

9. On the reception of the double-helix see Edward Yoxen, "Speaking About Competition: An Essay on *The Double Helix* as Popularization," in Terry Shims and Richard Whitley, eds., *Expository Science: Forms and Functions of Popularisation* (Dordrecht: D. Reidel, 1985), pp. 163–183.

10. K. V. Linderstrom-Lang, "How is a Protein Made?" *Scientific American, 189, No. 3* (1953), p. 105.

11. OSU, Pauling Papers, "Chemical Biology at the California Institute of Technology," 1953, pp. 25–26.

12. CIT, Delbrück Papers, Box 23.20, Delbrück to Watson, April 14, 1953.

13. G. S. Stent and R. Calendar, *Molecular Genetics*, pp. 229–231 (see Note 8); and Matthew Meselson and Franklin W. Stahl, "Demonstration of the Semiconservative Mode of DNA," in *Phage and the Origins of Molecular Biology*, pp. 246–251 (see Note 2).

14. G. S. Stent and R. Calendar, *Molecular Genetics*, pp. 245–249 (see Note 8); and Arthur Kornberg, "Biological synthesis of deoxyribonucleic acid," *Science, 131* (1960), p. 1503. See also Kornberg's later accounts in his *For the Love of Enzymes: The Odyssey of a Biochemist* (Cambridge: Harvard University Press, 1989), especially Ch. 4–7.

15. G. S. Stent and R. Calendar, *Molecular Genetics*, p. 470 (see Note 8). For a vivid account of the work on the genetic codes and related research see H. F. Judson, *The Eighth Day of Creation*, Part II (see Note 5).

16. G. S. Stent and R. Calendar, *Molecular Genetics*, p. 98 (see Note 8).

17. Ibid.; and H. F. Judson, *The Eighth Day of Creation*, Part II (see Note 15). See also, Francis Crick, *What a Mad Pursuit: A Personal View of Scientific Discovery* (New York: Basic Books, 1988), especially ch. 8.

18. G. S. Stent and R. Calendar, *Molecular Genetics*, Ch. 7 (see Note 8); *The Genetic Code*. XXXI Cold Spring Harbor Symposium on Quantitative Biology (New York: Cold Spring Harbor Laboratory of Quantitative Biology, 1966). For alternative views on the representational scope and limits of the genetice code see Lila L. Gatlin, *Information Theory and the Living System* (New York: Columbia University Press, 1972) and Susan Oyama, *The Ontogemy of Information*, (Cambridge University Press, 1985), especially ch. 5.

19. FFA, Report Section, Behavioral Sciences Program, Final Report, 1951–1957.

20. FFA, Report Section, Grant File PA 56-223 Reel 2741, Pauling to Heald, June 14, 1960 and Pauling to McDaniel, November 22, 1966 (letters acknowledging the receipt of grants).

21. CIT, Historical File, Box 88, Pauling File, "The Next Hundred Years," KRCA, Channel 4, December 13, 1958.

22. Ibid., pp. 10–12. Also OSU, Pauling Papers, "The Future of Science and Medicine," John P. Peters Memorial Lecture, Yale, November 17, 1958.

23. Linus Pauling, "A Molecular Theory of General Anesthesia," *Science 134* (1961), pp. 15–21, quote on p. 1.

24. FFA, Report Section, Grant File PA56–223, Reel 2741, "Orthomolecular Psychiatry," 1967, pp. 1–8; Linus Pauling, "Academic Address," in Max Pinkel, ed., *Biological Treatment of Mental Illness*, (New York: L. C. Page, 1962), pp. 31–37, Sect. 5, Attachments.

25. Gordon Wolstenholme, ed., *Man and His Future* (Boston: Little, Brown, 1963).

26. Ibid., Hermann J. Muller, "Genetic Progress by Voluntarily Conducted Germinal Choice," pp. 254–256.

27. Joshua Lederberg, "Biological Future of Man," in *Man and His Future*, pp. 264–265 (see Note 25).

28. Ibid., pp. 265–269. Lederberg, in fact, placed a great deal of emphasis on euphenics. He elaborated his stand in his "Experimental Genetics and Human Evolution," *The American Naturalist, 100* (1966), pp. 519–531, personal communications, Lederberg to Kay, January 4, and February 25, 1992.

29. Ibid., "Eugenics and Genetics," Discussion, pp. 286–289.

30. Ibid., pp. 284–285.

31. Linus Pauling, "Reflections on the New Biology," *UCLA Law Review, 15* (1968), p. 269. Quoted in Troy Duster, *Backdoor to Eugenics* (New York: Routledge, 1990), p. 46.

32. Robert Sinsheiner, "The Prospect of Designed Genetic Change," *Engineering and Science, 32* (1969), pp. 8–13; quoted in Evelyn Fox Keller, "Mastering the Master Molecule: Molecular

Biology Comes of Age," p. 5 (paper presented at the Mellon Workshop "Comparative Perspectives on the History and Social Study of Modern Life Science," MIT, April 5, 1991).

33. On the historical significance of this document and the "new eugenics" see P. Weingart, J. Kroll, and K. Bayertz, *Rasse, Blud und Gene: Geschichte der Eugenik und Rassenhygiene in Deutschland* (Frankfurt: Suhrkamp Verlag, 1988), pp. 646–652; and Daniel J. Kevles, *In the Name of Eugenics: Genetics and the Uses of Human Heredity* (Berkeley: University of California Press, 1985), Ch. 17.

Conclusion

Current discourse on genetic engineering technologies often characterizes these developments as a natural consequence of the theoretical research that took place during the 1950s, 1960s, and 1970s, a logical evolution from the pure to the applied. The lessons from this book imply the reverse: that from its inception around 1930, the molecular biology program was defined and conceptualized in terms of technological capabilities and social possibilities. Representations of life within the new biology were a priori predicated on interventions that, in turn, aimed from the start at reshaping vital phenomena and social processes. Constructed within the protein paradigm, these objectives were reformulated after 1953 around the concept of the DNA "master molecule." Molecular biology was mission-oriented basic research. The ends and means of biological engineering were inscribed into the Rockefeller Foundation's molecular biology program, and eugenic goals played a significant role in its design. The program, in turn, formed a key element in the Foundation's new agenda, "Science of Man," a cooperative venture between the natural, medical, and social sciences. This agenda sought to develop a comprehensive science of social control and a rational basis for human engineering. It was a scientific and a cultural enterprise shaped by the historical contingencies of the tumultuous era of the 1920s–1950s.

Caltech has offered an excellent vantage point for examining some of these developments, the capillary workings of power in the production of knowledge. Its institutional structures could most naturally accommodate the Rockefeller Foundation's cooperative enterprise. Caltech's research community, committed principally to physical science and engineering, provided an optimal setting for a technology-driven physicochemical biology. Leading schools of molecular biology had emerged there, that, in turn, attracted life scientists from around the world, extending Caltech's influence well beyond its temporal and spatial confines. As a primary site for implementing the Foundation's molecular biology program, the Institute was the beneficiary of enormous support, a vote of confidence in the formidable alliance between scientists and the private sector. Thus it offered a privileged vantage point from which to follow up-close the convergence of cognitive and social interests that inscribed genetic engineering ideals into the new biology and that led, by the 1960s, to the dominance of the molecular vision of life.

These historical lessons are not based on arguments of causality or even directionality. The Foundation did not impose on scientists a particular research agenda; the Foundation's officers depended at every turn on their scientific advisers—leading figures such as Morgan, Beadle, and Pauling—to inform them on cognitive priorities and technological capabilities. These scientists, in turn, sought to promote their own research goals and technologies. The rise of the new biology was a process of consensus formation in which the Rockefeller Foundation and an academic elite reinforced each other's interests, forming a hegemonic bloc sustained by a system of incentives and power sharing. Whether these scientists shared most of the Foundation's goals is secondary to the formation of consensus. What is of primary importance is that these particular representations of life supplied an instrumental rationality that legitimated and empowered both groups, both scientific and cultural projects.

What has been the significance of this disciplinary power? What have been the intellectual and social consequences of the dominance of the molecular vision of life? The history of science affords some valuable insights into these questions. Scientists, patrons, and public presentations of science have historically singled out certain research programs as "cutting edge"—a set of problems that signified and was promoted as the key to select mysteries of nature. Such research projects have been privileged both intellectually and socially, garnering resources, talent, and prestige. The flip side of this picture, the negative space, as it were, has been those areas bracketed out of the "vanguard," their modest resource bases and relative lack of prestige tending to discourage interest in their research problems. A circular process was set in motion. Lack of adequate support impeded the growth of less fashionable areas, whereas the targeting of massive resources to select fields accelerated their pace, creating a sense of rapid progress and public excitement, thus perpetuating mechanisms of knowledge claims and social authority.

Yet a look at the trajectories of a number of research programs within the mainstream of biology during the twentieth century (for example, cytology, embryology, evolutionary biology, ethology, and ecology) reveals that with the rise and decline of major research programs structural shifts have not mapped directly unto cognitive domains. Ascending fields have not necessarily directed their cognitive potency toward unsolved critical problems but, rather, redefined which problems were central. Important biological problems, such as differentiation, growth, organismic organization, selection, adaptation, and speciation, have remained unsolved for decades. Fields became marginalized without losing their intellectual validity.

The rise of molecular biology during the 1930s–1950s was such a process of redefinition of what counted as important in biology, followed by vigorous promotion of these select research fields; this targeting and amplification eventually led to wholesale redirection of biological research. There is no doubt that molecular biology has supplied deep insights into some of the fundamental mechanisms of life, notably reproduction and subcellular regulation; it has also supplied powerful tools for research in diverse areas in life science. Its very effectiveness, however, is grounded in its limits. By narrowing its epistemic domain, the new biology has bracketed out important animate phenomena from its discourse on life. There have been numerous scientific "truths" to be explored, many possible representations of life, and alternative visions of nature and nurture. That the architects of the Rockefeller Foundation's program—scientists and patrons—concentrated their resources on specific conceptions of nature and nurture, that they favored a molecular representation of life, reflected the special appeal of this particular kind of knowledge. The program expressed the perception that mechanisms of upward causation were necessary and sufficient explanations of life and the most productive path to biological and social control.

This view has persisted into the 1990s, backed by institutional and commercial interests that dwarf the millions of dollars of the Rockefeller Foundation. The seductive power of the molecular reading of life within a binary framework of DNA and proteins and the weight of genetic determinism have generally bracketed out of the consensus nonreductionistic explanations of life, health, disease, and behavior. Critical perspectives on nature and nurture have been neglected. Moreover, because social policy generally relies on dominant scientific fields, valuable input into social planning has been lost.

The disciplinary power of molecular biology, especially its expanding sphere of influence through the various human genome projects, displays some deep lines of continuity with the past. Today, just as half a century ago, there is a remarkable congruence between the cognitive and social realms, between our technocratic social policies and the technocratic approach to life, health, and disease. The enormous faith in the power of molecular genetics to explain human order and disorder has paralleled the enormous investments in genetic engineering in agriculture and medicine; the technological and cognitive realms drive and justify each other. This dialectical process of knowing and doing, empowered by a synergy of laboratory, boardroom, and federal lobby, has sustained the rise of molecular biology into the twenty-first century.

Key to Archival Sources

APS	American Philosophical Society, Philadelphia, Pennsylvania
CIT	California Institute of Technology, Pasadena, California
CUB	Columbia University, Butler Library, New York, New York
FFA	Ford Foundation Archives, New York, New York
LCM	Library of Congress, Manuscript Division, Washington, DC
OSU	Oregon State University, Corvallis, Oregon
PHS	Presbyterian Historical Society, Philadelphia, Pennsylvania
PPC	Pasadena Presbyterian Church, Pasadena, California
RAC	Rockefeller Archive Center, North Tarrytown, New York
YUB	Yale University, Beinecke Library, New Haven, Connecticut

Index

Printed in the United States
135353LV00005B/99/A

Printed in Great Britain
by Amazon.co.uk, Ltd.,
Marston Gate.